BIOLOGY OF VITRONECTINS
AND THEIR RECEPTORS

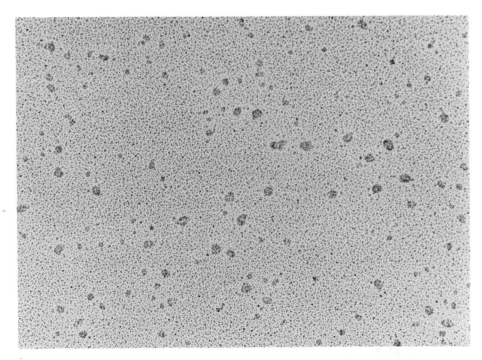

Illustration used for background hardcover: Electronic microscopical picture of rotary shadowed multimeric vitronectin (Preissner KT and Timpl R. 1993).

Illustration on the front of the cover: "Egg in the pan". A macroscopic vision of an adherent cell.

ACKNOWLEDGEMENT

We are indebted to the following organizations for the generous financial support they have provided to help ensure the success of this conference:
Amersham Buchler, Braunschweig (Germany); Behringwerke, Frankfurt-Marburg (Germany); Biotest Pharma GmbH, Dreieich (Germany); Hoffmann-La Roche, Basel (Switzerland); Luitpoldwerk, München (Germany); MAGV-GmbH, Rabenau (Germany); Merck, Darmstadt (Germany); Sandoz AG, Nürnberg (Germany); Schering AG, Berlin (Germany); Serbio, Gennevilliers (France); Serva Feinbiochemica, Heidelberg (Germany)

BIOLOGY OF VITRONECTINS AND THEIR RECEPTORS

Proceedings of the First International Vitronectin Workshop, Rauischholzhausen Castle, Marburg, Germany, 25–28 August, 1993

Editors:

KLAUS T. PREISSNER, SYLVIA ROSENBLATT, CHRISTINE KOST and JÖRG WEGERHOFF

Haemostasis Research Unit, Kerckhoff Klinik, Max-Planck Institut Bad Nauheim, Germany

DEANE F. MOSHER

Department of Medicine, University of Wisconsin, Madison, W.I., U.S.A.

 1993

EXCERPTA MEDICA AMSTERDAM – NEW YORK – LONDON – TOKYO

International Congress Series No. 1042
ISBN 0-444-81680-1

This book is printed on acid-free paper.

Published by:
Elsevier Science Publishers B.V.
P.O. Box 211
1000 AE Amsterdam
The Netherlands

Library of Congress Cataloging in Publication Data:

International Vitronectin Workshop (1st : 1993 : Marburg, Germany)
    Biology of vitronectins and their receptors : proceedings of the
First International Vitronectin Workshop, Rauischholzhausen Castle,
Marburg, Germany, 25-28 August, 1993 / editors, Klaus T. Preissner
... [et al.].
        p.   cm. -- (International congress series ; no. 1042)
    Includes bibliographical references and index.
    ISBN 0-444-81680-1
    1. Vitronectin--Congresses. 2. Vitronectin--Receptors-
-Congresses.   I. Preissner, Klaus T. II. Title.   III. Series.
QP552.C42I58   1993
612'.01575--dc20   93-33787
                                                       CIP

PREFACE

This proceedings volume contains a collection of chapters presented at the international workshop entitled 'The Biology of Vitronectins and Their Receptors', which was held in Rauischholzhausen Castle, Germany, on August 25-28, 1993. Eighty participants from eleven countries attended the meeting, which was meant to bring together scientists from different fields of biological/biomedical research whose activities concentrate on cellular adhesion related to a variety of physiological and pathophysiological conditions.

In this respect, the common players among diverse biological systems are the multifunctional adhesive protein, vitronectin and related components. These, together with the appropriate cell surface adhesion receptors, provide the basis for cellular communication relevant for development, proliferation, pericellular proteolysis, humoral defence mechanisms, and tumor metastasis. The discovery of vitronectin as a 'serum spreading factor' more than 25 years ago, and independent research activities at the Scripps Clinic Research Foundation, La Jolla, and the La Jolla Cancer Research Foundation in the seventies resulted in the demonstration of immune regulatory function devoted to 'complement S-protein', as well as to the description of an important fibronectin-independent cell adhesion factor, named vitronectin (in relation to its first isolation by adsorption to glass beads) (see scheme). In the mid-eighties comparison of cDNA clones and the functional activities of both factors resolved their relation, and demonstrated that serum spreading factor, S-protein and vitronectin all refer to the same protein. Meanwhile, vitronectin(-like) structures have been identified in a number of vertebrate and invertebrate organisms, and the molecule appears to be a member of the hemopexin-type super gene family. During the past decade we have witnessed the explosive development in the field of cellular adhesion, particularly related to the discovery of integrins and other cell surface adhesion receptors in a variety of biological systems. Several chapters in this volume are devoted to the role of vitronectin receptors in relation to different physiological settings.

The selection of presentations, which reflect the multidisciplinary nature of this field, was made on the basis of the high quality of the work of the individuals or groups involved. The opinion and views expressed in each contribution should be regarded as those of the author(s). The meeting in Rauischholzhausen Castle was only made possible by generous support and financial contributions from the Deutsche Forschungsgemeinschaft, Bonn-Bad Godesberg (Germany), and the Max-Planck-Gesellschaft, Munich (Germany), as well as by contributors listed in the Acknowledgement. The editors also wish to thank Angelika Püschel for her skilful secretarial assistance.

September 1993,

Klaus T. Preissner
Sylvia Rosenblatt
Christine Kost
Jörg Wegerhoff
Deane F. Mosher

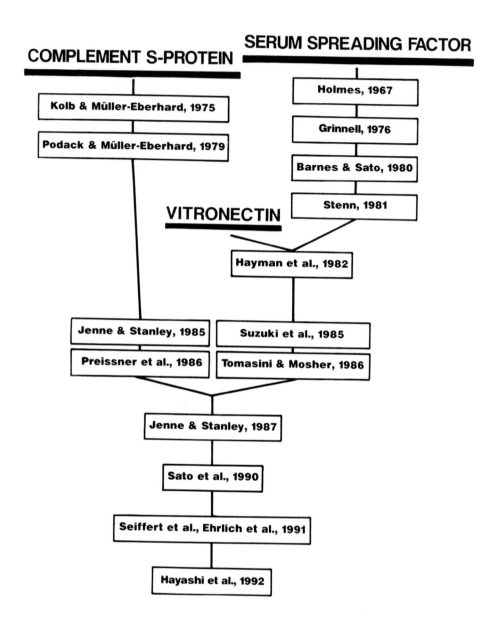

Selected references in a chronological setting

Selected references in alphabetical order

Barnes D, Sato G. Cell 1980; 22:649-655

Ehrlich HJ, Klein Gebbink R, Preissner KT, et al. J Cell Biol 1991; 115:1773-1781

Grinnell F. Exp Cell Res 1976; 102:51-60

Hayman EG, Engvall E, A'Hearn E, et al. J Cell Biol 1982; 95:20-23

Holmes R. J Cell Biol 1967; 32:297-308

Jenne D, Stanley KK. EMBO J 1985; 4:3153-3157

Jenne D, Stanley KK. Biochemistry 1987; 26:6735-6742

Kolb WP, Müller-Eberhard HJ. J Exp Med 1975; 141:724-735

Nakashima N, Miyazaki K, Ishikawa M, et al. Biochim Biophys Acta 1992; 1120:1-10

Podack ER, Müller-Eberhard HJ. J Biol Chem 1979; 254:9908-9914

Preissner KT, Heimburger N, Anders E, et al. Biochem Biophys Res Commun 1986; 134:951-956

Sato R, Komine Y, Imanaka T, et al. J Biol Chem 1990; 265:21232-21236

Seiffert D, Keeton M, Eguchi Y, et al. Proc Natl Acad Sci USA 1991; 88:9402-9406

Stenn KS. Proc Natl Acad Sci USA 1981; 78:6907-6911

Suzuki S, Oldberg A, Hayman EG, et al. EMBO J 1985; 4:2519-2524

Tomasini BR, Mosher DF. Blood 1986; 68:737-742

Participants of the International Vitronectin Workshop, Rauischholzhausen Castle, Germany

Contents

Vitronectin structure, biosynthesis, processing and distribution

Vitronectins: From vertebrates to invertebrates
    M. Hayashi   3
Purification of stabilized native vitronectin
    I. Hayashi and D.F. Mosher   13
Multimeric vitronectin: Structure and function
    S. Hess, A. Stockmann, W. Völker and K.T. Preissner   21
Expression of recombinant human vitronectin using the baculoviral system
    D.C. Sane and Y. Zhao   31
Vitronectin-integrin interactions: Analysis by site-directed mutations
    R.C. Cherny and P. Thiagarajan   39
A comparison of vitronectin and megakaryocyte stimulating factor
    D.M. Merberg, L.J. Fitz, P. Temple, J. Giannotti, P. Murtha,
    M. Fitzgerald, H. Scaltreto, K. Kelleher, K. Preissner, R. Kriz,
    K. Jacobs and K. Turner   45
Binding domain for β-endorphin and vitronectin
    S. Wöhner, D. Linder, H. Teschemacher and K.T. Preissner   55
Primary structure of vitronectins and homology with other proteins
    H.J. Ehrlich, B. Richter, D. von der Ahe and K.T. Preissner   59
Scanning microcalorimetry and equilibrium chemical denaturation studies of
human plasma vitronectin
    C.B. Peterson   67
Distribution of vitronection
    D. Seiffert, T.J. Podor and D.J. Loskutoff   75
Vitronectin expression in human germ cell tumours and normal mouse development
    S. Cooper, W. Bennett, S. Roach and M.F. Pera   83
Vitronectin association with skin
    K. Dahlbäck   91
The distribution of vitronectin in resting and thrombin stimulated platelets.
An ultra-immunocytochemical study
    E. Morgenstern and K.T. Preissner   99
Metabolism of vitronectin complexes
    H.C. de Boer, Ph. de Groot and K.T. Preissner   103
Receptor mediated endocytosis of vitronectin by fibroblast monolayers
    P.J. McKeown-Longo and T.S. Panetti   111

Vitronectin receptors and adhesive interactions

Platelets, extracellular matrix and integrins
    J.J. Sixma, H.K. Nieuwenhuis and Ph.G. de Groot   121
Megakaryocyte specific expression of the αIIb integrin
    G. Marguerie and G. Uzan   127

Vitronectin receptors in melanoma tumor growth and metastasis
    B. Felding-Habermann, B.M. Mueller, L.C. Sanders and D.A. Cheresh   135
Peptide and non-peptide antagonists of ß3 integrins
    B. Steiner, W.C. Kouns and T. Weller   141
Peptides from the second calcium binding domain of integrin chain αIIb inhibit
fibrinogen binding to αIIbß3 by direct interaction
    B. Diefenbach, B. Felding-Habermann, A. Jonczyk and F. Rippmann   149
Selectivity of the non-peptidic GPIIb/IIIa-receptor-antagonist BIBU 52 in
respect to other RGD-dependent integrin receptors using an *in vitro* cell
adhesion assay
    E. Seewaldt-Becker, F. Himmelsbach and Th.H. Müller   157
Tyrosine phosphorylation of a 38 kDa protein upon interaction of urokinase-type
plasminogen activator (u-PA) with its cellular receptor
    I. Dumler, T. Petri and W.-D. Schleuning   163
Modulation of the protein tyrosine kinase pp60^{c-src} upon platelet activation:
A protein kinase C inhibitor blocks translocation
    U. Liebenhoff and P. Presek   171
The role of αv integrins in adenovirus infection
    G.R. Nemerow, T.J. Wickham and D.A. Cheresh   177
* Biological properties of human immunodeficiency virus type-1 TAT protein:
  Angiogenic effects and adhesive interactions of extracellular TAT
    V. Fiorelli, G. Barillari, R.C. Gallo and B. Ensoli   351

Extracellular matrix interactions and alterations

Drosophila development and cellular adhesion
    J.H. Fessler, F.J. Fogerty, R.E. Nelson, D. Gullberg, T. Bunch
    and L.I. Fessler   187
Episialin modulates cell-cell and cell-matrix adhesion, promotes invasion in
matrigel and inhibits cytolysis by cytotoxic effector cells
    J. Hilkens, J. Wesseling, H.L. Vos, S.L. Litvinov, M. Boer,
    S. van der Valk, J. Calafat, E. van de Wiel van Kemenade and C. Figdor   193
Molecular organization and antiproliferative activity of arterial tissue
heparan sulfate proteoglycans
    E. Buddecke, A. Schmidt, P. Vischer and W. Völker   201
Thrombin interaction with the vascular system
    R. Bar-Shavit, Y. Eskohjido, M. Benezra and I. Vlodavsky   209
Basic fibroblast growth factor expression in human bone marrow cells and
phospholipase C release of biologically active growth factor-heparan sulfate
proteoglycan complexes
    G. Brunner and H. Nguyen   217
Vitronectin in inflammatory conditions: Localization in rheumatoid
arthritic synovia
    B.R. Tomasini-Johansson   223

---

* Due to circumstances beyond our control, this manuscript has been placed at the end of the
  volume.

Tenascin and DSD-1-proteoglycan: Extracellular matrix components involved
in neural pattern formation and remodeling
    A. Faissner, A. Clement, B. Götz, A. Joester, C. Mandl, C. Niederländer,
    O. Schnädelbach and A. Scholze   229

Immune defense and micro-organisms

Lymphocyte mediated cytotoxicity
    E.R. Podack   237
Structure and function of CD59
    P.J. Sims   243
Molecular mechanisms and therapeutical intervention strategies of the sepsis
syndrome: Induction-pattern and function of lipopoly-saccharide binding protein
    R.R. Schumann, C. Kirschning, N. Lamping, H.-P. Knopf, H. Aberle and
    F. Herrmann   249
Vitronectin binding surface proteins of staphylococci and *Helicobacter pylori*
    T. Wadström, A. Ljungh and J.-I. Flock   257
Bone sialoprotein binding to *Staphylococcus aureus*
    C. Rydén and A. Yacoub   265
Adhesive reactions between immobilized platelets and *Staphylococcus aureus*
    M. Herrmann, R.M. Albrecht, D.F. Mosher and R.A. Proctor   273

Vitronectin and pericellular proteolysis

Plasminogen activation: A multifunctional proteolytic cascade
    A.-P. Sappino and J.-D. Vassalli   283
Variants of the receptor for urokinase-type plasminogen activator
    G. Høyer-Hansen, H. Solberg, E. Rønne, N. Behrendt and K. Danø   289
Vitronectin and plasmin(ogen) in lesional skin of the bullous pemphigoid:
Colocalisation suggests binding interactions
    M.D. Kramer, H.M. Gissler, B. Weidenthaler-Barth and K.T. Preissner   295
The interaction of plasminogen activator inhibitor-1 with vitronectin
    D.A. Lawrence, M.C. Naski, M.B. Berkenpas, S. Palaniappan, T.J. Podor,
    D.F. Mosher and D. Ginsburg   303
Covalent modulation of vitronectin structure for the control of plasminogen
activation by PAI-1
    S. Shaltiel, I. Schvartz, Z. Gechtman and T. Kreizman   311
Novel approaches towards PAI-1 interactions
    H. Pannekoek, M. van Meijer, A.J. van Zonneveld, D.J. Loskutoff
    and C.F. Barbas   325
Clinical relevance of the plasminogen activator system in tumor invasion and
metastasis in breast cancer
    M. Schmitt, F. Jänicke, C. Thomssen, L. Pache, M. Kramer, J. Bläser,
    H. Tschesche, O. Wilhelm, U. Weidle and H. Graeff   331

xii

Low density lipoprotein receptor-negative fibroblasts internalize lipoprotein
(a) by the low density lipoprotein receptor-related protein/α_2-macroglobulin
receptor
>   A. Beckman, H. Scharnagl, B. Hertwig, R. Siekmeier, W. Schneider,
>   W. Groß and W. März   343

Index of authors   361

Keyword index   365

Vitronectin structure, biosynthesis, processing and distribution

Biology of Vitronectins and their Receptors
K.T. Preissner, S. Rosenblatt, C. Kost, J. Wegerhoff and D.F. Mosher, editors

Vitronectins: from vertebrates to invertebrates

M. Hayashi

Department of Biology, Ochanomizu University,
Bunkyo–ku, Tokyo 112, Japan

INTRODUCTION

Vitronectin (VN), also termed serum spreading factor or S–protein, is a glycoprotein existing in human blood plasma and connective tissues. It was originally purified from human serum by Hayman et al. [1] and Barnes and Silnutzer [2] in 1983, and the primary structure of human VN was deduced from cDNA analyses by Suzuki et al. [3] and Jenne and Stanley [4] in 1985.

At least three major roles of VN are known. Firstly, it promotes attachment and spreading of cultured animal cells on tissue culture substrate. VN rather than fibronectin is the principal cell–attachment protein of tissue culture serum (*e.g.* fetal bovine serum) [5, 6]. An Arg–Gly–Asp (RGD) sequence located close to the NH_2–terminus of VN is recognized by a transmembrane VN receptor protein, *e.g.* integrin $\alpha v\beta_3$, on the surface of cultured cells [7], and VN receptor mediates the signal from extracellular VN to intracellular cytoskeletal actin fibers [8]. Secondly, VN interferes with the cytolytic action of membrane attack complex of C5b–9 [9]. The principal site of this action exists at the heparin–binding site located close to the COOH–terminus of VN [10]. Thirdly, VN regulates blood coagulation by modulating the activities of thrombin–antithrombin III [11–13] and PAI–1 [14–17]. This action is exerted at the heparin–binding site [18] or at the NH_2–terminal 5 kD portion [19]. All these VN functions have been deduced only from knowledge of the *in vitro* properties of human VN and also to some extent fetal bovine serum VN.

In 1988, we developed an easy purification procedure of VN, which enabled us to obtain milligram quantities of VNs from many vertebrates. We subsequently characterized the structure and function of a variety of vertebrate VNs and have proposed a general model of VN structure. To understand the biological functions of VN in living organisms, we have recently extended our studies to more primitive organisms. This paper deals with the story of VN from vertebrates to invertebrates, focusing on the work done in my laboratory; it is not intended to provide a comprehensive review.

RAPID AND SIMPLE PURIFICATION OF VITRONECTIN

When we started to investigate the structure and function of VN in 1983, purification of VN was difficult. Barnes and Silnutzer [2] had reported a purification method using a conventional 4–step column chromatography procedure in 1983. In the same year, Hayman et al. [1] had reported another purification method using monoclonal antibody chromatography followed by heparin affinity chromatography. The monoclonal antibody

used by Hayman et al. [1] was raised against a crude VN preparation obtained according to the method by Barnes and Silnutzer [2].

However, we encountered serious difficulties with these purification procedures.
Barnes and Silnutzer [2] obtained an overall recovery of 2.48 mg of VN from 100 ml of
human serum, that is, about 8 % recovery. At best, it took us 3 weeks to get about 1 mg
of pure VN. Human serum is cheap and available in abundance, but we found their
procedure difficult to scale up. In particular, the glass bead and concanavalin A–

Table 1. Purification of Serum Vitronectin [24]; "Japanese Method"

---

1. Perform all procedures at room temperature unless otherwise specified. All volumes can
be scaled up proportionally.
2. If using (frozen) serum, start at step 3. If using fresh blood, coagulate for more than 1 h.
If using (frozen) plasma (anti–coagulated with acid citrate–dextrose but never with
heparin), coagulate by adding 2 ml of 1 M $CaCl_2$ to 100 ml of plasma in glassware and
incubate for more than 1 h at ambient temperature and then for 2 h (or overnight) at 4°C.
Serum after removal of fibronectin and probably other molecules can be used.
3. Centrifuge serum, blood, or plasma at 10,000 rpm for 10 min at 4°C.
4. To supernatant (100 ml) add 2.5 ml of 0.2 M EDTA and 0.5 ml of 0.2 M PMSF/ethanol
with vigorous stirring.
5. If a faster and simpler procedure is required, involving possible slight contamination, start
at step 9. This frequently results in similar VN purity, but sometimes yields an impure
product, probably depending on the preparation of serum used.
6. Prepare two columns in tandem at room temperature. The first should consist of 2 ml of
control Sepharose 4B, and the second should consist of 5 ml of heparin–Sepharose 4B.
Strictly, the volume of heparin–Sepharose 4B necessarily depends on the capacity of the
column.
7. Both columns should be prewashed with 25 ml of 2 M NaCl/PB–EDTA (10 mM Na–
phosphate buffer (pH 7.7)/5 mM EDTA), and then pre–equilibrated with 25 ml of 0.13 M
NaCl/PB–EDTA.
8. Pump the serum through the tandem columns, following with 25 ml of 0.13 M NaCl/ PB–
EDTA at a flow rate of roughly 100 ml/20 min. Discard the first 5 ml then collect the
flowthrough serum. When the serum has passed through the first column, wash the
column with 5 ml of 0.13 M NaCl/PB–EDTA. Disconnect the two columns and continue
washing the heparin–Sepharose 4B column with 20 ml of 0.13 M NaCl/PB–EDTA.
9. To the flowthrough serum (about 125 ml) or to serum (100 ml) when steps 6–8 are
skipped, add 96 g of powdered urea. Dissolve the urea at about 30°C or by incubating in a
boiling water bath. Make up to 200 ml with PB–EDTA, and allow to stand for more than
2 h. Alternatively, incubate the serum/urea in a boiling water bath for 5 min to save time.
10. During the 2 h, prepare 250 ml of 0.13 M NaCl/PB–EDTA/8 M urea, 25 ml of 0.5 M
NaCl/PB–EDTA/8 M urea, and 25 ml of 2 M NaCl/PB–EDTA/8 M urea.
11. The heparin–Sepharose 4B (5 ml) column at step 8 should be prewashed with 25 ml of 2
M NaCl/PB–EDTA/8 M urea, and then pre–equilibrated with 25 ml of 0.13 M NaCl/PB–
EDTA/8 M urea.

12. Pump the serum/urea at step 9 through the heparin–Sepharose column at a flow rate of 200 ml/2–4 h. A higher flow rate is apt to cause flow disruption. Without incomplete removal of fibrous materials at steps 2–3, the serum becomes very viscous within a few hours after the addition of 8 M urea. This causes low recovery of VN. If flow becomes severely restricted, it is better to restart from step 2.

13. Wash the column with 100 ml of 0.13 M NaCl/PB–EDTA/8 M urea, then with 10 ml of 0.13 M NaCl/PB–EDTA/8 M urea containing 7 μl of 2–mercaptoethanol. Halt for 2 h, or overnight.

14. Wash the column with 90 ml of 0.13 M NaCl/PB–EDTA/8 M urea containing 7 μl of 2–mercaptoethanol.

15. Elute the column with 25 ml of 0.5 M NaCl/PB–EDTA/8 M urea. Collect 1 ml fractions.

16. Read OD_{280} of the fractions and pool peak fractions, usually, into two lots of high and low VN concentrations. Usually discard the fractions showing less than 0.1 OD_{280}.

17. Extensively dialyze against a desired solution at 4°C, *e.g.* 0.13 M NaCl/10 ml Na–phosphate buffer (pH 7.2).

18. Centrifuge at 10,000 rpm for 10 min at 4°C.

19. Read OD_{280} of the supernatant. Calculate the protein concentration using OD_{280} value of 1.38 at 1–cm path length for 1 mg/ml. Usually, 3–6 mg VN is obtained from 100 ml of human serum.

20. Aliquot and freeze at – 80°C in plastic tubes until use.

21. The heparin–Sepharose 4B column can be used repeatedly. It should be washed with 10 ml of 0.13 M NaCl/PB–EDTA containing 0.02 % Na–azide and stored at 4°C.

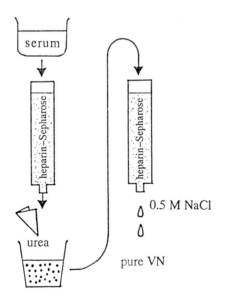

Figure 1. Flow chart of the vitronectin purification procedure by Yatohgo et al. [24].

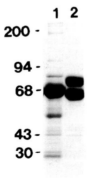

Figure 2. Protein staining of SDS–PAGE of starting human serum (1) and elute human vitronectin fraction (2) under reducing conditions.

Sepharose column chromatographies were difficult to reproduce. At that time, we did not have facilities to produce monoclonal antibody, and in any case we felt that the possible species specificity which might arise from the use of antibody for purification would be undesirable. Dahlbäck and Podack [20] and Preissner et al. [21] reported other conventional 5–step purification procedures in 1985, but they were also time–consuming. Therefore, we sought to develop a rapid, simple, and flexible purification method.

There was a discrepancy in the heparin–binding property between the VNs obtained by Barnes and Silnutzer [2] and by Hayman et al. [1]. The former VN did not bind to heparin, whereas the latter did. It was established that this was due to the use of 8 M urea in the purification procedure of Hayman et al. [1]; urea converts VN from a heparin–nonbinding to a heparin–binding form [22, 23]. This can be interpreted by assuming that the heparin–binding site is cryptic in the native form of VN and becomes exposed artificially at the molecular surface after 8 M urea treatment [22]. On the basis of this artificial conversion of the heparin–binding activity, we have developed a simple method (so–called Japanese method) to purify VN from human serum [24]. As illustrated in Fig. 1, human serum is passed through the first heparin–Sepharose column to remove endogenous heparin–binding proteins. The flowthrough portion of the serum is collected and powdered urea is added to give a concentration of 8 M. The urea activates the heparin–binding property of VN in the flowthrough serum and now the activated VN is applied to a second heparin–Sepharose column, to which it binds, and is eluted with 0.5 M NaCl (Fig. 2). This simple procedure provides milligram quantities of pure VN in 2–3 days with cheap reagents and ordinary column chromatography tools. A detailed protocol for this Japanese method is presented in Table 1.

The Japanese method purifies heparin–binding VN (hbVN), which has similar cell–spreading activity to heparin–nonbinding VN (hnbVN) [24]. However, the two species differ in some properties. For example, hnbVN is monomeric, but hbVN is a disulfide–bonded multimer [25–27]. Both hbVN and hnbVN bind to the cell layer of cultured fibroblasts, but only hbVN is incorporated intracellularly by endocytosis and degraded [27]. Furthermore, hbVN seems to be preferentially bound to collagen [28]. It is important to note that the heparin–binding form of VN is one of the endogenous forms of VN, and not always artificial. In fact, hbVN usually comprises 2 % of total VN in fresh unfrozen human plasma [25] and about 25 % in frozen stored blood plasma or serum [23]. So, it is likely that hbVN is a biologically active form *in vivo*, while hnbVN is a precursor that circulates in blood as an inactive form. The VN property of fast conversion to hbVN followed by rapid clearance by endocytosis and degradation seems to reflect the coupling of VN function with a transient biological phenomenon *in vivo*.

VITRONECTINS FROM A VARIETY OF LIVING ORGANISMS

By applying the Japanese method, we have purified blood VNs from human, rabbit, mouse, rat, hamster, guinea pig, dog, horse, porcine, bovine, goat, sheep, chicken, and goose [29, 30]. These species were practically all animals from which more than 20 ml of plasma or serum was readily available. Without exception, these VNs were purified as heparin–binding protein. Figure 3 shows the protein–staining of the SDS–PAGE pattern of these VNs. All the VNs have collagen–binding and cell–spreading activities, and show similar NH$_2$–terminal amino acid sequences and similar serological cross–reactivity. There are, however, four kinds of marked species–specific variations. The molecular weights of VNs vary from 59 to 78 kD depending on the species. The number and proportion of bands on SDS–PAGE seem to be species–specific. The amount of VNs purified from 100 ml of serum ranges from 0.5 to 9.4 mg, depending on species and probably also on the preparation of serum. The amount and composition of bound carbohydrate also vary species–specifically, *e.g.* human VN contains only *N*–linked saccharides and does not contain D–galactosamine, while bovine VN contains both *N*– and *O*–linked saccharides and 8 residues of D–galactosamine per VN molecule [29].

There is a discrepancy between the molecular weights of VN estimated from SDS–PAGE and cDNA analysis, even after consideration of bound carbohydrates, bound phosphate amounting to about 2.5 mol [31], and bound sulfate amounting to about 2 mol [32] per mol of VN. The molecular weight of the core polypeptide estimated from the cDNA sequence is 52 kD for human [3, 4], rabbit [33], and mouse VNs [34]. On the other hand, molecular weights from SDS–PAGE are 76/68 kD for human VN, 68 kD for rabbit VN, and 71 kD for mouse VN [30]. These discrepancies are due to abnormal behavior of VN on SDS–PAGE and the presence of bound carbohydrate. Estimated molecular weights of human VN depend on the concentration of acrylamide, *e.g.* 79/72 kD at 6 % acrylamide, but 74/61 kD at 15 % acrylamide. Ferguson plot analysis of deglycosylated VNs on SDS–PAGE gave far more consistent results [30]. Based on the above data, we present a general model of the functional structure of animal VN in Fig. 4 [30].

VN in human plasma is selectively phosphorylated by protein kinase [31, 35]. On the basis of this specific phosphorylation, Korc–Grodzicki et al. [36] suggested that phosphorylated plasma proteins of 135 kD in rabbit, 75 kD in guinea pig, 140 kD, 116 kD, and 75 kD in rat, and 90 kD in mouse are probably VNs. Proteins larger than 75 kD, however, may not be VNs, since their molecular weights are much larger than those of purified VNs estimated from SDS–PAGE, although phosphorylated pure VNs have not been examined by SDS–PAGE.

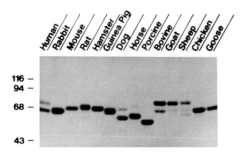

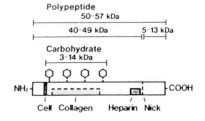

Figure 3. SDS–PAGE of purified vitronectins from a variety of animal sera under reducing conditions. Reproduced from Nakashima et al. [30].

Figure 4. A general model of the functional structure of animal vitronectin. Reproduced from Nakashima et al. [30].

8

Little work has been done on the purification and characterization of VNs from non–plasma/serum sources. The first such report described a partial purification of VN from human fetal membrane by Hayman et al. [1]. Fetal membrane VN is similar in size on SDS–PAGE and in cell–spreading activity to the blood homologue. A similar VN was purified from human yolk sac carcinoma by Cooper and Pera [37]. On the other hand, yolk VN, which we recently purified from chick egg yolk, is distinct [38]. Yolk VN is similar in cell–spreading activity and immunocross–reactivity but distinct in size and structure from the chicken blood homologue. Yolk VN separates into two truncated forms of 54 and 45 kD bands on SDS–PAGE, while the blood homologue shows a single major band of 70 kD. Both 54 and 45 kD yolk VNs lack the COOH–terminal heparin–binding domain. The 45 kD yolk VN also lacks both the NH_2–terminal 49 amino acids and cell–spreading activity.

Immunoassay of VN in animal tissues indicates that VN exists in loose connective tissues of skeletal muscle, embryonal lung, kidney, and skin [1, 39]. Human platelets contain VN and secrete VN after stimulation [40]. Among cultured cells, Hep G2 human hepatoma cells synthesize and secrete VN into culture medium [41]. Some non–hepatic cells, *e.g.* HeLa P3#6 cells from human cervix cancer and A549 cells from human lung

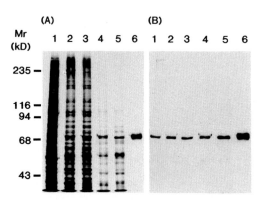

Figure 5. Purification of *Physarum* vitronectin–like protein. Protein staining (A) and immunoblotting (B) of SDS–PAGE step were done on starting *Physarum* homogenate (1), insoluble cell debris (2), Triton X–100 extract (3), anti–vitronectin Sepharose chromatography eluate (4), heparin–Sepharose chromatography eluate (5), and preparative SDS–PAGE product (6). Reproduced from Miyazaki et al. [47].

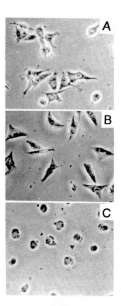

Figure 6. BHK cell–spreading activity of *Physarum* vitronectin–like protein (A), human plasma vitronectin (B), and bovine serum albumin (C). Reproduced from Miyazaki et al. [47].

cancer, also synthesize and secrete VN into culture medium [42]. VN mRNA is, however, detected only in liver and slightly in brain [34, 43]. Therefore, almost all animal VNs in tissues seem to be derived from blood plasma VN, which is produced in the liver and circulates in blood.

Although most of our knowledge of VN has been obtained from human plasma VN, for experimental analysis of biological functions *in vivo*, VNs from small laboratory animals and invertebrates should be analyzed, because they are obviously much more suitable than human VN for manipulation. Immunoblotting with anti–VN antibody can detect VNs from a variety of animals [30]. Strongly stained bands have been observed in quail serum (69 kD), finch serum (3 bands between 69–74 kD), whole *Physarum* (70 kD), whole *Drosophila* adults (80 kD and 85 kD) and whole nematode (38 kD and 17 kD). Weakly stained bands were detected in newt serum (96 kD), Japanese terrapin serum (51 kD and 79 kD), *Xenopus* serum (58–63 kD), goldfish serum (86 kD), sea bream serum (66 kD, 68 kD, 86–96 kD), hagfish serum (64 kD), loach serum (51–55 kD and 84 kD), blue crab serum (66 kD), and prawn serum (70 kD). No bands have been detected in sera from octopus, clam, abalone, lamprey, and ascidian, or in whole extracts of planaria, *Hydra*, sponge, *Tetrahymena*, yeast, *Dictyostelium*, 3rd instar larvae of *Drosophila*, and sea urchin (sperms, eggs, and embryos at the stages of blastula and gastrula). The absence of bands in the immunoblot does not necessarily mean the absence of VN in the species, but merely shows the absence of immunologically cross–reactive protein. The presence or absence of protein reacting with anti–VN antibody does not correlate with phylogeny in the animal kingdom.

Surprisingly, plants also contain proteins reacting with anti–VN antibody. Anti–human VN antibody cross–reacts with a 55 kD protein in leaf and root protein extracts from lily, broad bean, soybean, and tomato [44]. Genomic DNAs from these plants hybridize with human VN cDNA [44]. A 62 kD protein reacting with anti–human VN antibody was also detected in and purified from a brown alga, *Fucus* [45]. These plant VNs exist extra-cellularly and seem to function in cell adhesion.

PHYSARUM VITRONECTIN–LIKE PROTEIN

The most primitive VN–producing organism so far found is slime mold, *Physarum polycephalum* [30]. *Physarum* was grown in a shaking incubator at 24°C, as a suspension in a sterilized solution containing 1.5 % bactotrypton, 0.22 % yeast extract, and 1.5 % glucose supplemented with several inorganic ions and hemin [46]. A protein reacting with anti–bovine VN antibody was detected in an immunoblot of *Physarum* homogenate (Fig. 5, lane 1) [47]. The protein, *Physarum* VN–like protein, migrates at 70 kD on SDS–PAGE under reducing conditions. *Physarum* VN–like protein is not secreted from *Physarum* and is recovered in the insoluble fraction of *Physarum* homogenate. It was purified by anti–bovine VN affinity chromatography followed by heparin affinity chromatography from Triton X–100 extracts of *Physarum* homogenate (Fig. 5), and thus is a heparin–binding protein. Purified *Physarum* VN–like protein shows a typical VN–type cell–spreading activity for BHK cells, both in terms of specific activity and cell–spreading morphology (Fig. 6) [47]. The cell–spreading activity is specifically inhibited by GRGDSP peptide, but not GRGESP peptide, with similar dose–dependency to that of authentic human plasma VN.

On the other hand, the NH_2–terminal sequence and some internal sequences of *Physarum* VN–like protein are, surprisingly, distinct from those of vertebrate VNs [48]. Computer–assisted homology search revealed significant sequence homology with mitochondrial dihydrolipoamide acetyltransferase, a component of pyruvate dehydrogenase complex. Therefore, *Physarum* VN–like protein may be a chimeric molecule of dihydrolipoamide transferase and VN. A solution to this problem must await complete sequencing of *Physarum* VN–like protein.

The localization of *Physarum* VN–like protein is also unique [48]. All the VNs so far found exist in extracellular matrix or body fluids in animals and plants. On the other hand, dihydrolipoamide acetyltransferase is a mitochondrial enzyme. Immunohisto-chemistry for *Physarum* VN–like protein with both anti–bovine VN and anti–rat pyruvate dehydrogenase complex antibodies indicates that the staining is distributed evenly in cytoplasm and nucleus of *Physarum* [48]. In recent preliminary experiments, we have found some intracellular localization of VN in particular cells of mouse and rat tissues [Tsuchiya et al., Kitami et al., Sawada et al., unpublished results]. Panetti and McKeown–Longo [27] also reported VN endocytosis by cultured fibroblasts. Therefore, intracellular VN may be present in many living organisms, and not specific for *Physarum*.

Studies on VNs from typical laboratory experimental organisms should lead to new insights into the nature and dynamics of VN function *in vivo*.

ACKNOWLEDGEMENTS
We thank Ms. K. Hayashi for her secretarial assistance and Mr. W. R. S. Steele for his linguistic advice during preparation of the manuscript.

REFERENCES
1 Hayman EG, Pierschbacher D, Öhgren Y, Ruoslahti E. Proc Natl Acad Sci USA 1983; 80: 4003–4007.
2 Barnes DW, Silnutzer J. J Biol Chem 1983; 258: 12548–12552.
3 Suzuki S, Oldberg A, Hayman EG, Pierschbacher MD, Ruoslahti E. EMBO J 1985; 4: 2519–2524.
4 Jenne D, Stanley KK. EMBO J 1985; 4: 3153–3157.
5 Fath KR, Edgell C–JS, Burridge K. J Cell Sci 1989; 92: 67–75.
6 Underwood PA, Bennett FA. J Cell Sci 1989; 93: 641–649.
7 Pytela R, Pierschbacher MD, Ruoslahti E. Proc Natl Acad Sci USA 1985; 82: 5766–5770.
8 Burridge K, Fath K, Kelly T, Nuckolls G, Turner C. Ann Rev Cell Biol 1988; 4: 487–525.
9 Podack ER, Kolb WP, Müller–Eberhard HJ. J Immunol 1978; 120: 1841–1848.
10 Tschopp J, Masson D, Schäfer S, Peitsch M, Preissner KT. Biochemistry 1988; 27: 4103–4109.
11 Ill CR, Ruoslahti E. J Biol Chem 1985; 260: 15610–15615.
12 Jenne D, Hugo F, Bhakdi S. Thromb Res 1985; 38: 401–412.
13 Podack ER, Dahlbäck B, Griffin JH. J Biol Chem 1986; 261: 7387–7392.

14 Declerck PJ, De Mol M, Alessi M–C, Baudner S, et al. J Biol Chem 1988; 263: 15454–15461.
15 Wiman B, Almquist A, Sigurdardottir O, Lindahl T. FEBS Lett 1988; 242: 125–128.
16 Mimuro J, Loskutoff DJ. J Biol Chem 1989; 264: 936–939.
17 Salonen EM, Vaheri A, Pöllänen J, Stephens R, et al. J Biol Chem 1989; 264: 6339–6343.
18 Kost C, Stüber W, Ehrlich HJ, Pannekoek H, Preissner KT. J Biol Chem 1992; 267: 12098–12105.
19 Seiffert D, Loskutoff DJ. J Biol Chem 1991; 266: 2824–2830.
20 Dahlbäck B, Podack ER. Biochemistry 1985; 24: 2368–2374.
21 Preissner KT, Wassmuth R, Müller–Berghaus G. Biochem J 1985; 231: 349–355.
22 Hayashi M, Akama T, Kono I, Kashiwagi H. J Biochem 1985; 98: 1135–1138.
23 Barnes DW, Reing JE, Amos B. J Biol Chem 1985; 260: 9117–9122.
24 Yatohgo T, Izumi M,. Kashiwagi H, Hayashi M. Cell Struct Funct 1988; 13: 281–292.
25 Izumi M, Yamada KM, Hayashi M. Biochim Biophys Acta 1989; 990: 101–108.
26 Høgåsen K, Mollnes TE, Harboe M. J Biol Chem 1992; 267: 23076–23082.
27 Panetti TS, McKeown–Longo PJ. J Biol Chem 1993; 268: 11988–11993.
28 Ishikawa M, Hayashi M. Biochim Biophys Acta 1992; 1121: 173–177.
29 Kitagaki–Ogawa H, Yatohgo T, Izumi M, Hayashi M, et al. Biochim Biophys Acta 1990; 1033: 49–56.
30 Nakashima N, Miyazaki K, Ishikawa M, Yatohgo T, et al. Biochim Biophys Acta 1992; 1120: 1–10.
31 McGuire EA, Peacock ME, Inhorn RC, Siegel NR, et al. J Biol Chem 1988; 263: 1942–1945.
32 Jenne D, Hille A, Stanley KK, Huttner WB. Eur J Biochem 1989; 185: 391–395.
33 Sato R, Komine Y, Imanaka T, Takano T. J Biol Chem 1990; 265: 21232–21236.
34 Seiffert D, Keeton M, Eguchi Y, Sawdey M, Loskutoff DJ. Proc Natl Acad Sci USA 1991; 88: 9402–9406.
35 Korc–Grodzicki B, Tauber–Finkelstein M, Chain D, Shaltiel S. Biochem Biophys Res Commun 1988; 157: 1131–1138.
36 Korc–Grodzicki B, Chain D, Kreizman T, Shaltiel S. Anal Biochem 1990; 188: 288–294.
37 Cooper S, Pera M. Development 1988; 104: 565–574.
38 Nagano Y, Hamano T, Nakashima N, Ishikawa M, et al. J Biol Chem 1992; 267: 24863–24870.
39 Dahlbäck K, Löfberg H, Dahlbäk B. Acta Derm Venereol 1986; 66: 461–467.
40 Preissner KT, Holzhüter S, Justus C, Müller–Berghaus G. Blood 1989; 74: 1989–1996.
41 Barnes DW, Reing J. J Cell Physiol 1985; 125: 207–214.
42 Yasumitsu H, Seo N, Misugi E, Morita H, et al. In Vitro Cell Dev Biol (in press).
43 Solem M, Helmrich A, Collodi P, Barnes D. Mol Cell Biochem 1991; 100: 141–149.
44 Sanders LC, Wang C–S, Walling LL, Lord EM. Plant Cell 1991; 3: 629–635.
45 Wagner VT, Brian L, Quatrano RS. Proc Natl Acad Sci USA 1992; 89: 3644–3648.
46 Daniel JW, Rusch HP. J Gen Microbiol 1961; 25: 47–59.
47 Miyazaki K, Hamano T, Hayashi M. Exp. Cell Res 1992; 199: 106–110.
48 Miyazaki K, Miyata T, Sawada H, Matuda S, Hayashi M. (in preparation).

Biology of Vitronectins and their Receptors
K.T. Preissner, S. Rosenblatt, C. Kost, J. Wegerhoff and D.F. Mosher, editors

Purification of stabilized native vitronectin

I. Hayashi and D.F. Mosher

Department of Medicine, University of Wisconsin, 1300 University Avenue, Madison, WI 53706 USA

A modified method of Dahlbäck and Podack (1) has been developed to purify native vitronectin from large volumes of human plasma. The modifications simplify the preparation of the 9-20% polyethylene glycol precipitate, minimize use of protease inhibitors, handle labile cysteinyl residues as intrachain disulfides rather than attempting to keep them reduced with glutathione, improve the purity of the final product by a separation on heparin-Sepharose, allow the purification to be done efficiently over a 2-week period, and result in a product that is stable when stored as an ammonium sulfate precipitate.

MATERIALS AND METHODS

Materials. The following chemicals were purchased from Sigma (St. Louis, MO): urea, sodium dodecyl sulfate (SDS), 5,5'-dithionitrobenzoic acid (DTNB), and soybean trypsin inhibitor. DEAE-Sephacel, Blue-Sepharose, heparin-Sepharose, and Sephacryl S200 HR were purchased from Pharmacia LKB Biotechnology Inc. (Uppsala, Sweden). Electrophoresis reagents were obtained from BioRad (Richmond, CA). Citrated fresh frozen human plasma was obtained from Badger Red Cross (Madison, WI) or during therapeutic plasmapheresis of patients with paraproteins.

Method. Plasma, 4.5 or 91, was thawed at 37°C, a crude grade of soybean trypsin inhibitor added to 10 µg/ml, put into one or two 5-1 polypropylene beakers, and taken to a 4°C cold room. Subsequent steps were done in the cold room or in a centrifuge cooled to 4°C. To each stirred beaker containing 4.5-1 plasma, 360 ml of 1 M barium chloride (80 ml/l) was added followed by 438 gm (90 gm/L final concentration) polyethylene glycol 8000. After 1 hr of stirring, the precipitates were removed by centrifugation. The supernatant was put in fresh 5-1 beaker(s), and to each stirred beaker containing 5-1 of supernatant, 550 gm (to give a final total concentration of 200 gm/L) polyethylene glycol was added. Stirring was continued for 15-30 min until the precipitate coalesced and adhered to the side of the beaker. The beakers were left overnight without stirring, during which time the precipitate settled to the bottom of the beaker and formed a gummy brown cake.

Each beaker, containing its adherent cake, was rinsed twice with a small volume of 20% polyethylene glycol in 20 mM Tris, 150 mM sodium chloride, pH 7.4. The cakes were dissolved over 3-4 hr with gentle stirring in 2 1 20 mM sodium phosphate, pH 7.0, containing 1 mM DTNB. This solution was prepared just before use by mixing 1 part of 5 mM DTNB (2 mg/ml) in 50 mM sodium phosphate, pH 8.0, with 4 parts of 12.5 mM sodium phosphate, pH 6.4. As the precipitate dissolved, the solution became bright yellow due to formation of thionitrobenzoate anion.

The solution was clarified by centrifugation and pumped at a rate of 180 ml/hr onto a 5 (i.d.) x 40 cm column of DEAE-Sephacryl equilibrated in 20 mM sodium phosphate, pH 7.0. The column was developed with 100 ml of 1 mM DTNB in 20 mM sodium phosphate, pH 7.0; 1 l of 20 mM sodium phosphate, 0.1 mM EDTA, pH 7.0; and a 4.4 l gradient of sodium chloride, 0 to 500 mM in 20 mM sodium phosphate, 0.1 mM EDTA, pH 7.0. Fractions were assayed for absorbance at 280, 320, and 600 nm; conductivity; fluorescence when irradiated with near ultraviolet light; vitronectin concentration by electroimmunoassay (2); and protein composition by polyacrylamide gel electrophoresis in sodium dodecyl sulfate (SDS-PAGE) (3) and agarose gel electrophoresis (4). Vitronectin was found in fractions containing between 130 and 220 mM sodium chloride (Figure 1). These fractions extended from just prior to the blue ceruloplasmin-containing fractions that absorb at 600 nm to just prior to the fluorescent retinol-binding protein-containing fractions that absorb at 320 nm. Thus, the progress of the separation could be monitored easily by inspection of the column and fractions with white and near ultraviolet light.

Fractions from the DEAE-Sephacryl column with vitronectin concentrations > 500 μg/ml were pooled, and solid ammonium sulfate, 472 gm/l (to 70% saturation), was added with stirring. The precipitate was collected by centrifugation, washed with 70% saturated ammonium sulfate, and stored as a slurry waiting the next steps.

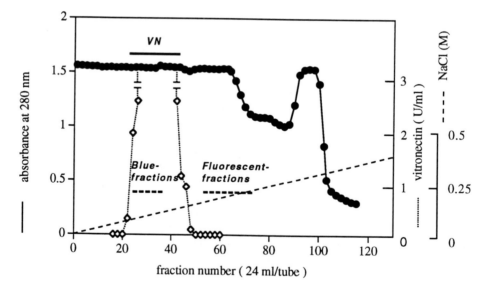

Figure 1. DEAE-Sephacel column chromatography of the 9-20% polyethylene glycol precipitate from human plasma.

The fractions containing celuroplasmin that adsorb at 600 nm and the fractions containing retinol-binding protein that adsorb at 320 nm are indicated as Blue-fractions and Fluorescent-fractions with dashed bars, respectively. The fractions indicated by a solid bar were pooled, and proteins were precipitated by 70% ammonium sulfate. A unit of vitronectin is approximately 100 μg.

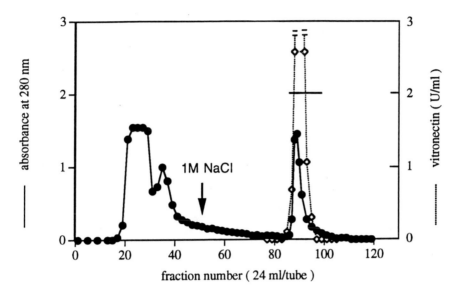

Figure 2. Blue-Sepharose column chromatography of vitronectin fraction from DEAE-Sephacryl column.

The stepwise elution with 1M sodium chloride was started at the fractions indicated by an arrow. Fractions containing vitronectin indicated by a solid bar were collected and pooled. Only traces of vitronectin were present in passthrough fractions measured in other purifications.

The precipitate was collected by centrifugation and dissolved in 50 mM Tris, 150 mM sodium chloride, 0.1 mM EDTA, pH 7.4. Dissolved proteins were dialyzed against 3 changes (4-12 hr each) of this buffer. The solution was clarified by centrifugation and pumped at a rate of 180 ml/hr onto a 5 x 35 cm column of Blue-Sepharose equilibrated with the buffer. The column was developed with 1 l of the buffer and then with 50 mM Tris, 1 M sodium chloride, 0.1 mM EDTA, pH 7.4. Vitronectin eluted with the 1 M salt (Figure 2).

Fractions from the Blue-Sepharose column containing vitronectin were pooled, concentrated by precipitation with ammonium sulfate as described above, and dialyzed against 20 mM Tris, 20 mM sodium chloride, 0.1 mM EDTA, pH 7.4. The solution was clarified by centrifugation and pumped at a rate of 60 ml/hr onto a 2.6 x 20 cm column of heparin-Sepharose equilibrated with the same buffer. The column was developed with 100 ml of starting buffer followed by a 1200 ml gradient of sodium chloride, 20 to 1000 mM in 20 mM Tris, 0.1 mM EDTA, pH 7.4. Vitronectin, along with variable amounts of a 40 kDa degradation product of C3 (5), eluted in fractions containing between 100 and 200 mM sodium chloride (Figure 3). Contaminating albumin (which constituted ~12% of applied protein) did not bind to the column, and other contaminating proteins (a mixture that constituted ~33% of applied protein) bound more strongly and eluted with >200 mM sodium chloride.

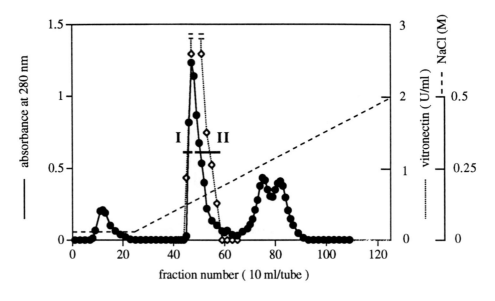

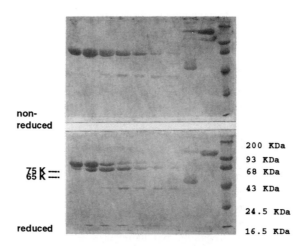

Figure 3. Heparin-Sepharose column chromatography of vitronectin fraction from Blue-Sepharose column.

SDS-polyacrylamide gel electrophoresis of fractions was done using 10% gel under non-reduced (upper panel) and reduced (lower panel) conditions. The right lane (Mr) shows molecular weight marker proteins. Vitronectin-containing fractions indicated by solid bars were pooled (pool I and II) and concentrated by precipitation with 70% ammonium sulfate. Note that pool I contained most 1-chain, 75 kDa vitronectin, whereas later tubes pooled in pool II contained more and more 2-chain form.

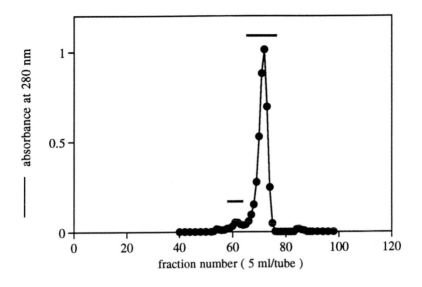

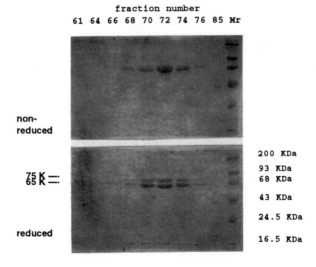

Figure 4. Gel filtration of pool-II on a column of Sepharcryl S-200.

Only the pool II fractions had contaminating 40 kDa protein and required further separation by gel filtration. SDS-polyacrylamide gel electrophoresis of fractions was done using 10% gel under non-reduced (upper panel) and reduced (lower panel) conditions. The right lane (Mr) shows molecular weight marker proteins. Fractions containing vitronectin indicated by a solid bar were pooled and dialyzed against 70% ammonium sulfate to precipitate the protein.

Vitronectin-containing fractions from the heparin-Sepharose column were pooled and concentrated by dialysis against 70% saturated ammonium sulfate. The precipitate was dissolved in 3-4 ml of 20 mM Tris, 150 mM sodium chloride, 0.1 mM EDTA, Ph 7.4, and applied to twin 2.6 x 90 cm columns of Sephacryl S-200 connected in series (so the buffer ran from top to bottom in the first and bottom to top in the second). These columns, pumped at a rate of 30 ml/hr, separated the proteins into 3 fractions: traces of vitronectin oligomers (<5% of applied protein), vitronectin monomer, and the 40 Kda protein (Figure 4). Vitronectin-containing fractions were dialyzed against 70% saturated ammonium sulfate, and the precipitated protein was stored as a slurry (concentration 40-50 mg/ml) at 4°C.

A measured portion and percentage of the slurry was dissolved, dialyzed against Tris-buffered saline (0.15 M sodium chloride, 10 mM Tris, Ph 7.4), and analyzed for protein concentration and purity by SDS-PAGE and agarose gel electrophoresis. The content of free sulfhydryl was assayed with 1 mM DTNB in 20 mM sodium phosphate, Ph 8.0 (6). The amount of incorporated thionitrobenzoate was assessed by spectroscopy after treatment of the protein with 1 mM 2-mercaptoethanol in 3M guanidine. Vitronectin protein concentrations were quantitated in the presence or absence of SDS with a micro protein assay (Pierce, Rockford, IL) using human IgG as a standard.

RESULTS AND DISCUSSION

The purification was carried out over a 2-week period and reproduced by several different laboratory workers. The final product was pure as assessed by SDS-PAGE and agarose gel electrophoresis (5). A key step was heparin affinity chromatography (Figure 3). Most heparin-binding proteins were removed in prior steps, and native vitronectin bound readily to heparin-Sepharose in 20 mM sodium chloride at 4°C and remained bound until the salt concentration was raised to 100-150 mM. Albumin and a number of more strongly binding proteins were thus removed, leaving only a 40 kDa fragment of C3 as a contaminant. The C3 fragment was removed by gel filtration (Figure 4). Inasmuch as the 40 kDa C3 fragment contaminated vitronectin in only the later eluting fractions and these fractions contained mainly the 2-chain form of vitronectin, we routinely do the gel filtration step only on pooled later eluting proteins, giving us 3 pools of purified vitronectin-- mostly 1-chain, mostly 2-chain, and the small amount of mostly 2-chain protein that was multimeric (Table I).

Table 1
Yields from 11 liters of human plasma

Fractions	Amount	Description
Fraction-I (monomer)	32 mg	80% 1-chain
Fraction-II (multimer)	2 mg	68% 2-chain
Fraction-II (monomer)	24 mg	65% 2-chain

The yields in term of percentage of starting plasma vitronectin have been low- 5-to 10%. We have not tried systematically to increase the yields. The yields in term of protein have been substantial--60 to 100 mg per preparation--and could be scaled up further easily.

The final product was native as judged by its inability to bind heparin at physiological ionic strengths and temperature and poor reactivity with the conformationally sensitive 8E6 anti-vitronectin monoclonal antibody (5). Treatment with >2.5M urea induced heparin and 8E6 binding activities and multimerization (5). Native vitronectin, but not urea-treated vitronectin, accelerated the inactivation of thrombin by plasminogen activator inhibitor 1 (7). Conversely, urea-treated, but not native vitronectin, was taken up and degraded by cultured cells (8).

The original intention of the DTNB treatment was to handle free sulfhydryls as mixed disulfides. Native vitronectin purified by the modified method had <0.1 mol/mol free thiol as opposed to the 1 to 4 free thiols found in vitronectin purified in the presence of reduced glutathione (9). There was no evidence, however, of release of thionitrobenzoate anion when the purified protein was incubated with 2-mercaptoethanol and guanidine (5). Unless steric factor interfere, sulfhydryl-disulfide exchange reactions between disulfide compounds with sulfur directly attached to aromatic groups (as in DTNB) and simple alkymercaptans (protein sulfhydryls) should go to completion (6); this apparently occurs during the modified purification of native vitronectin. Despite the possible introduction ofan intrachain disulfide that may not exist in circulating plasma vitronectin, the vitronectin, as described above, has the same conformational lability as vitronectin in plasma when exposed to urea.

Vitronectin purified by the modified methods is stable for >1year when stored as a concentrated slurry in 70% saturated ammonium sulfate and can be shipped inexpensively and safely. We hope, therefore, that our purification will make native vitronectin more accessible for experiments elucidating the interesting differences between native and altered forms.

ACKNOWLEDGEMENTS

These studies were supported by the National Institutes of Health, Bethesda, MD through grant HL29586 and a Fogarty International Fellowship. The purification scheme was developed during a sabbatical by DFM in the Department of Clinical Chemistry, University of Lund, Malmö, Sweden with the kind assistance of Johan Stenflo, Bjorn Dahlbäck, Carl-Bertil Laurell, and Kathy Mosher.

REFERENCES

1.   Dahlbäck B, Podack ER. Biochemistry 1985; 24: 2368-2374.
2.   Laurell CB. Anal Biochem 1966; 15: 45-52.
3.   Laemmli EK. Nature 1970; 227: 680-685.
4.   Jeppson JO, Laurell CB, Franzén B. Clin Chem 1979; 25: 629-638
5.   Bittorf SV, Williams EC, Mosher DF. 1993; J Biol Chem (in press).

6.   Habeeb AFSA. In: Hirs CHW, Timasheff SN, eds. Methods in Enzymology. New York: Academic Press, 1972; 25: 457-464.
7.   Naski MC, Lawrence DA, Mosher DF, Podor TJ, Ginsburg D. J Biol Chem 11993; 268: 12367-12372.
8.   Panetti TS, McKeown-Longo PJ. J Biol Chem 1993; 268: 11988-11993.
9.   Tomasini BR, Mosher DF. Progress Hemostas Thromb 1991; 10: 269-305.

Biology of Vitronectins and their Receptors
K.T. Preissner, S. Rosenblatt, C. Kost, J. Wegerhoff and D.F. Mosher, editors

Multimeric vitronectin: structure and function

S. Hess[a], A. Stockmann[a], W. Völker[b], K. T. Preissner[a]

[a]Max-Planck-Institut, Haemostasis Research Unit, Kerckhoff-Klinik,
D-61231 Bad Nauheim (Germany)

[b]Institute for Arteriosclerosis Research at the University of Münster,
D-48149 Münster (Germany)

INTRODUCTION

Vitronectin (VN) is a multifunctional, adhesive glycoprotein with a Mr of 75 kDa in the single chain form and 65 + 10 kDa in the two chain form due to an endogenous proteolytic cleavage site. VN shows a high degree of conformational flexibility which enables the molecule to respond in multifunctional processes in immune defense, blood coagulation, fibrinolysis and pericellular proteolysis. Exposure of cryptic binding sites which may lead to homo- or hetero-type protein interactions is the prerequisite for certain functions of VN. The intention of our study was to investigate in more detail the properties of the reactive (heparin-binding) form of VN by using chemical denaturation with urea, which resulted unexpectedly in the formation of non-covalently associated VN multimers.

LOCALIZATION OF DIFFERENT FORMS OF VITRONECTIN IN VIVO

VN is present in plasma and serum at a concentration of 200-400 μg/ml and the major site of synthesis is the liver. Also platelets, megakaryocytes, monocytes/ macrophages (for review see [1]) and some tumor cell lines [2] contain an immunological identical protein. The majority of VN found in the circulation is present in the plasma form in a "closed" conformation which serves as a large pool of soluble, latent VN. A minor fraction of circulating VN is in the "opened", heparin-binding, reactive form [3]. This form may either be complexed to other VN molecules to form VN multimers or it is associated with other ligands such as the C5b-9 complex or the thrombin-antithrombin (AT III) complex and thereby serves a scavenger function. Another type of heparin-binding form of VN exists when it is associated with extracellular matrix sites such as in skin or with normal and diseased vessels (for review see [1,4]). This insoluble form of VN functions as adhesion, spreading and migration factor for cells and as a stabilizing component for plasminogen activator inhibitor-1 (PAI-1) (Table 1).

Table 1
Distribution and destinations of vitronectin

Origin/destination	Molecular form	Process/functional role
Liver	Plasma form	Biosynthesis
Circulation	Major: plasma form (closed, latent) Minor: heparin-binding form (opened, reactive):	Plasma pool
	a) multimers	Clearance
	b) complement complexes	Humoral defense
	c) ternary complexes	Haemostasis and clearance
Extravascular spaces (e.g. skin, vessel wall)	Matrix bound, heparin - binding (multimeric?) form	Adhesion, spreading, migration, PAI-1 stabilization

TRANSITION FROM THE "CLOSED" TO THE "OPENED" FORM

Two differently charged regions in VN, a highly acidic portion close to the amino-terminus and a cluster of basic amino acids, the "heparin-binding domain", close to the carboxy-terminus are believed to internally stabilize the whole molecule in the latent, plasma form in which most of the binding sites for other ligands (such as heparin) are cryptic. A transition to a partly opened or totally opened form can be achieved when VN reacts with ligands such as the C5b-9 complex, the thrombin-ATIII-complex, when it is coated to surfaces or when it is treated with denaturants such as 6 M urea. The totally opened form of VN has the potential to self-associate and to form non-covalently associated multimers [5]. On polyacrylamide gel electrophoresis in the *absence* of SDS, plasma VN which was isolated under non-denaturing conditions [6] consists primarily of monomer, dimer and little trimer, whereas treatment of the plasma form with 6 M urea for 1 hr at 37°C shows a multimeric pattern consisting of 3-18 VN molecules with an estimated Mr >1,200 kDa. Gel electrophoresis in the *presence* of SDS under *non-reducing* conditions revealed that urea treatment also resulted in *intra-molecular* disulfide bridge rearrangement resulting in dissociation of the 10 kDa light chain from the 65 kDa heavy chain of two-chain VN (Fig. 1).

It is important to note, that different methods to denature VN yield different types of VN multimers. VN purified according to Yatohgo et al [7] which includes denaturation with urea and *additionally* treatment with reductants resulted in *inter-molecular* rearrangement of disulfide bonds, leading to covalently linked highly *aggregated* multimers. Similarly, SDS-stable aggregated multimers were obtained by boiling plasma VN for 3 min (not shown).

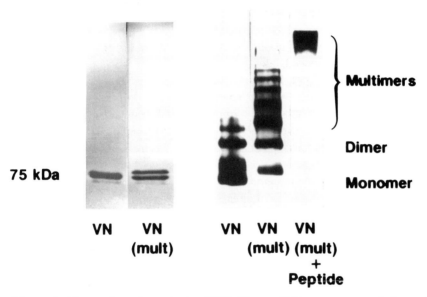

Figure 1. Electrophoretic analysis of VN. Plasma VN or urea-treated plasma VN (= multimeric VN) were subjected to polyacrylamide gel electrophoresis in the presence (left panel) or absence of SDS (right panel) under non-reducing conditions. Additionally, VN was incubated with 40-fold molar excess of VN peptide # 2 for 1 hr at 37°C prior to analysis.

The conformational transition of plasma VN to multimeric VN was analyzed in parallel by (a), a competitive Elisa [8] using 13H1 (a conformation-dependent antibody which only recognizes the opened form of VN, kindly provided by Dr. Declerck, Leuven, Belgium); (b), by gel electrophoresis in the absence of SDS, measuring multimers; and (c), by analyzing direct binding of heparin (Fig. 2). Exposure of cryptic epitopes for 13H1 preceeded multimerization and heparin binding. A critical urea concentration of > 2 M was needed to induce multi-merization. Multimerization was irreversible under native conditions and dependent on the concentration of VN (>100 µg/ml). Quantitative analysis of the data revealed a non-linear, cooperative polymerization.

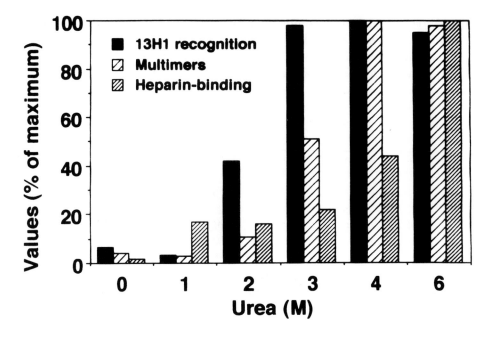

Figure 2. Conformational transition of VN: Correlation between 13H1 recognition, multimer formation and heparin-binding. Purified plasma VN treated for 1 hr with the indicated urea concentrations was subsequently analyzed in parallel by competitive ELISA (monoclonal antibody 13H1 recognition), by non-denaturing gel electrophoresis (multimer formation), and for direct heparin-binding. Values are given as % of maximum from a typical experiment (from [5]).

STABILIZATION OF MULTIMERS BY COVALENT LINKAGES

To further characterize the structure-dependent reactivities of plasma VN and multimeric VN we utilized tissue transglutaminase, an enzyme which is widely distributed in tissues and body fluids and which has been shown to mediate covalent cross-linking between protein chains of VN [9]. In the presence of tissue transglutaminase only minor portions of plasma VN were *inter-molecularly* crosslinked into dimers and some trimers, whereas multimeric VN was covalently crosslinked into *SDS-stable multimers*. The same results were obtained with the chemical crosslinker dimethyl suberimidate. These findings indicate, that the reactive, multimeric form of VN is by far more suceptible to covalent linkages, which may lead to stable high molecular weight VN complexes either in the circulation, on the cell surface or in the extracellular matrix.

INVOLVEMENT OF THE HEPARIN-BINDING DOMAIN IN THE MULTIMERIZATION PROCESS

Since multimerization of plasma VN by urea resulted in exposure of the heparin-binding site we wanted to investigate whether this domain is involved in the multimerization process. Synthetic VN peptide # 2 (residues 348-361) from the core region of the heparin-binding site was reacted with plasma VN. Gel electrophoresis in the absence of SDS showed that peptide # 2 at 10-40-fold molar excess (50-200 μM) was able to induce multimerization of plasma VN (not shown) and to induce a further multimerization of multimeric VN (5 μM) (Fig. 1), whereas a synthetic peptide representing a flanking region of this domain (residues 371-383) was ineffective. Additionally, Western-blot analysis showed that the peptide itself was incorporated into the multimers, indicating that this highly basic peptide may react with a complementary acidic domain at the amino-terminus, leading to destabilization of the VN molecule. To further determine the role of the heparin-binding domain in the multimerization process, this domain was removed from intact VN by proteolysis of plasma VN with plasmin. Limited proteolysis resulted initially in the formation of a 61 kDa VN fragment, which lacked the heparin-binding-domain and which was unable to multimerize upon urea treatment. Likewise, in a time-dependent fashion, plasmin-proteolysis prevented urea-induced formation of VN multimers, but resulted in the formation of fragments. These data show that the heparin-binding site in VN not only functions to bind glycosaminoglycans but that this motif is also involved in the self-assembly of VN multimers. Moreover, electron microscopy of VN multimers revealed specimens of 15-28 nm in diameter with an overall globular shape [5]. We propose at least two populations of "opened" form of VN molecules in multimers: those which form the core region of these "particles" by interacting with complementary binding sites and those which provide exposed binding sites for other ligands such as heparin.

BINDING OF SOLUBLE PLASMA VITRONECTIN AND MULTIMERIC VITRONECTIN TO ADHERENT CELLS

VN is commonly used as an insoluble substrate for adhesion studies, and the integrins αvβ1, αvβ3 and αvβ5 on various cell types have been shown to recognize *immobilized* VN (for review see [1]). We wanted to elucidate the binding characteristics of *soluble* plasma and multimeric VN to adherent porcine endothelial cells (EC) to mimic the interaction of fluid phase VN with luminal receptor sites and to determine the fate of both forms of VN. Confluent porcine EC monolayers were incubated at 37°C for up to 6 hr with either radiolabeled plasma or multimeric VN at 220 ng/ml. While plasma VN showed only moderate binding to the cells, the multimeric form of VN demonstrated after 6 hr incubation 4 - 5 fold more efficient association with the cell layer, indicating that the multimeric form has additional sites exposed or is avidly recognized due to its multivalent nature. These conclusions were supported by cross-competition experiments demonstrating that only excess multimeric VN competed binding of radiolabeled multimers.

Moreover, in competition assays it was established that heparin at 50 and 500 μg/ml did not compete for the binding of the plasma form of VN to cell layers, whereas there was a concentration-dependent inhibition of binding for the multimeric form of VN. Also heparan sulfate and VN peptide # 2 were effective in competing for the multimeric form of VN, while there was no effect of chondroitin sulfate or other heparin-binding proteins such as antithrombin III or fibrinogen. RGD peptides did not compete for binding of the plasma or the multimeric form of VN, indicating that integrins are either not present or not functional at the *luminal* side of cultured porcine EC or that the RGD-domain in soluble VN is not recognized by adherent cells. Previous studies with human umbilical vein EC in suspension demonstrated an RGD-dependent binding of plasma VN, indicating that integrins have the capability to bind soluble VN [10]. Scalise-Panetti and McKeown-Longo report that the initial binding of soluble VN (which was isolated by the authors according to Yatohgo et al [7], thus representing an highly aggregated form of VN) to fibroblasts in monolayer could not be competed for by RGD, whereas integrin αvß5 was involved in internalization and degradation of this form of VN [11, and this issue].

In order to examine the binding capacity of matrix, cells were gently removed by detergent treatment, and radiolabelled multimeric VN (3 nM) was incubated with the isolated matrix. The isolated matrix showed 3-4 fold higher capacity to bind multimeric VN, as compared to cell binding (Fig. 3A). Binding was specific and could be drastically reduced by 4 μM heparin. Plasma VN showed hardly any binding to isolated matrix and this binding could not be competed for by heparin (Hess and Preissner, in preparation).

Uptake and translocation of only the multimeric form of VN could be demonstrated when binding was compared at 37°C and at 4°C (Fig. 3B). After 6 hr incubation, binding was discriminated between the cell phase and the underlying matrix. Cells bound multimeric VN at 37°C and at 4°C, but matrix associated VN was only found when cells were metabolically active indicating that translocation had occured.

To corroborate the binding data with radiolabelled ligands we used ultrastructural analysis to study the fate of multimeric VN. Plasma VN was coupled to colloidal gold particles (17 nm) which yielded VN-gold conjugates that resembled the multimeric form in size and functional properties. Platinum/carbon replicas of subconfluent EC showed VN-gold conjugates on the surface of EC in close association with proteoglycans, which were stained with cuprolinic blue and to a

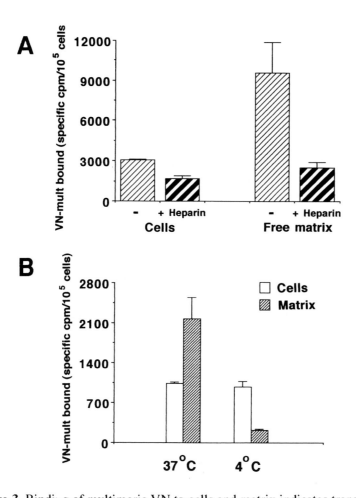

Figure 3. Bindi:.g of multimeric VN to cells and matrix indicates translocation. Confluent porcine endothelial cells in 48-well dishes were incubated overnight in serum-free medium. **A.** Cell monolayers or isolated matrix, which was obtained by gently detaching cells with 0.2 M NH_4OH/1% Triton X, were incubated at 37°C for 4 hr with 3 nM ^{125}I-multimeric VN in albumin-containing serum-free medium in the absence or presence of 4 μM heparin. Binding was determined by solubilizing the material with 1 M NaOH, followed by counting. **B.** Cell monolayers were incubated at 37°C or at 4°C with 3 nM ^{125}I-multimeric VN in albumin-containing serum-free medium. After 6 hr incubation the cell layer was washed three times and the cell fraction was solubilized with 0.2 M NH_4OH/1% Triton X, while the remaining matrix was detached with 1 M NaOH and both fractions were counted. Values (mean +/- SD, n = 3) are corrected for non-specific binding which was determined in the presence of 200-fold molar excess of unlabeled multimeric VN.

higher degree at exposed matrix. In the presence of heparin, VN-gold decoration to cells and to free matrix sites was drastically reduced or even inhibited. Incubation of EC with VN-gold conjugates for 1 hr at 4°C and post-incubation for 2 hr at 37°C resulted in disappearance of gold marker from the cell surface, indicating that uptake has taken place. After 3 hr incubation the gold marker was seen in lysosomes and also at the basolateral side of the cells in association with matrix structures [12].

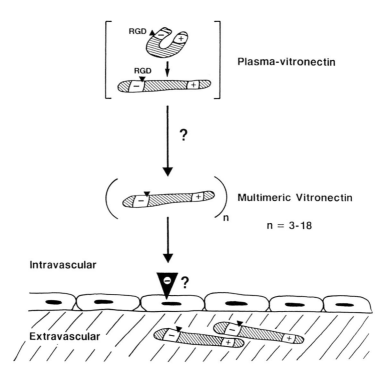

Figure 4. Hypothetical model of the conformational transition of plasma VN to multimeric VN which may result in clearance of this highly reactive form from the circulation and subsequent deposition into the subendothelial matrix.

CONCLUSIONS

The multimeric form of VN which represents the prototype conformer of the reactive heparin-binding form not only binds glycosaminoglycans in the circulation, but also interacts via its heparin-binding domain with presumably poly-anionic structures at cell surfaces resulting in deposition of multimeric VN in the subendothelium or partial proteolysis in lysosomal compartments (Fig. 4). These data are in agreement with the studies of de Boer et al [13] who showed that VN, when associated with thrombin-ATIII also binds via the heparin-binding domain to endothelial cells, followed by translocation of the complex. The different reactivities of plasma VN and the heparin-binding form of VN with endothelial cells appear to be crucial for maintaining only a minor fraction of reactive VN in the circulation and not allowing VN or other adhesion proteins to simply attach to the intact endothelium. Moreover translocation of multimeric VN serves an important role in the normal and diseased vessel wall and may be relevant for cellular responses such as migration and proliferation and maintanence of PAI-1 stabilization.

REFERENCES

1   Preissner KT. Annu Rev Cell Biol 1991; 7: 275-310.
2   Yasumitsu H, Seo N, Migugi E, Morita H et al. In Vitro Cell Dev Biol 1993; 29A: 403-407.
3   Izumi M, Yamada KM, Hayashi M. BBA 1989; 990: 101-108.
4   Tomasini B, Mosher DF. Prog Hemostasis Thromb 1990; 10: 269-305.
5   Stockmann A, Hess S, Declerck P, Timpl R, et al. J Biol Chem 1993 (in press).
6   Preissner KT, Wassmuth R, Müller-Berghaus G. Biochem J 1985; 231: 349-355.
7   Yatohgo T, Izumi M, Kashiwagi H, Hayashi M. Cell Struct Funct 1988; 13: 281-292.
8   Tomasini B, Mosher DF. 1988; Blood; 72: 903-912.
9   Sane DC, Moser TL, Pippen AMM, Parker CJ, et al. Biochem Biophys Res Commun 1988; 157: 115-120.
10   Preissner KT, Anders E, Grulich-Henn J, Müller-Berghaus G. Blood 1988; 71: 1581-1589.
11   Scalise Panetti T, McKeown-Longo P. J Biol Chem 1993; 268: 11492-11495.
12   Völker W, Hess S, Vischer P, Preissner KT. J Histochem Cytochem 1993 (in press).
13   de Boer HC, Preissner KT, Bouma BN, de Groot PG. J Biol Chem 1992; 267: 2264-2268.

© Elsevier Science Publishers B.V. All rights reserved.
Biology of Vitronectins and their Receptors
K.T. Preissner, S. Rosenblatt, C. Kost, J. Wegerhoff and D.F. Mosher, editors

EXPRESSION OF RECOMBINANT HUMAN VITRONECTIN USING THE BACULOVIRAL SYSTEM [1]

D. C. Sane [2] and Y. Zhao

Duke University Medical Center, Durham NC   27710   USA

INTRODUCTION

The gene structure [1], cDNA sequence [2,3], and domain map [4] of VN [3] have been published. Considerable preliminary structural information has been derived about binding sites for the numerous ligands of VN. Nevertheless, there are several unresolved issues concerning VN structure and function. The PAI-1 binding site of VN remains controversial, with evidence supporting the somatomedin B domain (residues 1-44, [5]), residues 115-121 [6] and the heparin binding domain [7,8]. Furthermore, the functional significance of the multiple post translational modifications of VN (Table 1) remains almost totally obscure. To address these and other structural questions, we have expressed wild type and mutant forms of VN in insect cells using the baculoviral system.

Table 1
Post-translational modifications of VN

Modification	Residues modified	Possible functions
N-linked glycosylation	N67,N150,N223	unknown
Tyrosine sulfation	Y56,Y59	Site of thrombin interaction; enhances intramolecular binding to GAG
Serine phosphorylation	S378	Decreases heparin affinity
Proteolysis	R379	Creates 2-chain form with increased heparin affinity

OVERVIEW OF THE BACULOVIRAL SYSTEM

Autographa californica nuclear polyhedrosis virus (AcNPV or "wild type baculovirus") is the prototype virus from the family *Baculoviridae* , which infect insect cell hosts including Sf9 cells used in our studies. During viral replication, a 29-kDa polyhedron protein, which protects the viral genome, is produced in large quantity. Baculoviral expression of a foreign protein requires construction of a recombinant baculovirus in which the polyhedron gene is replaced with the gene of interest, so that expression is driven by the strong polyhedron promoter. Because the baculovirus genome is very large, the cDNA of interest cannot be directly inserted into the viral DNA. Instead, the first step is to create a transfer vector in a plasmid containing the gene of interest attached to the polyhedron promoter and flanked by regions of viral DNA. The recombinant baculovirus is produced by co-transfecting Sf9 cells with wild type linearized viral DNA and the transfer vector. The flanking viral DNA sequences in the transfer vector promote homologous recombination, producing the desired integration of the cDNA of interest. Cells infected with the wild type virus contain clusters of the polyhedron protein and are recognizable using light microscopy as "occlusion positive". Cells infected with the recombinant baculovirus, which lack the polyhedron protein are "occlusion negative" but are rather difficult to identify by light microscopy. To aid in identification of the recombinant baculovirus, we have used transfer vectors (pBlueBac and pBlueBacHis, Invitrogen) that allow co-expression of β-galactosidase. The recombinant baculovirus is isolated from blue, occlusion negative plaques by plaque purification.

ADVANTAGES AND DISADVANTAGES OF BACULOVIRAL EXPRESSION

There are several advantages of the baculovirus system over expression in bacterial, yeast, and mammalian systems. First, high levels (up to 500 mg/l) of functional protein are expressed. Second, the Sf9 cells are relatively easy to grow and do not require a CO_2 incubator. Third, most mammalian proteins expressed in insect cells have undergone similar post-translational modifications to those in mammalian cells. Recombinant proteins expressed in Sf9 insect cells are proteolytically processed (including correct cleavage of mammalian signal peptide), secreted, phosphorylated, N-glycosylated, O-glycosylated, myristylated, and palmitylated. Phosphorylation has been reported at the correct serine and tyrosine residues [9,10] . A cAMP-dependent protein kinase has been reported in Sf9 cells and recognizes the correct target residue of synapsin IIa [11]. Nevertheless, other post-translational modifications may not occur in individual proteins; human complement protein C1s, for example is not β-hydroxylated and is incompletely glycosylated when expressed in insect cells [12]. In addition, high mannose type N-linked carbohydrates are retained on some expressed proteins [13]. Tyrosine sulfation had not previously been studied in this system (see below). We have found the process of plaque purification to be relatively slow, with each round requiring approximately 6 days, and 3-4 rounds of purification needed to rid the recombinant virus of contaminating wild type virus.

CONSTRUCTION OF THE TRANSFER VECTOR, pBLUEBAC-HVN

A cDNA for human VN (pGEMhVN) was kindly provided by Dr. Erkki Ruoslahti, La Jolla Cancer Research Foundation. A start codon as well as BamHI and NheI restriction sites were added to pGEMhVN using PCR. The PCR product was digested with BamHI and subcloned into M13, where the entire PCR product was sequenced to exclude PCR-induced erros. The M13 insert was then digested with NheI and cloned into the unique NheI site of pBlueBac, producing the transfer vector, pBlueBac-HVN.

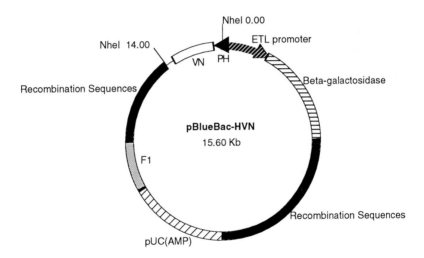

FIGURE 1 pBlueBac-HVN. The transfer vector used for the production of the recombinant baculovirus expressing the nonfusion rVN is shown. The cDNA for VN is cloned into the unique NheI site of pBlueBac, under the control of the polyhedron promoter (PH). The expression of β-galactosidase is driven by the ETL promoter. The recombination sequences flanking the cDNA for VN and β-galactosidase promote allelic replacement of the polyhedron protein in the wild type baculovirus genome. The transfer vector is large (15.6 kb), reducing transfection efficiency. Smaller transfer vectors (eg. pBlueBac II and III) with additional cloning sites are now available.

CONSTRUCTION AND PLAQUE PURIFICATION OF RECOMBINANT BACULOVIRUS

Sf9 cells were co-transfected with linearized wild type baculovirus DNA and pBlueBac-HVN with the cationic liposome method. Infected cells were overlaid with Sea plaque agarose in Grace's medium containing X-gal. At 4-7 days, blue polyhedron negative plaques were picked and subjected to two additional rounds of plaque purification. A high titer recombinant virus stock was then established.

EXPRESSION AND CHARACTERIZATION OF WILD TYPE RECOMBINANT VN AS A NONFUSION PROTEIN

Sf9 cells were grown in serum free medium (EX-CELL 401™, JRH Biosciences) to eliminate contamination of rVN with bovine VN from FBS. Western blotting demonstrated a single band at ~ 70 kDa. Maximum rVN levels of 33.9 mg/ml were obtained at 48 hrs post-infection. Cells were metabolically labeled with [^{32}P]-Pi, then rVN was immunoprecipiated with anti-VN (Quidel). The [^{32}P]-labeled rVN was then subjected to phosphoaminoacid analysis, revealing a phosphoserine residue.

Plasma VN is sulfated on tyrosine residues 56 and 59. Based upon the similarity of the sulfated site in VN with that in other thrombin-binding sulfated proteins [14], we postulated that Y(56)-SO4 and Y(59)-SO4 promote the interactions of VN with thrombin. In order to study the functional effects of sulfation we first had to determine whether tyrosine sulfation occurred in the insect cell host. In the first analysis, we performed TPST assays on Sf9 cell homogenates since the presence of TPST activity is a pre-requisite for tyrosine sulfation. TPST is a Golgi membrane enzyme that catalyzes the following reaction:

$$\text{Protein-Tyr-OH + PAPS} \xrightarrow{\text{TPST}} \text{Protein-Tyr-SO3- + PAP,}$$

where PAPS is the sulfate donor, 3' phosphoadenosine, 5'phosphosulfate and PAP is 3', 5' adenosine diphosphate. TPST had not previously been demonstrated in the Sf9 insect cells . The TPST assay of Rens-Domiano and Roth [15] and Niehrs and Huttner [16] was used with minor modifications. The insect cells homogenates had a specific activity of 0.340 pmol EAY-S04 product formed/min/mg. This compares with 0.405 pmol/min/mg for homogenates of human platelets and 0.472 pmol/min/mg for bovine liver microsomes [17]. Thus, TPST activity is present in Sf9 insect cells demonstrating the potential for tyrosine sulfation of recombinant vitronectin expressed in these cells.

In the second assay, Sf9 cells were metabolically labeled with [^{35}S]-SO4, then lysed. An immunoprecipitation with anti-VN revealed that rVN was sulfated by the Sf9 cells, but this assay does not differentiate between sulfation of tyrosine or carbohydrate residues. Treatment of the [35S]-SO4 labeled rVN with N-glycanase reduced its apparent Mr from 70 kDa to 60 kDa, without diminishing the sulfate label. This experiment demonstrates the presence of N-linked oligosaccharides in rVN, and is consistent with sulfation of tyrosine residues rather than carbohydrates (no O-linked oligosaccharides could be detected using O-glycanase, Genzyme). To further verify the presence of tyrosine sulfation, [35S]-SO4 labeled rVN was alkaline hydrolyzed with Ba(OH)2, then subjected to 2-D electrophoresis. Autoradiography revealed a spot that co-migrated with tyrosine sulfate standard.

These studies verify the presence of tyrosine sulfate in rVN expressed in Sf9 cells, and represent the first time this modification has been demonstrated in this system. Further studies are underway to determine whether Y56 and Y59 are the sulfated residues . A recombinant baculovirus with Y56 and Y59 mutated to phenylalanines has been constructed to examine the significance of this modification.

Recombinant VN expressed in Sf9 cells possessed many of the same modifications of plasma VN as previously described [18]. Interestingly, rVN was expressed as a single chain form, despite the presence of a threonine at residue 381 in the cDNA used for expression. This suggests that Sf9 cells lack the protease responsible for this cleavage. The rVN retained cell and PAI-1 binding capacity. The baculoviral system appears to be useful for further structure-function studies of VN.

EXPRESSION OF WILD TYPE AND MUTANT RECOMBINANT VN AS FUSION PROTEINS

The wild type recombinant VN expressed as a nonfusion protein was easily purified using mono Q anion exchange chromatography. Because the deletion of significant protions of the VN molecule could alter the binding to mono Q, mutant forms of VN were expressed as fusion proteins to facilitate purification. A 1401 bp DNA fragment with an new unique Bam H1 site containing the whole sequence corresponding to VN was created using the polymerase chain reaction and subcloned into M13. Oligonucleotide-directed mutagenesis was performed using T7-GEN to produce RAD-rVN, RGE-rVN, ΔH-rVN (deletion of residues 343-376) and ΔS-rVN (deletion of residues 1-44). The mutated VN cDNAs and wild type VN cDNA were purified, sequenced, and cloned into the baculovirus transfer vector pBlueBacHis B, which contains the polyhistidine affinity tag coding sequence and enterokinase cleavage sequence located just 5' to the polylinker insertion site (Figure 2). The recombinant baculoviruses were produced and plaque-purified as before.

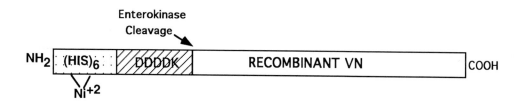

FIGURE 2. Recombinant VN fusion protein, expressed in cells infected by a recombinant baculovirus constructed using the transfer vector pBlueBacHisB. The polyhistidine tag endows the fusion protein with affinity for Ni^{+2}. An enterokinase cleavage sequence (DDDDK) is aminoterminal to rVN, allowing purification of the protein of interest.

Purification of Recombinant VN fusion protein.
The purification scheme used for the fusion proteins has been described in detail [19]. Briefly, Sf9 cells were harvested at 72 hours post-infection, sonicated, then incubated with Ni^{2+}-charged Sepharose resin. The resin was packed into a column, and washed sequentially with Buffers A, B, C, and D (Buffer A: 8M urea, 20 mM Na phosphate, pH 7.8, 500 mM NaCl; Buffer B: 8M urea, 20 mM Na phosphate, pH 5.9, 500 mM NaCl; Buffer C: Buffer B with 5 mM imidazole; Buffer D: Buffer B with 20 mM imidazole). Recombinant VN was eluted using 8M urea, 20 mM Na phosphate, pH 6.3, 500 mM NaCl, 300 mM imidazole. Recombinant VN-containing fractions were pooled and dialysed against 50 mM , Tris·HCl, pH 7.8, 1mM EDTA, 1mM DTT.

Cell attachment studies.
Wild-type, RAD-rVN, RGE-rVN, ΔH-rVN and ΔS-rVN were coated on microtiter wells and the ability of EAhy.926 cells to attach and spread was measured. EAhy.926 cells bound equally well to wild type, ΔH-rVN and ΔS-rVN, but binding to RAD-rVN and RGE-rVN was inhibited by more than 90%. These findings demonstrate the critical role of the RGD sequence of VN in cell attachment. Under the assay conditions there was no evidence of a secondary cell binding site in the heparin or somatomedin B domains. Cherny et al. have also found that the RGD sequence of VN is critical for its binding to platelets, HUVEC, and Panc-1 cells [20].

CHOICE OF INSECT CELL HOSTS FOR VN EXPRESSION.

The expression of VN as fusion and nonfusion proteins has been performed using Sf9 cells, derived from *Spodoptera frugiperda*. Recent reports indicate that higher levels of expression can be achieved using cells derived from *Trichoplusia ni* , the cabbage looper. These cells also grow better in serum free medium.

Unfortunately, these cells proved to be unsuitable for VN expression, since a heparin-binding 125 kDa protein recognized by a monoclonal anti-VN antibody was discovered in both egg cells (High Five™) and midgut cells (MG1™) derived from this organism. Sf9 cells did not contain this VN-like protein. The discovery of a VN-like protein in insect cells adds to the growing list of organisms with a protein homologous to VN [21-24] and points to the need for careful selection of host cells for expression of recombinant VN.

FOOTNOTES

1   Supported by a James A. Shannon Director's Award, a FIRST Award (R29HL46993) and an American Society of Hematology Scholar's Award.
2   Current address: The Bowman Gray School of Medicine of Wake Forest University, Section of Cardiology, Medical Center Boulevard, Winston-Salem, NC 27157 USA
3.   Abbreviations: (r)VN: (recombinant) vitronectin; PAI-1: plasminogen activator inhibitor type-1; SmB: somatomedin B domain; GAG: glycosaminoglycan binding domain; Sf9: Insect cells from *Spodoptera frugiperda* ; PCR: polymerase chain reaction; TPST: tyrosylprotein sulfotransferase;ΔH-rVN: heparin binding domain deletion mutant lacking residues 343-376; ΔS-rVN: SmB deletion mutant lacking residues 1-44.

REFERENCES

1   Jenne D, Stanley KK. Biochemistry 1987; 26: 6735-6742.
2   Suzuki S, Oldberg Å, Hayman EG, Pierschbacher MD, Ruoslahti E. EMBO J 1985; 4: 2519-2524.
3   Jenne D, Stanley KK. EMBO J 1985; 4: 3153-3157.
4   Suzuki S, Pierschbacher MD, Hayman EG, Nguyen K, Öhgren Y, Ruoslahti E. J Biol Chem 1984; 259: 15307-15314.
5   Seiffert D, Loskutoff D J Biol Chem 1991; 266: 2824-2830.
6   Mimuro J, Muramatsu S, Kurano Y, Uchida Y, Ikadai H, Watanabe S, Sakata Y. Biochemistry 1993; 32: 2314-2320.
7   Kost C, Stüber W, Ehrlich HJ, Pannekoek H, Preissner KT J Biol Chem 1992; 267: 12098-12105.
8   Gechtman Z, Sharma R, Kreizman T, Fridkin M, Shaltiel S. FEBS Lett 1993; 315: 293-297.
9   Marais RM, Hsuan JJ, McGuigan C, Wynne J, Treisman R. EMBO J 1992: 11: 97-105.
10   Frappier L, O'Donnell M. J Biol Chem 1991; 266: 7819-7826.
11   Siow YL, Chilcote TJ, Benfenati F, Greengard P, Thiel G. Biochemistry 1992; 31: 4268-4275.

12	Luo C, Thielens NM, Gagnon J, Gal P, Sarvari M, Tseng Y, Tosi M, Zavodsky P, Arlaud GJ, Schumaker VN. Biochemistry 1992; 31: 4254-4262.
13	Chen WY, Shen QX, Bahl OP. J Biol Chem 1991; 266: 4081-4087.
14	Hortin GL. Blood 1990; 76: 946-952.
15	Rens-Domiano S, Roth JA. J Biol Chem 1989; 264: 899-905.
16	Niehrs C, Huttner WB. EMBO J 1990; 9:35-42.
17	Sane DC, Baker MS. Thromb Haemostas 1993; 69: 272-275.
18	Zhao Y, Sane DC. Arch Biochem Biophys August 1993 (in press).
19	Zhao Y, Sane DC. Biochem Biophys Res Commun 1993; 192: 575-582.
20	Cherny RC, Honan MA, Thiagarajan P. J Biol Chem 1993; 268: 9725-9729.
21	Nakashima N, Miyazaki K, Ishikawa M, Yatohgo T, Ogawa H, Uchibori H, Matsumoto I, Seno N, Hayashi M. (1992) Biochim Biophys Acta 1120: 1-10.
22	Miyazaki K, Hamano T, Hayashi M. Exp. Cell Res. 1992; 199: 106-110.
23	Wagner VT, Brian L, Quatrano RS Proc. Natl. Acad. Sci. USA 1992; 89: 3644-3648.
24	Sanders LC, Wang C-S, Walling LL, Lord E M Plant Cell 1991; 3: 629-635.

Biology of Vitronectins and their Receptors
K.T. Preissner, S. Rosenblatt, C. Kost, J. Wegerhoff and D.F. Mosher, editors

Vitronectin–Integrin Interactions: Analysis by Site–directed Mutations

R.C. Cherny and P. Thiagarajan

Department of Medicine, University of Washington, Seattle WA, 98195

INTRODUCTION

Vitronectin is a major cell adhesion glycoprotein found in plasma and in the subendothelial matrix (1-2). Many cell adhesion molecules, including VN, exhibit cell adhesion properties by virtue of the tripeptide sequence, Arg-Gly-Asp (RGD), in the cell binding domain (3). The RGD-containing cell adhesion molecules interact with heterodimeric integral membrane glycoproteins --termed integrins, consisting of noncovalently linked α and β -chains (4). Three distinct integrin receptors of varying specificity are known to interact with VN, namely $\alpha_{IIb}\beta_3$ expressed on the surface of stimulated platelets (5,6), $\alpha_v\beta_3$ present on endothelial cells (7,8) and $\alpha_v\beta_5$ identified on several pancreatic and lung carcinoma cell lines (9,10). Despite the fact all these receptors interact with the RGD sequence, they are still able to distinguish among the various ligands. Thus, fibronectin receptor does not interact with vitronectin. The mechanism of receptor specificity remains unclear. One hypothesis suggests that amino acid sequences adjacent to the RGD site confer a specific secondary structure thereby imparting specificity. A second explanation for the receptor specificity despite a common RGD site in each ligand, is that there is a second binding site unique to each ligand. The goal of these experiments was to determine, by direct means, what role RGD in VN plays in determining integrin receptor specificity. Additionally, we sought to identify other independent binding sites that determine integrin specificity. We performed site-directed mutagenesis of the RGD sequence in the VN cDNA and expressed and isolated the wild type and mutant proteins from a mammalian expression system. We also generated fragments of vitronectin by formic acid digestion and by limited proteolysis by plasmin. Interactions of the recombinant VNs and the proteolytic fragments were tested in cell adhesion assay.

EXPERIMENTAL PROCEDURES

Specific oligonucleotides for DNA sequencing, site-directed mutagenesis and the polymerase chain reaction were synthesized using a Applied Biosystems 380B DNA Synthesizer. $Na_2{}^{51}CrO_4$ was obtained from Amersham (Arlington Heights, IL). Standard laboratory reagents were purchased from Sigma Chemical Company (St. Louis, MO).

Cell lines: MG63 osteogenic sarcoma cells and Panc-1 pancreatic carcinoma cells were purchased from American Tissue Culture Collection (Rockville, MD). These cell lines were maintained in Dulbecco's modified Eagle's medium supplemented with 5% fetal calf serum (Hyclone, Utah), 2mM L-glutamine, 50 ug/ml penicillin, 100 ug/ml neomycin and 50 ug/ml streptomycin. Human umbilical vein endothelial cells (HUVEC) were grown as described before (11).

Expression of Recombinant Vitronectin: BHK cell lines were transfected with mammalian expression vector pZEM229R provided by Dr. Eileen Mulvihill of ZymoGenetics (Seattle, WA), containing the cDNA for vitronectin and the stable transfectants of BHK cells (BHK-VN) were selected by the addition of 1 uM MTX. Recombinant vitronectin was isolated from serum-free medium as decribed in detail before (11).

Site-directed mutagenesis: Single point mutations in the wild-type cDNA of VN were introduced according to the methods of Vandeyar and co-workers (12) using the United States Biochemical Corp. T7GEN^{TM} *In Vitro* Mutagenesis Kit, resulting in a substitution of Asp^{47} with Glu^{47} (RGE-VN), Gly^{46} to Ala^{46} (RAD-VN), and Arg^{45} to Leu^{45} (LGD-VN). The BHK cells were transfected, selected and serum free tissue culture supernatants were collected from Cell Factory cultures and the recombinant vitronectin was isolated by Heparin-agarose chromatography (13).

VN from human plasma was also isolated by the same procedure. Following isolation, the recombinant VNs and plasma-derived VN were reduced and S-carboxy methylated. Protein concentrations were determined as described before (9).

Preparation of vitronectin fragments: Ten mg of plasma-derived vitronectin is incubated in 70% formic acid for 16 hours at 37^OC. Following the cleavage of Asp^{217} and Pro^{218}, the vitronectin fragments were lyophilized and subjected heparin-agarose chromatography and the non-binding fraction was isolated. In seperate experiment 10 mg of plasma-derived vitronectin was incubated with plasmin (0.1 units) for 4 hours at 37^OC (14) and subjected to heparin-agarose chromatography as above.

Adhesion Assays: Substrates for adhesion studies were prepared in 96 well polystyrene multiwell microtiter plates (Nunc, Denmark). A 100 ul solution containing various amounts of either plasma VN, its fragment or recombinant VN in 10 mM Tris (pH 7.4), 0.15 M NaCl was added to the wells and incubated overnight at 4^OC. After incubation, the solution was aspirated from the wells and blocked with 3% bovine serum albumin for 2 hours at room temperature prior to the addition of ^{51}Cr-labelled cells or platelets. Control wells were filled with bovine serum albumin. Platelets were isolated from peripheral blood by differential centrifugation and labelled with ^{51}Cr as described before (5). Confluent 75 cm^2 flasks of HUVEC or Panc-1 cells were washed in D-PBS. Cells were harvested using trypsin digestion. The cells were resuspended in serum-containing medium for 15 minutes and then centrifuged 1,000 x g for 4 minutes. The cell pellet was resuspended in 300-500 ul of medium and labeled with 100 uCi $Na_2^{51}CrO_4$ for 1 hour. The cells were then washed thrice in serum-free medium and aliquots of 100 ul were added to each well and incubated for 1 hour at 37^OC for HUVEC and Panc-1 cells and for 20 minutes at room temperature for thrombin-activated platelets. The degree of binding was determined by incubating the wells twice with 2% SDS and the extracts were combined and the ^{51}Cr contents were determined.

RESULTS

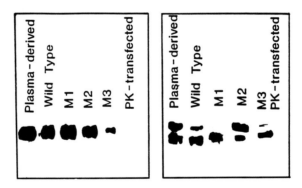

Figure 1. Immunoblot of Vitronectin. Serum-free medium from transfected BHK cells were subjected to SDS-PAGE under reducing and non-reducing conditions, transferred to nitrocellulose paper and probed with a polyclonal antibody to vitronectin.

Expression of Recombinant Vitronectin: Tranfection of BHK-570 cell line with pZEM229R containing the cDNA for VN and its mutants resulted in several cell lines secreting recombinant VN into the tissue culture medium. Despite amino acid substitutions at the critical cell binding motif, mutant VN molecules were readily synthesized, processed and secreted similar to wild-type VN. The nontransfected parental BHK cell line and BHK cells transfected in an identical fashion with the cDNA for prekallikrein did not produce detectable VN (Figure 1). As with plasma-derived VN, rVN

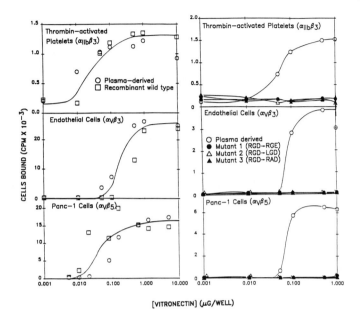

Figure 2: Adhesion of cells to vitronectin: Polystyrene plates were coated with increasing amount of vitronectin. ^{51}Cr-labelled thrombin activted platelets, endothelial cell or Panc-1 cells were added. Adhesion was determined by residual radioactivity after washing.

present in the supernatant of transfected BHK cells migrated both as a single-chain form (75 kDa) and a two-chain form (65 and 10 kDa) linked by a disulfide bond in SDS-PAGE.

Interaction of vitronectin receptors with recombinant vitronectin: Cell adhesion function for isolated recombinant wild-type VN was compared directly to that of plasma-derived VN. Figure 1 shows closely overlapping dose-response curves using thrombin activated platelets ($\alpha_{IIb}\beta_3$), HUVEC ($\alpha_v\beta_3$) and Panc-1 cells ($\alpha_v\beta_5$). These experiments show that recombinant wild-type VN synthesized from BHK cells behaves similarly to plasma-derived VN for all three integrins tested and that this mammalian expression system represents a viable approach to the expression of recombinant VN forms for further structure/function analysis. The recombinant mutant VNs were tested for cell adhesion and compared to plasma-derived VN. Figure 1 demonstrates no appreciable binding regardless of the integrin tested. Even at 10 ug of substrate per well with, well above the saturating amounts, adhesion binding of activated platelets or cells was no greater than background levels. It is possibile that the decresed cell adhesion to the mutant proteins was due to defective adhesion to the tissue culture plate. To test this possibilty, 1 ug of each of the recombinant VN was coated to tissue cuture plate. After washing and blocking the amount of VN bound was quantitated by the monoclonal antibody H3.68, which reacts with all three rVNs using an elisa procedure. These results show adhesion of various recombinant VNs are similar to plasma-derived VN and the absence of adhesion with rVN-RGE and rVN-RAD is not due to failure to attach to the tissue culture plates. However rVN-LGD showed markedly decreased adhesion to the plates compared to wild-type vitronectin.

Interaction vitronectin receptor with vitronectin fragments: 70 % formic acid cleves the peptide bond between Asp^{217} and pro^{218} resulting in the formation of an N-terminal RGD containing fragment and a C-terminal fragment containing the glycosaminoglycan site. The N-

42

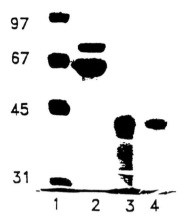

97

67

45

31

1   2   3   4

Figure 3: Isolation of the N-terminal fragment: Acid hydrolysis of vitronectin. Vironectin was incubated with 70% formic acid overnight at 37°. Lane 1, Mol wt std. Lane 2, vitronectin. Lane, 3 Acid hydrolysed vitronectin, Lane 4, N-terminal fragment.

terminal fragment retains the cell adhesion activity while the C-terminal fragment does not promote cell adhesion. When compared with intact vitronectin in dose response isotherm, the N-terminal fragment decreased cell adhesion activity especially towards Panc-1 cells, suggesting the N-terminal fragment may be involved in cell cooperativity with RGD site in promoting optimum cell adhesion activity.

DISCUSSION

BHK cells synthesize, process and secrete wild-type VN and mutant forms of VN. Isolation of all forms of VN, plasma-derived and recombinant alike, was accomplished in analogous fashion thereby allowing direct comparison of results obtained in functional cell adhesion assays. The ability of VN to maintain cell adhesion function even after exposure to 8 M urea was critical and allowed the use of heparin-agarose chromatography for isolation of all forms of VN. Direct comparison of recombinant wild-type VN to plasma-derived VN showed similar dose-response curves in cell adhesion assays to all three integrins tested. These results demonstrate that mammalian cell expression, as described, is a viable experimental approach to address questions of structure and function relationships. Furthermore, our results show complete loss of cell adhesion activity when one of the critical amino acids involved in cell binding is substituted via mutagenesis techniques. Even a conservative change of Asp^{47} to Glu^{47} or Gly^{46} to Ala^{46} resulted in complete loss of binding activity. These results provide direct confirmatory evidence that the consensus cell adhesion sequence RGD is critical for interaction of all three known integrins mediating cell adhesion. Despite the fact that all the integrin receptors for VN react with the RGD site, they differ considerably in their ligand specificity. Both $\alpha II\beta 3$ and $\alpha v\beta 3$ interact with fibrinogen, fibronectin and von Willebrand protein, in addition to VN (5,15). In contrast, $\alpha v\beta 5$ does not interact with fibrinogen, fibronectin or von Willebrand protein (9,10). In addition, $\alpha IIb\beta 3$ also interacts with the sequence HHLGGAKQAGDV in the carboxy terminus of the γ-chain of fibrinogen, while none of the other two integrins react with this site (17). The mechanism of receptor specificity remains unclear.

Our experiments with RGD VN mutants clearly show that the tripeptide, RGD, must be present for cell adhesion VN by all three known integrins of VN and that there is no apparent independent binding site as described for fibronectin or von Willebrand protein. It is likely that additional sites operate only in the presence of an obligatory RGD cell adhesion site. Our experiments suggest that at least in $\alpha v\beta 5$ additional sites are involved in the C-terminal region for optimum interaction.

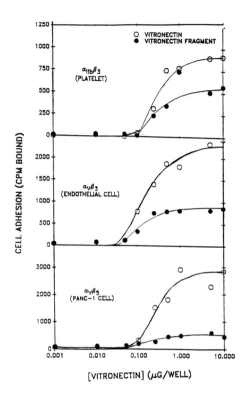

Figure 4. Adhesion of cells to vitronectin fragment. Polystyrene plates were coated with vitronectin or its fragment and cell adhesion was determined as in figure 2.

In this regard Orlando *et al.* (18) have noticed the interaction of $\alpha_v\beta_3$ with VN leads to subsequent stabilized interactions. While the initial interaction was inhibited almost completely by peptides containing the RGD sequence, these subsequent interactions were not inhibited by these peptides and were independent of cytoskeletal reorganization. Such interactions may be involved in ligand specificity. Experiments are currently in progress with site-directed and deletional mutations to address these possibilities.

REFERENCES

1. Preissner KT. (1989). Blut: 59: 419-431.
2. Tomasini BR and Mosher DF: Vitronectin. Progress in Thrombosis and Hemostasis. 10: 269-305, 1990.
3. Hynes RO.(1987) Cell 48: 549-554.
4. Ruoslahti E. (1991) J Clin Invest 87: 1-5, 1991.
5. Thiagarajan P, Kelly K. (1988) Throm. Haemost. 60: 514-517, 1988.
6. Thiagarajan P, Kelly KL. (1988) J. Biol. Chem. 263: 3035-3038, 1988.
7. Thiagarajan, P., Shapiro, S. S., Levine, E., DeMarco, L., and Yalcin, A. (1985) J. Clin. Invest. 75, 896-901
8. Fitzgerald, L. A., Charo, I.F., and Phillips, D. R. (1985) J. Biol. Chem. 260, 10893-10896
9. Cheresh, D. A., Smith, J. W., Cooper, H. M., Quaranta, V. (1989) Cell 57, 59-69

44

10. Smith, J. W., Vestal, D. J., Irwin, S. V., Burke, T. A., Cheresh, D. A. (1990) J. Biol. Chem. 265, 11008-11013
11. Cherny RC, Honan MA and Thiagarajan P. 1993. J Biol Chem 268: 9725-9729.
12. Vandeyar M, Weinwe MP, Hutton CJ, Batt CA (1988) Gene 65: 129-133.
13. Kitagaki-Ogawa, H., Yatogho, T., Izumi, M., Hayashi, M., Kashiwagi, H., Matsumoto, I., Seno, N. (1990) Biochim. Biophys. Acta 1033, 49-56
14. Chain D, Kreizman-T, Shapira H and Shaltiel S, FEBS letters 1991 285: 251-256.
15. Pytela, R., Pierschbacher, M. D., Ginsberg, M. H., Plow, E. F., Ruoslahti, E. (1986) Science 231, 1559-1562
16. Hawiger, J., Kloczewiak, M., Bednarek, M. A., Timmons, S. (1989) Biochemistry 28, 2909-2914
17. Chen C and Hawiger J: (1991) Blood 77: 6121-6126.
18. Orlando, R. A., Cheresh, D. A. (1991) J. Biol. Chem. 266, 19533-19550

Biology of Vitronectins and their Receptors
K.T. Preissner, S. Rosenblatt, C. Kost, J. Wegerhoff and D.F. Mosher, editors

A Comparison of Vitronectin and Megakaryocyte Stimulating Factor

David M. Merberg, Lori J. Fitz, Patty Temple, Joanne Giannotti, Pat Murtha, Mike Fitzgerald, Heidi Scaltreto, Kerry Kelleher, Klaus Preissner*, Ron Kriz, Ken Jacobs, and Katherine Turner.

Genetics Institute, Cambridge, MA U.S.A.

* Max Planck Institut, Bad Nauheim, Germany.

Introduction

Megakaryocyte stimulating factor (MSF) was first detected in and subsequently purified from urine obtained from bone marrow transplant patients during the period of acute thrombocytopenia. In murine bone marrow fibrin clot cultures, urinary MSF stimulates the growth of both pure and mixed megakaryocyte colonies with a specific activity of greater than 5×10^7 units per milligram protein. Purified urinary MSF is a cysteine-rich protein with molecular weights of 28-35KD and 20-25KD under non-reducing and reducing conditions, respectively (1).

We have recently cloned MSF, and determined the genomic and cDNA sequences. The DNA sequence of the full length MSF cDNA has revealed that recombinant MSF is first synthesized as a 400,000 Dalton glycosylated precursor, and that the urinary form of the protein is located at the N-terminus of the unprocessed molecule. Analysis of the precursor sequence has shown that several regions are strikingly similar to domains of vitronectin. Vitronectin, a member of a group of adhesive plasma proteins which includes fibrinogen, fibronectin and von Willebrand factor, is a multifunctional regulator of both complement and coagulation systems (2). To learn more about the structure and function of MSF, we have undertaken a comparative structural analysis of the two molecules.

Comparative analysis of the cDNA and deduced amino acid sequences of MSF and vitronectin.

The full length MSF cDNA is transcribed from a 4.5 kb mRNA and encodes a polypeptide of 1404 amino acids. The high molecular weight of the precursor form is partly due to an extensively O-linked glycosylated region in the central portion of the protein. Apart from the N-terminal secretory leader sequence, hydropathy analysis of MSF does not reveal any membrane-spanning regions in the molecule (data not shown). Whereas a single mRNA transcript exists for vitronectin, variably spliced MSF mRNA species have been observed in several sources such as peripheral blood mononuclear cells. The observation that 10 of the 12 exons present in the MSF gene are separated by introns of identical phase suggests that MSF may be a mosaic protein as described by Patthy (3). In contrast to vitronectin, which exists in plasma at concentrations of 200-400ug/ml, MSF is present at a much lower concentration (<1ng/ml in serum and urine).

The nucleotide and amino acid sequences of MSF were compared to the GenBank, EMBL, PIR, SwissProt, and GenPept sequence databases using the FASTA program (4,5). The FASTA algorithm identified a region of MSF from amino acids 1140 to 1300 that is 40% identical to amino acids 150 to 310 of vitronectin. Because both MSF and vitronectin are proteins with multiple domains, we sought to determine whether other regions of these sequences might also be similar. Accordingly, we compared each of the 12 protein sequences encoded by the 12 exons of MSF to the entire vitronectin sequence using a global similarity algorithm (Table 1). This analysis revealed that there are indeed several regions of MSF which are similar to vitronectin. Exons 2 and 3 of MSF are each approximately 45% identical to the somatomedin-B domain of vitronectin encoded within exon 2. MSF exons 7, 8, and 9 are also clearly similar to exons 3, 4, and 5 of vitronectin, respectively. Interestingly, there are segments of unrelated sequence interspersed between the homologous regions of these two proteins. For example, exon 6 of MSF, which encodes the 940 amino acid glycosylated domain, has little similarity to any region of vitronectin.

MSF DOMAIN	VITRONECTIN DOMAIN	% IDENTITY	DOMAIN TYPE
2	2	44	Somatomedin B
3	2	45	Somatomedin B
4	3	24	
5	5	26	
6	*	15	
7	3	54	
8	4	42	Hemopexin
9	5	43	Hemopexin
10	3,4	26	
11	5	27	
12	8	22	

Table 1. Sequence similarity between MSF and vitronectin. Sequences of peptides encoded by exons 2 through 12 of MSF (exon 1 encodes 25 amino acids residues containing the signal peptide) were compared to the entire vitronectin protein sequence via the program Gap (4). Gap identified the segment of vitronectin most similar to the MSF exon and reported the percentage of identical residues. Since exon 6 of MSF is longer than the entire vitronectin sequence, it was not aligned with a specific region. Sequences that are 30% or more identical are likely to have a common structure and/or function.

Given the very different lengths of MSF and vitronectin, and the discontinuous regions of similarity between them, the possibility that they evolved from a common ancestral protein was explored. We attempted to align the full length protein sequences of the MSF precursor and vitronectin. Pairwise alignment algorithms are often unsuccessful in comparing sequences which have homologous regions separated by long dissimilar regions. To circumvent this difficulty, we aligned the sequence of MSF with those of human, mouse, and rabbit vitronectin simultaneously, using the programs Pileup and LineUp (4). Inclusion of the sequences of vitronectin from several species served to weight important residues (those conserved through evolution) more heavily. The resulting alignment of MSF and human vitronectin is shown in Figure 1. For simplicity, the murine and rabbit vitronectins sequences are omitted. Many of the exon boundaries are conserved; for example, the boundaries of MSF exons 3, 8, 9, and 12 are aligned with those of exons 2, 4, 5, and 8 of vitronectin, respectively.

Exon 2 of the vitronectin gene encodes the somatomedin-B portion of the molecule. Somatomedin-B, produced by proteolytic cleavage of vitronectin, circulates in serum and was originally believed to have growth promoting activity, but this activity was later found to result from contamination by EGF (6). The similarities between exons 2 and 3 of MSF and the somatomedin-B domain of vitronectin are particularly noteworthy because several other proteins contain similar domains. Figure 2 shows the sequence comparison of exons 2 and 3 of MSF with the somatomedin-B-like domains of the following proteins: Vitronectin, PC-1 (a plasma cell transmembrane glycoprotein (7,8)), Tcl-30 (a T cell specific protein expressed in immature thymocytes (9)), and PP11 (a human placental-specific protein (10)). It is clear from this comparison that eight cysteines and other noted key residues are conserved among these domains. The presence of two similar, but not identical, somatomedin-B-like domains in MSF suggests that a duplication event occurred in the evolution of the MSF gene.

The lack of sequence homology between MSF and the amino region of the peptide encoded by exon 3 of vitronectin is interesting because this section of vitronectin contains an "RGD" (Arg-Gly-Asp) sequence. This motif, found immediately adjacent to the somatomedin-B domain, has been observed in several adhesive glycoproteins, and has been shown to be an important cell adhesion sequence for a subset of integrin receptors.

Exon 4 of the MSF precursor confers heparin binding activity to the full length precursor, a function that has been mapped by mutational analysis (LFitz, KBean, HFinnerty; unpublished data). This domain has a very high density of basic residues (30% His + Lys + Arg) as is characteristic of heparin binding domains. The capacity of the heparin binding region of MSF to interact with proteoglycans and glycosaminoglycans has not yet been evaluated. The urinary form of MSF is unable to bind heparin, suggesting that the C-terminus of the processed form is either upstream of the heparin-binding domain in exon 4 or is present but in a conformation unable to bind heparin. The heparin binding domain of vitronectin is located much closer to the C-terminus of the protein (encoded by exon 7). Conformational flexibility of the heparin-binding domain of vitronectin exon 7 has been reported (11). It is interesting to note that MSF contains a potential cAMP-dependent protein kinase phosphorylation site "KKKTK" at the carboxyl terminus of the heparin binding domain. Vitronectin contains a similar site at the carboxyl terminus of its heparin binding domain, which has been shown to be phosphorylated (12).

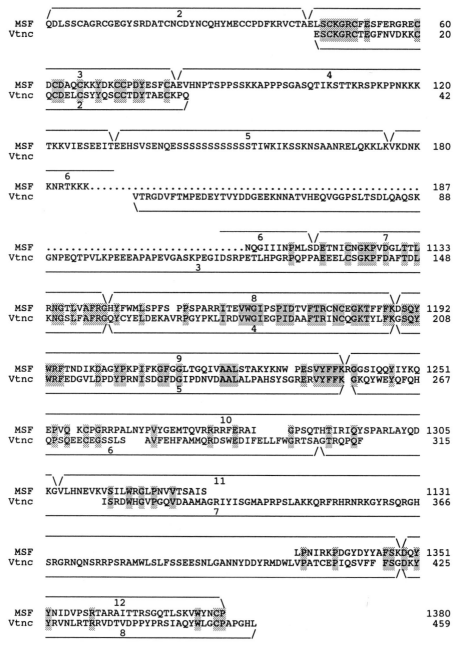

```
         /                 2                 \/
MSF   QDLSSCAGRCGEGYSRDATCNCDYNCQHYMECCPDFKRVCTAELSCKGRCFESFERGREC   60
Vtnc                                              ESCKGRCTEGFNVDKKC   20
                                                   _____

         _____3_____\/_____4_____
MSF   DCDAQCKKYDKCCPDYESFCAEVHNPTSPPSSKKAPPPSGASQTIKSTTKRSPKPPNKKK  120
Vtnc  QCDELCSYYQSCCTDYTAECKPQ                                       42
      _____2_____/

      _____\/_____5_____\/_____
MSF   TKKVIESEEITEEHSVSENQESSSSSSSSSSSSSSTIWKIKSSKNSAANRELQKKLKVKDNK  180
Vtnc

         _____6_____
MSF   KNRTKKK.........................................................  187
Vtnc              VTRGDVFTMPEDEYTVYDDGEEKNNATVHEQVGGPSLTSDLQAQSK       88
                  _____

                                    _____6_____\/_____7_____
MSF   ..............................NQGIIINPMLSDETNICNGKPVDGLTTL 1133
Vtnc  GNPEQTPVLKPEEEAPAPEVGASKPEGIDSRPETLHPGRPQPPAEEELCSGKPFDAFTDL  148
      _____3_____

      _____\/_____8_____\/_____
MSF   RNGTLVAFRGHYFWMLSPFS PPSPARRITEVWGIPSPIDTVFTRCNCEGKTFFFKDSQY 1192
Vtnc  KNGSLFAPRGQYCYELDEKAVRPGYPKLIRDVWGIEGPIDAAFTRINCQGKTYLFKGSQY  208
      _____/_____4_____/\___

      _____9_____\/_____
MSF   WRFTNDIKDAGYPKPIFKGFGGLTGQIVAALSTAKYKNW PESVYFFKRGGSIQQYIYKQ 1251
Vtnc  WRFEDGVLPDYPRNISDGFDGIPDNVDAALALPAHSYSGRERVYFFK GKQYWEYQFQH  267
      _____5_____/_____

      _____10_____
MSF   EPVQ KCPGRRPALNYPVYGEMTQVRRRRFERAI     GPSQTHTIRIQYSPARLAYQD 1305
Vtnc  QPSQEECEGSSLS    AVFEHFAMMQRDSWEDIFELLFWGRTSAGTRQPQF         315
      _____6_____/_____

      _\/_____11_____
MSF   KGVLHNEVKVSILWRGLPNVVTSAIS                     1131
Vtnc                ISRDWHGVPGQVDAAMAGRIYISGMAPRPSLAKKQRFRHRNRKGYRSQRGH  366
                    _____7_____

                                    _____\/____
MSF                               LPNIRKPDGYDYYAFSKDQY 1351
Vtnc  SRGRNQNSRRPSRAMWLSLFSSEESNLGANNYDDYRMDWLVPATCEPIQSVFF FSGDKY  425
      _____/\____

      _____12_____\
MSF   YNIDVPSRTARAITTRSGQTLSKVWYNCP                 1380
Vtnc  YRVNLRTRRVDTVDPPYPRSIAQYWLGCPAPGHL           459
      _____8_____/
```

Figure 1. Sequence alignment of MSF precursor and vitronectin (VTNC). Sequences of MSF and vitronectin were aligned as described in the text. The section of MSF encoded by exon 6 is not significantly similar to any region of vitronectin; most of this sequence (residues 188 to 1105) has been omitted for clarity. The exon from which each portion of the peptide is derived is shown above and below the alignment for MSF and Vitronectin, respectively. Identical residues are shaded. Numbering of each sequence begins at the predicted signal cleavage site.

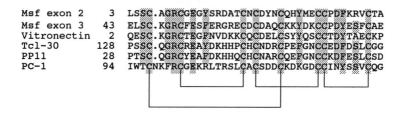

```
Msf exon 2     3   LSSC.AGRCGEGYSRDATCNCDYNCQHYMECCPDFKRVCTA
Msf exon 3    43   ELSC.KGRCFESFERGRECDCDAQCKKYDKCCPDYESFCAE
Vitronectin    2   QESC.KGRCTEGFNVDKKCQCDELCSYYQSCCTDYTAECKP
Tcl-30       128   PSSC.QGRCREAYDKHHPCHCNDRCPEFGNCCEDFDSLCGG
PP11          28   PTSC.QGRCYEAFDKHHQCHCNARCQEFGNCCKDFESLCSD
PC-1          94   IWTCNKFRCGEKRLTRSLCACSDDCKDKGDCCINYSSVCQG
```

Figure 2. Sequence comparison of somatomedin-B type Domains. Sequences of the somatomedin-B type domains of MSF, vitronectin, Tcl-30 PP11, and PC-1 were aligned using the program PileUp (Devereux). Predicted disulfide bonding pattern, based on computer modelling of vitronectin is shown (K. Preissner, unpublished results).

The most distinguishing feature of the peptide encoded by exon 5 of the MSF gene is a run of 12 consecutive serine residues. Vitronectin does not contain a similar serine rich region. As previously mentioned, exon 6 of MSF encodes the highly glycosylated central portion of the precursor molecule. The composition of this domain is predominated by five amino acids: 26.9% threonine, 22.3% proline, 13.5% lysine, 9.0% alanine, and 8.5% glutamic acid. There are 54 occurrences of the degenerate (i.e. allowing one mismatched and/or omitted amino acid) pattern "KEPAPTT". In vitronectin exon 3, there is a single occurrence of the somewhat similar pattern "KPEEEAPAP", but it is unclear whether these two motifs share any structural or functional properties.

Exon 7 of MSF is remarkably similar to the carboxyl end of exon 3 of vitronectin (15/27 identical residues), considering that there is little similarity between the amino end of vitronectin exon 3 and any portion of MSF. The large proportion of conserved amino acids suggests that these sequences have an important biological property which is common to both MSF and vitronectin. Thus far, these sequences are unique to MSF and vitronectin; no related sequences were found in the protein sequence databanks.

Exons 8 and 9 of MSF are clearly homologous to exons 4 and 5 of vitronectin, sharing many common residues and exon boundaries. These are typical hemopexin domains, with central hydrophobic motifs ("IDTVF" and "IVAAL" in MSF, "IDAAF" and "VDAAL" in vitronectin) and proximal aromatic residues. There is less similarity between exons 10 and 11 of MSF and exons 6 and 7 of vitronectin. Exon 7 of vitronectin contains the heparin binding region of this molecule, the counterpart in MSF being much closer to the amino terminus of the molecule. Although the percentage of identical residues between exon 12 of MSF and exon 8 of vitronectin is not high, these sequences may have similar functions because the residues which are identical in both include cysteine, tyrosine, tryptophan and proline, amino acids that are often conserved between homologous domains.

Figure 3 provides a schematic display of the overall relationship between the MSF and vitronectin sequences. The homologous somatomedin-B domains are found at the amino terminus of both proteins. These are followed by non-homologous regions of very different lengths. The shared hemopexin domains are found more than 1000 residues downstream in the MSF sequence. Again, these are followed by stretches of sequence that appear to be unrelated. Finally, the last 25 residues of these proteins may also be homologous. It is particularly interesting that MSF and vitronectin both have a heparin binding domain, although it is located in different regions in each protein.

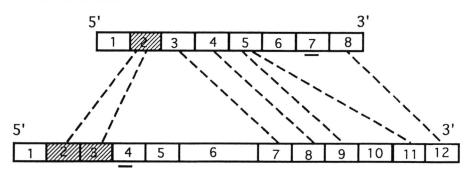

VITRONECTIN

MSF

Figure 3. Structural comparison of MSF precursor and vitronectin. Homologous exons of MSF and vitronectin are connected with a dashed line. The somatomedin-B domain of vitronectin and the homologous exons of MSF precursor are indicated by hatched boxes. The heparin binding domains of vitronectin and MSF precursor are underlined.

Discussion

This study describes the structural domains of MSF, a complex glycoprotein with intriguing structural homology with vitronectin. The arrangement of the domains of MSF is shown in Figure 4. The full length rMSF precursor is not efficiently processed in COS and CHO cells (data not shown) and the pathway by which the MSF precursor is converted to the mature form is not well understood. Since vitronectin is known to associate with plasminogen and thrombin, it will be important to determine whether these proteases play a role in MSF processing. The actual C terminus of the mature form of MSF has not been identified, but tryptic mapping experiments indicate that it is located in exon 4. Thus the somatamedin-B domains are a major component of the urinary form.

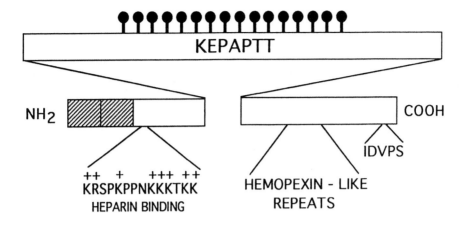

Figure 4. MSF precursor domain structure. The central cluster of potential O-glycosylation sites is indicated by knobs located on exon 6 surrounded by the somatomedin-B-like (hatched box), heparin binding, hemopexin-like, and IDVPS domains, respectively.

MSF is the fifth protein in which a cysteine rich somatomedin-B domain has been found. The somatomedin-B domain of vitronectin has been implicated in binding plasminogen activator inhibitor-1 (PAI-1)(13). The function of this domain in other proteins is unknown but it has been suggested to provide strucural stability. Both PC-1 and Tcl-30 are class II transmembrane proteins which have a reverse orientation in the membrane, exposing the C-terminus and the somatomedin B-like domains extracellularly. Recently PC-1, which is expressed on the surface of antibody-secreting plasma cells, has been shown to be a threonine-specific protein kinase and to be identical to human fibroblast nucleotide pyrophosphatase (14,15). The presence of somatomedin-B domains in a form of MSF which has growth stimulating effects on megakaryocyte progenitors raises the possibility that this domain has proliferative effects. However, this function remains to be shown definitively.

Another characteristic feature shared by MSF and vitronectin is homology with the "hemopexin" repeats first described in human hemopexin, a heme binding plasma protein (16,17). Both hemopexin and vitronectin can undergo conformational changes resulting in exposure of different binding sites. Vitronectin occurs in at least two conformational forms: a native form that does not bind heparin and a heparin binding form which occurs when vitronectin is treated with urea, low pH or adsorbed to surfaces. The cryptic location of the heparin and glycosaminoglycan binding domains of vitronectin suggest that vitronectin is conformationally labile (18). Preliminary experiments indicate MSF may share this property. Addition of only 5-10 amino acids of the heparin binding domain to the C-terminus of truncated rMSF, engineered to terminate upstream of the heparin binding site, restores heparin binding but completely alters the conformation of the protein as assessed by SDS PAGE (data not shown).

In addition, MSF may share some of the adhesive properties of vitronectin. Although rMSF is secreted by COS and CHO cells, preliminary studies have shown that MSF can bind vitronectin and can associate with the extracellular matrix (K. Preissner, L. Fitz, unpublished results). Unlike vitronectin, MSF does not contain the "RGD" motif and therefore probably does not function as a ligand for the α_{IIb}-β_3 or the $\alpha_v\beta_3$ integrin receptors. However, MSF exon12 contains the pentapeptide "IDVPS" (see Fig. 1). This sequence is similar to the "LDVPS" motif found in the CS-1 peptide of one of the alternatively spliced forms of fibronectin. In fibronectin, these residues are responsible for binding to the VLA-4 integrin receptor and have been implicated in hematopoietic stem cell-microenvironment interactions (19).

The long, highly glycoslylated domain encoded by exon 6 of MSF is unlike any region of vitronectin. However, similar domains do occur in growth factor-like peptides belonging to the "trefoil" family (20). This family includes: (i) the pS2 peptide expressed by the breast cancer cell line MCF-7, (ii) hSP, a tandem duplication of the pS2 domain expressed by normal stomach epithelium, and (iii) spasmolysin , a protein containing two pS2 domains which has been identified in frog skin but whose function is unknown (21,22,23). In spasmolysin the highly glycosylated repetitive domain is located in the central region of the protein, making it most similar in overall structure to the MSF precursor.

Although vitronectin is best known for its procoagulant, complement, and adhesion related activities, it is also suspected of influencing platelet development and responding to platelet signals. For example, the tetrapeptide RGDS and a monoclonal antibody specific to the α_v β_3 vitronectin receptor were found to inhibit guinea pig proplatelet formation *in vitro* (24). In addition, phosphorylation of ser 378 in vitronectin has been proposed to occur *in vivo* by a protein kinase released from platelets or injured cells activated during hemostasis (25). Further study of both MSF and vitronectin, and elucidation of the functions of their various domains, is likely to lead to a greater understanding of platelet biology.

REFERENCES

1. Turner K, Fitz L, Temple P, Jacobs K. Blood 1991; 78:279a.
2. Preissner KT, Jenne D. Thrombosis and Haemostasis 1991; 66:189-194.
3. Patthy L. Current Opinion in Structural Biol 1991; 1:351-361.
4. Devereux J, Haeberli P, Smithies O. Nucleic Acids Research 1984; 12:387-395.
5. Pearson WR. Methods in Enzymology, R.F. Doolitte, ed 1990; 183:63-98.
6. Heldin C-H, Wasteson A, Fryklund L, Westermark B. Science 1981; 213:1122-1123.
7. van Driel IR, Goding JW. J Biol Chem 1987; 262:4882-4887.
8. Buckley, MF, Loveland, KA, McKrinstry, WJ, Garson, OM, Goding, JW. J Biol Chem 1990; 265:17506-17511.
9. Baughman G, Lesley J, Trotter J, Hyman R, Bourgeois S. J Immunol 1992; 149:1488-1496.
10. Grundmann U, Römisch J, Siebold B, Bohn H, Amann E. DNA and Cell Biol; 1990; 9:243-250.

11. Tomasini BR, Mosher DF. Blood 1988; 72:903-912.
12. McGuire EA, Peacock ME, Inhorn RC, Siegel NR, Tollefsen DM. J Biol Chem 1988; 263:1942-1945.
13. Seiffert D, Loskutoff DJ. J Biol Chem 1991; 266:2824-2830.
14. Oda Y, Kuo M, Huang S, Huang J Biol Chem 1991; 266:16791-16795.
15. Funakoshi I, Kato H, Horie K, Yano T. Arch of Biochem and Biophys 1992; 295:180-187.
16. Jenne, D, Stanley KK. Biochem 1987; 26:6735-6742.
17. Hunt LT, Barker WC, Chen HR. Protein Seq Data Anal 1987; 1:21-26.
18. Panetti TS, McKeown-Longo PJ. J Biol Chem 1993; 268:11988-11993.
19. Williams DA, Rios M, Stephens C, Patel VP. Nature 1991; 352:438-441.
20. Thim L. FEBS Letters 1989; 250: 85-90.
21. Masiakowski P, Breathnach R, Bloch J, Gannon F, Krust A, Chambon P. Nuc Acids Res 1982; 10:7895-7903.
22. Tomasetto C, Rio M-C, Gautier C, Wolf C, Hareuveni M, et al. EMBO 1990; 9:407-414.
23. Hoffman W. J Biol Chem 1988; 263:7686-7690.
24. Leven RM, Tablin F. Exp Hematol 1992; 20:1316-1322.
25. Korc-Grodzicki B, Tauber-Finkelstein M, Chain D, Shaltiel S. Biochem and Biophys Res Commun 1988; 157:1131-1138.

Biology of Vitronectins and their Receptors
K.T. Preissner, S. Rosenblatt, C. Kost, J. Wegerhoff and D.F. Mosher, editors

Binding domain for ß-endorphin on vitronectin

S. Wöhner[a], D. Linder[b], H. Teschemacher[a] and K.T. Preissner[c]

[a]Rudolf-Buchheim-Institut für Pharmakologie der Justus-Liebig-Universität, Frankfurter Str. 107, D-35392 Gießen, FRG

[b]Biochemisches Institut des FB Humanmedizin der Justus-Liebig-Universität, Friedrichstr. 24, D-35392 Gießen, FRG

[c]Kerckhoff-Klinik, Forschungsgruppe Hämostaseologie der Max-Planck-Gesellschaft, Sprudelhof 11, D-61231 Bad Nauheim, FRG

SUMMARY

Recently human vitronectin has been demonstrated to exhibit specific non-opioid binding sites for human ß-endorphin (β_H-endorphin) upon interaction with heparin or surfaces. The present study was performed in order to obtain information about the location of the β_H-endorphin binding site on the vitronectin molecule.

In a first approach, microtiter plates were coated with vitronectin and a polyclonal antiserum against vitronectin or monoclonal antibodies directed against various vitronectin epitopes were tested for their capability to inhibit β_H-endorphin binding to vitronectin. Although β_H-endorphin binding proved to be inhibited by the polyclonal antiserum, none of the monoclonal antibodies was able to influence the β_H-endorphin interaction with vitronectin. In a further type of investigation β_H-endorphin (27-31), which represents the β_H-endorphin fragment primarily responsible for the β_H-endorphin binding to vitronectin, was radioactively labeled and allowed to bind to vitronectin; subsequently it was covalently coupled to the vitronectin molecule and the conjugate was subjected to endoproteinase ArgC digestion. SDS-PAGE of the digest revealed several radioactively labeled vitronectin cleavage products which were blotted over to a PVDF membrane; they were localized on the membrane by autoradiography and were analysed for their position within the vitronectin molecule by amino acid sequencing. Evidence was obtained for one or several sites of β_H-endorphin interaction with vitronectin, which are apparently related to the middle region of the vitronectin amino acid sequence.

INTRODUCTION

Opioidergic systems, consisting of opioid receptors and their endogenous ligands, the opioid peptides, are located in the nervous, in the endocrine and in the immune system of the mammalian organism. All opioid peptides demonstrated so far origin from three precursor proteins, proenkephalin, prodynorphin and proopiomelanocortin (POMC). Although a lot of information has been collected about biosynthesis, structure, location or effects of the opioid peptides during the last two decades, really conclusive evidence for their physiological significance has remained meagre [1].

In search of the functional relevance of ß-endorphin (a POMC processing product)

as far as found in human blood, we recently succeeded in demonstrating a possible physiological target found in human blood as well: Vitronectin. As a protein of multifunctional significance [2], vitronectin would serve as a particular attractive candidate for a ß-endorphin target molecule of functional relevance.

Human ß-endorphin has been shown to bind to vitronectin in a highly specific manner [3]. Binding does not occur via its N-terminus, which is able to interact with opioid receptors, but via its C-terminus indicating that $ß_H$-endorphin might represent a functional cross-linker between vitronectin and opioidergic systems. Native vitronectin does only bind very small amounts of $ß_H$-endorphin [3]; binding, however, is strongly increased upon steric vitronectin conversion, e.g. (i) upon interaction with certain compounds such as heparin [3] (ii) upon integration into macromolecules such as the terminal complement complex SC5b-9 [4] or (iii) upon attachment to surface structures such as nitrocellulose [3].

We tried to localize $ß_H$-endorphin binding sites on vitronectin by testing monoclonal antibodies directed against various vitronectin epitopes for their capability to inhibit ß-endorphin binding to vitronectin. In a further type of investigation, the C-terminal fragment of $ß_H$-endorphin, $ß_H$-endorphin (27-31), (primarily responsible for the $ß_H$-endorphin/vitronectin interaction) was radioactively labeled and allowed to bind to the specific $ß_H$-endorphin binding sites; subsequently it was covalently coupled to vitronectin using disuccinimidyl suberate as a cross-linker. Upon enzymatic digestion of the conjugate the cleavage products were separated by gel electrophoresis and were subsequently blotted onto a PVDF membrane; the radiolabeled ones were visualized by autoradiography and were finally analysed by amino acid sequencing for their position in the vitronectin molecule.

METHODS

The study has been performed using methods as, in principle, described previously [3, 5].

Testing for inhibition of $ß_H$-endorphin binding to vitronectin (VN) by a polyclonal antiserum and monoclonal antibodies against vitronectin epitopes, in brief, has been conducted as follows: Microtiter plates were coated with VN for 14 hours at 4 C. After saturation with bovine serum albumin for 1 hour at room temperature a polyclonal antiserum against vitronectin or monoclonal antibodies against certain VN epitopes were allowed to bind to the immobilized VN for 10 min at room temperature. Then ^{125}I-$ß_H$-endorphin was added, and the samples were incubated for 4.5 hours at 4 C. After washing, the radioactivity bound in each well was measured in a gamma counter.

The localization of ß-endorphin binding sites on vitronectin by analysis of vitronectin fragments marked by radiolabeled $ß_H$-endorphin (27-31) was achieved as follows: In brief, VN was activated by incubation with heparin and ^{125}I-$ß_H$-endorphin (27-31) was allowed to bind to VN for 3 hours at 4 C. Then, the radiolabeled peptide was covalently linked to its binding sites by incubation with disuccinimidyl suberate for 10 min at room temperature. After its purification and concentration, the conjugate was subjected to enzymatic digestion using endoproteinase ArgC. After various digestion periods the samples were subjected to SDS-PAGE.

The radiolabeled and non-labeled cleavage products of the ß$_H$-endorphin/vitronectin conjugate as separated by SDS-PAGE were blotted from the gel to a PVDF-membrane and there VN-fragments covalently linked to the radiolabeled peptide were visualized by autoradiography. The respective areas were cut off the membrane and subjected to amino acid sequencing.

RESULTS AND DISCUSSION

A polyclonal antiserum against vitronectin and monoclonal antibodies directed against various epitopes on the N-terminal, the C-terminal and the middle region of vitronectin were tested for their capability to prevent ß$_H$-endorphin binding to vitronectin in a solid phase assay. Whereas the polyclonal antiserum inhibited ß$_H$-endorphin binding by about 50 %, none of the monoclonal antibodies did influence ß$_H$-endorphin binding to vitronectin. Information about the ß$_H$-endorphin binding site location would have been obtained in case of a positive, i.e. an inhibitory effect observed for one or several of the monoclonal antibodies. The negative result as obtained, however, did not provide us with any information.

In a further series of experiments, we tried to localize the ß$_H$-endorphin binding sites on vitronectin by covalent coupling of radiolabeled ß$_H$-endorphin (27-31) to the binding site and subsequent search for the radioactivity signal on the vitronectin molecule. Therefore, the ß$_H$-endorphin/vitronectin conjugate was subjected to enzymatic digestion by endoproteinase ArgC, and the cleavage products were separated by SDS-PAGE; after blotting them onto a PVDF membrane, the radiolabeled ones were visualized by autoradiography and analyzed for their position within the vitronectin molecule by amino acid sequencing.

Control gels were run with radiolabeled vitronectin , which, however, had not been subjected to ArgC digestion. A 75 000 Da and 65 000 Da band appeared on the gel, corresponding to the two forms of vitronectin usually observed [3].Enzymatic degradation of the radiolabeled ß$_H$-endorphin/vitronectin conjugate with ArgC endoproteinase, however, resulted in a complete loss of the labeled 75 000 Da band and in the generation of a more stable fragment of about 55 000 Da. Further labeled fragments were produced , in addition. N-terminal amino acid sequencing of the material contained in the 55 000 Da band revealed the sequence G D V F T M P E - E Y .which is exactly corresponding to position 45-55 of the primary structure of vitronectin, starting within the well-known RGD sequence.

Thus, the labeled and N-terminally identified 55 000 Da fragment should include the vitronectin amino acid residues 45 to, very roughly, 60 000. In conclusion, the ß$_H$-endorphin (1-31) binding site should be rather located on the middle region than on the C- or on the N-terminal fragment of vitronectin. Binding of ß$_H$-endorphin to the middle region could take place simultaneously to a number of vitronectin interactions with functionally relevant ligands, e.g. from the coagulation or the complement system, without great risk of steric hindrance: all of those so far identified ligands interact with N- or C-terminal vitronectin fragments. Thus, ß$_H$-endorphin, known to be bound to vitronectin through its C-terminus and known to elicit various effects on immune cells via its N-terminus, might, in fact, represent a functional link between immune cells and vitronectin-related systems [6].

REFERENCES

1 Opioids I. Handb. Exp. Pharm. 104/I. Ed. A. Herz, Springer, Berlin 1993.
2 Preissner KT, Jenne D. Thromb Hemostas 1991; 66: 123-132.
3 Hildebrand A, Preissner KT, Müller-Berghaus G, Teschemacher H. J Biol Chem 1989; 264: 15429-15434.
4 Schweigerer L, Bhakdi S, Teschemacher H. Nature 1982; 296: 572-574.
5 Hildebrand A, Schweigerer L, Teschemacher H. J Biol Chem 1988; 263: 2436-2441.
6 Teschemacher H, Koch G, Scheffler H, Hildebrand A, Brantl V. Ann NY Acad Sci 1990; 594: 66-77.

Biology of Vitronectins and their Receptors
K.T. Preissner, S. Rosenblatt, C. Kost, J. Wegerhoff and D.F. Mosher, editors

Primary structure of vitronectins and homology with other proteins

H. J. Ehrlich[*], B. Richter, D. von der Ahe and K. T. Preissner

Haemostasis Research Unit, Kerckhoff-Klinik, Max-Planck-Institut, Sprudelhof 11, D-61231 Bad Nauheim, Germany

[*]Present address: Sandoz AG, Deutschherrnstrasse 15, D-90429 Nürnberg, Germany

INTRODUCTION

Isolation of vitronectins by heparin-Sepharose in the presence of 8 M urea according to Yatohgo et al. [1] allowed to obtain purified protein preparations from different species in sufficient quantities for further structural analysis. Yet, for conformational/functional analysis of vitronectins the more tedious isolation procedures to yield the intact plasma form of the protein have to be used [2,3, see also Hayashi and Mosher, this issue]. Comparison among 14 different vitronectins showed considerable amino-terminal sequence homology and sizes of polypeptide chains were found to be in the apparent M_r range of 50-57 kDa [4]. Depending on the site and nature of attached carbohydrate chains via N- and/or O-glycosylation, apparent M_r of vitronectins vary between 59 and 80 kDa. While these (liver-derived) vitronectins were isolated from plasma or serum, compared to adult chicken vitronectin egg yolk contains a vitronectin homologue with lower M_r which lacks heparin-binding function [5].

At present the cDNA-derived amino acid sequences of three mammalian vitronectins have been identified, indicating that 80-86% identity exists between the vitronectins from human [6,7], rabbit [8], and mouse [9,10] origin (Table 1). Furthermore, the complete genomic sequences of human [11] and mouse (Seiffert et al., personal communication) vitronectin have been completed, demonstrating identical intron-exon boundary structures.

Due to high homology among vitronectins from different sources, it is generally accepted that the following domain structure (based mostly on experiments with human vitronectin) exists (from the amino- to the carboxy-terminus): "Somatomedin B"-domain [PAI-1 binding]; RGD-epitope [cell attachment site, recognition by vitronectin receptors]; collagen-binding domain; cross linking site(s) for factor XIIIa/tissue transglutaminase [Gln-93]; plasminogen-binding site [330-349]; heparin-binding site [348-361]; PAI-binding site [~355-370]; endogenous cleavage site [Arg-379/Ala-380]. In addition, putative sulfation sites are at Tyr-56 and Tyr-59 and the cAMP-dependent protein kinase dependent phosphorylation site is at Ser-378. An established polymorphism of the human protein exists resulting in homozygote as well as heterozygote distribution of vitronectin with Thr or Met at position 381 [for review see 12,13].

Table 1

Comparison of amino acid sequences from human (H), rabbit (R), and mouse (M) vitronectins [6-10].

```
H-19 M A P L R P L L I L A L L A W V A L A/D Q E S C K G R C T E
R-19 * * * * * * I F T * * * * L * * V * */* * * * * D * * * *
M-19 * * * * * * F F * * * V * * * S * */* * * * * * * * * Q

H 12 G F N A N R K C Q C D E L C S Y Y Q S C C T D Y T A E C K P
R 12 * * * * * * * * * * * * * * * * * * * A * * A * * * * *
M 12 * * M * S K * * * * * * * * T * * * * * A * * M E Q E * *

H 42 Q V T R G D V F T M P E D E Y T V Y D - D G E E K N N A T V
R 42 * * * * * * * * * * * * * * * G P * * Y I E Q T * D * * S *
M 42 * * * * * * * * * * * * * * * W S * * * V E * P * * * T N T

H 71 H E Q V G G P S L T S D L Q A Q S K G N P E Q T P V L K P E
R 72 * A * P E S * T V G Q E - P T L * P D L Q T E G G A E P T H
M 72 G V * P E N T * P P G * * N P R T D G T L K P * A F * D * *

H101 E E A P A P E V G A S K P E G I D S R P E T L H P G R P Q P
R101 * V P L E * * M E T L R * * * E * L Q A G * T E L * T S A S
M102 * Q P S T * A P K V E Q Q * E * - L * * D * T D Q * T * E F

H131 P A E E E L C S G K P F D A F T D L K N G S L F A F R G Q Y
R131 * * * * * * * * * * * * * * * * * * * * * * * * * * * * *
M131 * - * * * * * * * * * * * * * * * * * * * * * * * * * * *

H161 C Y E L D E K A V R P G Y P K L I R D V W G I E G P I D A A
R161 * * * * * * T * * * * * * * * * * Q * * * * * * * * * * * *
M160 * * * * * * T * * * * * * * * * * Q * * * * * * * * * * * *

H191 F T R I N C Q G K T Y L F K G S Q Y W R F E D G V L D P D Y
R191 * * * * * * * * * * * * * * * * * * * * * * * * I * * * * *
M190 * * * * * * * * * * * * * * * * * * * * * * * * * * * * G *

H221 P R N I S D G F D G I P D N V D A A L A L P A H S Y S G R E
R221 * * * * * E * * S * * * * * * * * F * * * * * * * * * * * *
M220 * * * * * E * * S * * * * * * * * F * * * * * R * * * * * *

H251 R V Y F F K G K Q Y W E Y Q F Q H Q P S Q E E C E G S S L S
R251 * * * * * * D K * * * * * * * Q * * * * * * * * * * * * * *
M250 * * * * * * * * * * * * E * * Q * * * * * * * * * * * * * *

H281 A V F E H F A M M Q R D S W E D I F E L L F W G R T S A G T
R281 * * * * * * * * L H * * * * * * * K * * * * * * P * G * A
M280 * * * * * * * L L * * * * * * * N * * * * * * * S * D * A

H311 R Q P Q F I S R D W H G V P G Q V D A A M A G R I Y I S G M
R311 * * * * * * * * * * * * * * * * K * * * * * * * * * * * * L
M310 * E * * * * * * N * * * * * * * K * * * * * * * * * V T * S
```

Table 1 (continued)

```
H341 A P R P S L A K K Q R F R H R N R K G Y R S Q R G - - H S R
R341 T * S * * - * * * * K S * R * S * * R * * * R Y * R G R - -
M340 L S H S A Q * * * * K S K R * S * * R * * * R * * R G * R *

H369 G R N Q N S R R P S R A T W L S L F S S E E S N L G A N N Y
R366 S Q - - * * * * L * * S I S R L W * * * * * V S * * P Y * *
M370 S Q S S * * * * * S * * S I * F * * * * * * * * G * * T Y * N

H399 D D Y R M D W L V P A T C E P I Q S V F F F S G D K Y Y R V
R396 E * * E T S * * K * * * S * * * * * * Y * * * * * * * * * *
M400 Y * * D * * * * * * * * * * * * * * * * Y * * * * * * * * * *

H429 N L R T R R V D T V D P P Y P R S I A Q Y W L G C P A P G H
R426 * * * * Q * * * * * N * * * * * * * * * * * * * * * * * * * G
M430 * * * * * * * * S * N * * * * * * * * * * * * * * * * * T S E -

H459 L
R456 Q
M459 K
```

Identical amino acid residues to the human homologue are indicated by *; gaps (-) are included to yield maximal overlap of sequences. Note the high homology among all three sequences, except for the connecting segment (residues 70-130). The site of proteolytic processing prior to secretion is indicated by /; the entire basic region (containing the heparin-binding site) is marked by overlining.
(Methods: Using a human vitronectin probe (kindly provided by Dr. D. Jenne), a cDNA liver cell library from mouse was screened and two positive clones containing the 5' and 3' region of mouse vitronectin were identified. Utilizing polymerase chain reaction technology a combined full-length clone was obtained which was sequenced by standard procedures.)

HOMOLOGY BETWEEN DOMAINS IN VITRONECTIN AND OTHER PROTEINS

"Somatomedin B"-domain. The amino-terminal portion of the mature vitronectin molecule, the so called "somatomedin B"-region was the first part of the molecule whose primary structure had been identified [14] before the entire polypeptide primary structure was reported. This 44 amino acids long polypeptide appears to be a highly folded nodule, since 8 cysteine residues are connected within 4 disulfide bridges whose exact coordination has not been established yet. Interestingly, a number of otherwise non-related proteins contain one or more repeats of this characteristic cysteine pattern with additional homology to the overall "somatomedin B"-domain. This indicates that the independently - folded

Human PC-1   M D V G E E P L E K A

PP 11   M R A C

Human PC-1   A R A R T A K D P N T Y K V L S V C V L T T I L G C I F G L K P S C A K E V K S

PP 11   I S L V L A V L C G L A W A E D H K E S E P L P Q L E E T E E A L A S N L Y S A P T S

MSF-Precursor   M A W K T L P I Y L L L L S V F V I Q Q V S S Q D L S S

Human Vitronectin   M A - I - P L R P L L I H L A L L - A W V A L A - I - I - - D Q E S
Rabbit Vitronectin M A - I - P L R P I F T L A L L - L L W V V L A - I - I - - D Q E S
Mouse Vitronectin   M A - I - P L R P F F I L A L - V A W V S L A - I - I - - D Q E S

Human PC-1 I   C - K G R C F E - R T F G N - C R C D A A C V E L G N C C L D Y Q E T C I E P E H I W T
Human PC-1 II   C N K F R C G E K R L T R S L C A C S D D C K D K G D C I N Y S S V C Q G E K S W V E

PP 11   C Q - G R C Y E A F D K H H Q C H C N A R C Q E F G N C C K D F E S L C S D H E V S H S

MSF-Precursor I   C A - G R C G E G Y S R D A T C N C D Y N C Q H Y M E C C P D F K R V C T A E L S - I - T S
MSF-Precursor II   C - K G R C F E S F E R G R E C D C D A Q C P D Y E S F C A E V H N P T S

Human VN   C - I K G R C T E G F N A N R K C Q Q C D E L C S Y Y Q Q S C C T D Y T A E C K P Q V T R G D
Rabbit VN   C - I K D R C T E G F N A N R K C Q Q C D E L C S Y Y Q Q S C C A D Y A A E C K P Q V T R G D
Mouse VN   C - I K G R C T Q G F M A S K K C Q C D E L C T Y Y Q S C C A D Y M E Q E K P Q V T R G D

Figure 1. Sequence comparison between vitronectins, PC-1, placental protein 11 (PP 11) and megakaryocyte stimulating factor (MSF) precursor in their "somatomedin B" type repeats.

"somatomedin B"-domain which is flanked by phase I-type introns may have undergone exon-shuffling during evolution. Examples are the plasma cell-expressed surface protein PC-1 [15], the placental protein 11 [16], and the precursor of megakaryocyte-stimulating factor (K. Turner et al., unpublished observations). The comparison of primary structure within the "somatomedin B"-type repeats of these proteins is shown in figure 1.

RGD-dependent attachment site. The RGD motif in a number of extracellular (matrix) proteins has been demonstrated clearly to be one of the recognition signals for heterodimeric integrin receptors [17]. This tripeptide motif appears to be necessary but not sufficient to function as binding site for integrin receptors in a divalent metal-dependent manner. At least in part the flanking sequences of this tripeptide have some influence on the selectivity and specificity of ligand-receptor interaction [18]. Based on the three dimensional structure of RGD embedded within the sequence of the thenth type III repeat of fibronectin [19] as well as using site-directed mutagenesis [20], a number of linear and cyclic peptidic or non-peptidic derivatives have been produced, which may be used as competitors in cell attachment studies. Comparison between RGD flanking regions from different adhesion proteins is shown in table 2.

Table 2
Sequence comparison between vitronectin and other adhesion glycoproteins around their attachment site region (for complete sequences see [6,7,20-24])

VN	(36)	T	A	E	C	K	P	Q	V	T	R	G	D	V	F	T	M	P	E	D	E	Y	T	V	Y
BSP	(280)	T	Y	D	E	N	N	G	E	P	R	G	D	T	Y	R	A	Y	E	D	E	Y	S	Y	Y
OP	(34)	I	A	P	T	V	D	V	P	D	R	G	D	S	L	A	Y	G	L	R	S	K	S	R	S
FN	(543)	T	I	T	V	Y	A	V	T	G	R	G	D	S	P	A	S	S	K	P	I	S	I	N	Y
vWF	(1744)	A	C	E	V	V	T	G	S	P	R	G	D	S	Q	S	S	W	K	S	V	G	S	Q	W
PT	(563)	A	G	Y	K	P	D	E	F	K	R	G	D	A	C	E	G	D	S	G	G	P	F	V	M

VN: vitronectin; BSP: bone sialo glycoprotein; OP: osteopontin; FN: fibronectin; vWF: von Willebrand factor; PT: prothrombin

Hemopexin repeats. Computer-assisted structural analysis of vitronectin revealed distant homology to hemopexin-type repeat containing proteins as in hemopexin itself or in certain matrix metalloproteinases such as strmelysin or transin [11,25] (Fig. 2). Although hemopexin, a major heme-binding and transporting protein in the circulation, undergoes a unique conformational change upon ligand binding, the conformational flexibility of vitronectin may not be necessarily related to the presence of hemopexin-type repeats.

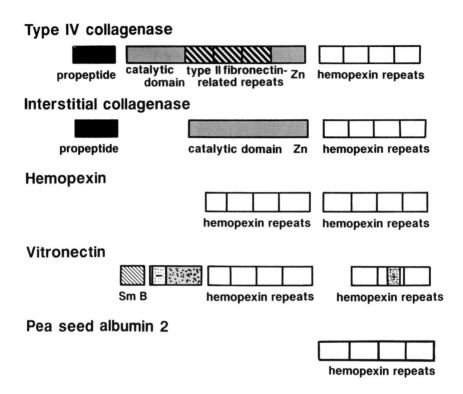

Figure 2. Hemopexin family of proteins which present homology with respect to the organization of hemopexin-type repeats [for details see 11,26,27]

Heparin-binding region. A feature in several heparin-binding proteins is a particular consensus site sequence of the type X-B-B-X-B-X or X-B-B-B-X-X-B [28] serving as poly-anionic binding motif (B: basic residues). Interestingly, within the heparin-binding site of vitronectin two adjacent consensus sites of this type exist, and using synthetic peptides this particular region has been demonstrated to be the heparin interaction domain [29]. Part of this region and adjacent basic portions of the carboxy-terminus of vitronectin interact also with other protein ligands such as plasminogen activator inhibitor-1 [30,31] or collagen and other matrix proteins (Rosenblatt and Preissner, unpublished observations). It is believed that this basic region is also involved in intra- as well as intermolecular stabilization of vitronectin monomers or multimers, respectively [32].

REFERENCES

1 Yatohgo T, Izumi M, Kashiwagi H, Hayashi M. Cell Struct Funct 1988; 13: 281-292.
2 Preissner KT, Wassmuth R, Müller-Berghaus G. Biochem J 1985; 231: 349-355.
3 Dahlbäck B, Podack ER. Biochemistry 1985; 24: 2368-2374.
4 Nakashima N, Miyazaki K, Ishikawa M, Yatohgo T, et al. Biochim Biophys Acta 1992; 1120: 1-10.
5 Nagano Y, Hamano T, Nakashima N, Ishikawa M, et al. J Biol Chem 1992; 267: 24863-24870.
6 Suzuki S, Oldberg A, Hayman EG, Pierschbacher MD, et al. EMBO J 1985; 4: 2519-2524.
7 Jenne D, Stanley KK. EMBO J 1985; 4: 3153-3157.
8 Sato R, Komine Y, Imanaka T, Takano TJ. J Biol Chem 1990; 265: 21232-21236.
9 Seiffert D, Keeton M, Eguchi Y, Sawdey M, et al. Proc Natl Acad Sci USA 1991; 88: 9402-9406.
10 Ehrlich HJ, Richter B, von der Ahe D, Preissner KT. 1991; EMBL accession No. X63003.
11 Jenne D, Stanley KK. Biochemistry 1987; 26: 6735-6742.
12 Tomasini BR, Mosher DF. Prog Hemostas Thromb 1990; 10: 269-305.
13 Preissner KT. Ann Rev Cell Biol 1991; 7: 275-310.
14 Fryklund L, Sievertsson H. FEBS Lett 1978; 87: 55-60.
15 Buckley MF, Loveland KA, McKinstry WJ, Garson OM. J Biol Chem 1990; 265: 17506-17511.
16 Grundmann U, Römisch J, Siebold B, Bohn H, et al. DNA Cell Biol 1990; 9: 243-250.
17 Hynes RO. Cell 1992; 69: 11-25.
18 Ruoslahti E, Pierschbacher MD. Science 1987; 238: 491-497.
19 Main AL, Harvey TS, Baron M, Boyd J, et al. Cell 1992; 71: 671-678.
20 Aota S, Nagai I, Yamada KM. J Biol Chem 1991; 266: 15938-15943.
21 Oldberg A, Franzen A, Heinegard D. J Biol Chem 1988; 263: 19430-19432.
22 Oldberg A, Franzen A, Heinegard D. Proc Natl Acad Sci USA 1986; 83: 8819-8823.
23 Sadler JE, Shelton-Inoles BB, Serace JM, Harlan JM, et al. Proc Natl Acad Sci USA 1985; 62: 6394-6398.
24 Mann KG, Blion J, Butkowski RJ, Downing M, et al. Meth Enzymology 1981; 80: 286-302.
25 Hunt LT, Barker WC, Chen HR. Prot Seq Data Anal 1987; 1: 21-26.
26 Stanley KK. FEBS Lett 1986; 199: 249-253.
27 Jenne D. Biochem Biophys Res Commun 1991; 176: 1000-1006.
28 Cardin AD, Weintraub HJR. Arteriosclerosis 1989; 9: 21-31.
29 Preissner KT, Grulich-Henn J, Ehrlich HJ, Declerck P, et al. J Biol Chem 1990; 265: 18490-18498.

30   Kost C, Stüber W, Ehrlich H, Pannekoek H, et al. J Biol Chem 1992; 267: 12098-12105.
31   Gechtman Z, Sharma R, Kreizman T, Fridkin M, et al. FEBS Lett 1993; 315: 293-297.
32   Stockmann A, Hess S, Declerck P, Timpl R, et al. J Biol Chem 1993; in press.

Biology of Vitronectins and their Receptors
K.T. Preissner, S. Rosenblatt, C. Kost, J. Wegerhoff and D.F. Mosher, editors

Scanning microcalorimetry and equilibrium chemical denaturation studies of human plasma vitronectin.

Cynthia B. Peterson

Department of Biochemistry, University of Tennessee, Knoxville, M407
Walters Life Science Building, Knoxville, TN 37996

INTRODUCTION

Features of proteins which confer specificity in interaction with other proteins, small molecular weight ligands, or macromolecules are of central importance for regulatory processes in biology.   In many cases, conformational adaptability enables a given protein to interact with different biomolecules depending on environmental conditions. This type of conformational flexibility is proposed to account for the different intermolecular interactions of the human glycoprotein, vitronectin. Vitronectin is an adhesive glycoprotein that interacts with a wide variety of macromolecules within the circulation and within the extracellular matrix, and it has been proposed to have a regulatory role in diverse physiological processes, including control of coagulation and fibrinolysis, regulation of the immune response, control of pericellular proteolysis, and cell attachment (reviewed in 1-5). Vitronectin exists predominantly in human plasma in a "folded" or "closed" form which is stabilized by intramolecular interactions. However, once incorporated into the extracellular matrix, the protein assumes an elongated conformation. Denaturation of the protein is thought to convert vitronectin to an "extended" or "open" form which may mimic the conformation of the protein in the extracellular matrix.

Interestingly, binding of heparin to purified vitronectin was shown to increase significantly upon chemical denaturation with urea or guanidine (6, 7) or heat treatment (7). Evidence from the chemical denaturation experiments indicating a cryptic binding site for heparin led to a model incorporating different conformations of the protein which have different ligand binding properties. The model is that vitronectin is stabilized by interactions between the positively charged heparin-binding region near the C-terminus and a negatively charged region near the N-terminus, in effect sequestering the heparin-binding region (6). The action of denaturants is presumably to interrupt noncovalent interactions which stabilize the "closed" form of the protein, converting it to an "open" form which can interact with heparin. Support for this model involving at least two conformational forms of the protein has been since added from demonstrations that other macromolecular ligands bind preferentially to the "denatured" or "open" form of vitronectin (7-9). Furthermore, it is increasingly apparent that the conformation of vitronectin markedly influences its biological properties, including binding of the terminal complement complex (10), interaction with PAI-1 (11), and receptor-mediated endocytosis of the protein (12).

The rapidly emerging evidence for different conformational forms of vitronectin which have distinct functional properties necessitates further characterization of native and denatured protein using biophysical techniques.   In order to more fully characterize the native and unfolded forms of the protein, a detailed treatment of the denaturation and renaturation behavior of vitronectin as a "protein folding" problem was initiated using spectroscopic and calorimetric methods.

MATERIALS AND METHODS

Materials -- Vitronectin was purified from human plasma essentially as described by Dahlback and Podack (13). Purity was assessed by denaturing polyacrylaminde gel electrophoresis (PAGE) in the presence of sodium dodecyl sulfate and β-mercaptoethanol (14). Multimeric vitronectin was prepared by denaturing protein in 8M urea for 2 hr, followed either by affinity chromatography on heparin-agarose (7) or by dialysis into PBS. Formation of vitronectin multimers was analyzed by non-reducing PAGE (15). Western blotting was performed with a rabbit polyclonal serum against vitronectin, a gift from Deane Mosher at the University of Wisconsin and a goat-anti-rabbit IgG conjugated with horseradish peroxidase which was purchased from Vector Labs. Heparin (average molecular weight = 15,000) was obtained from Sigma Chemical Company. Ultrapure-guanidine hydrochloride was purchased from Bethesda Research Laboratories. Urea was a product of Schwartz-Mann.

Differential scanning calorimetry -- Scanning microcalorimetry was performed using a Microcal MC-2D instrument interfaced with an IBM-AT computer for data collection and analysis using software provided by the manufacturer. The instrument was operated at a scan rate of 53°C per hr in the temperature range of 30 to 90°C, with data points collected at 15 sec intervals. A pressure of 30 psi was maintained on the cells during calorimetry experiments to prevent boiling inside the sample cell at high temperature. Buffer for the experiments was 40 mM potassium phosphate buffer, pH 7.4, containing 0.15 M NaCl (PBS). Protein samples were prepared by dialysis overnight into PBS, and both protein and dialysate were filtered through a 0.45 µm sterile filter and degassed for 10 min before loading into the calorimeter. Sample concentration was 2 mg•ml^{-1} vitronectin. For the run with added heparin, the ligand was added from a concentrated stock to give a final concen- tration of 1 x 10^{-4}M. Normalization of the curves to baseline levels empirical, selecting points before and after the curves at which the instrument response deviated from the linear. The total calorimetric enthalpy, ΔH_{cal}, was calculated as the integrated area of the endotherm in the ΔC_p vs. temperature plot. The t_m was assigned to the point at which ΔC_p was maximum.

Equilibrium denaturation studies -- Unfolding of vitronectin was induced by guanidine hydrochloride (GdnHCl) or urea and monitored by changes in intrinsic protein fluorescence. Buffer used for the experiments was 0.1 M sodium phosphate buffer, pH 7.5, containing 0.15 M NaCl and 0.04% Tween-80. Urea was dissolved to a final concentration of 10 M in the same buffer, and Gdn-HCl was dissolved to a final concentration of 8 M in the buffer. Buffer and denaturant solutions were filtered through a 0.45 µm membrane prior to use. Vitronectin was initially dialyzed into the buffer, and filtered through a 0.45 µm membrane to remove any particulate material prior to the fluorescence measurement. Aliquots from the concentrated protein stock were diluted in varying combinations of buffer and denaturant solution.

Denaturation was performed at room temperature until equilibrium was achieved approximately 10 hr after mixing of the protein and denaturant. Protein fluorescence measurements were made on a Perkin-Elmer Model LS-5B fluorimeter at 25 °C. Emission spectra were recorded from 320 to 450 nm using an excitation wavelength of 290 nm.

Observed fluorescence (F_{obs}) at a given wavelength was normalized to relative fluorescence (f) using the relationship $f = (F_{obs}-F_u)/(F_n-F_u)$, where F_u and F_n represent the fluorescence at that wavelength for unfolded and native protein, respectively. The fluorescence unfolding data were fit to a two-state model to calculate a Gibbs free energy change, ΔG, between native and denatured protein by non-linear regression of the equation:

$$f = ((Y_n + M_n{*}D) + ((Y_u + M_u{*}D) {*}\exp(-(\Delta G_{H2O})/RT + MG{*}D/RT)))/$$
$$(1 + \exp(-(\Delta G_{H2O}/RT + MG{*}D/RT)))$$

where R is the gas constant, T in the temperature in °K, ΔG_{H2O} is the free energy change between native and unfolded protein in the absence of denaturant, D is denaturant concentration, MG is the slope describing the dependence of ΔG on denaturant concentration, Y_n and Y_u are the concentration independent optical values and M_n and M_u are the slopes of the pre- and post-transition regions, respectively (16).

Iodide quenching of fluorescence -- Quenching of tryptophan fluorescence was performed by addition of sodium iodide as described previously (17). Sodium iodide was dissolved in PBS to a final concentration of 5 M. For quenching measurements on native protein, vitronectin was diluted to a concentration of 0.1 mM in PBS, and 2 ml of the protein sample was placed in the cuvette. Multimeric vitronectin for quenching studies was prepared in the same way. For quenching measurements on fully denatured vitronectin, protein was diluted to 0.1 mM in a solution of 8 M GdnHCl prepared as described above. Small volume aliquots from the sodium iodide stock were added and protein fluorescence was measured immediately. The excitation wavelength was 290 nm. Emission was measured at the maximum wavelength observed in the emission spectra for native, denatured, and renatured protein, equal to 338nm, 354nm, and 350nm, respectively.

RESULTS

Thermal denaturation of vitronectin -- With the current focus in the literature on protein stability and denaturation, it was of prime interest to evaluate the thermodynamics of denaturation and effects that ligands have on denaturation of a protein that appears to carry out some of its physiological functions in a "denatured" form. Preliminary studies on vitronectin have been implemented using differential scanning calorimetry to monitor thermal denaturation of the protein. The calorimetric scan observed for vitronectin in 40 mM sodium phosphate, pH 7.4, containing 0.15 M sodium chloride are shown in Figure 1a. A relatively symmetric endotherm is observed, with a calorimetric enthalpy (ΔH_{cal}) for the scan of 620 Kcal•mol^{-1}, while the van't Hoff (ΔH_{vh}) enthalpy calculated is

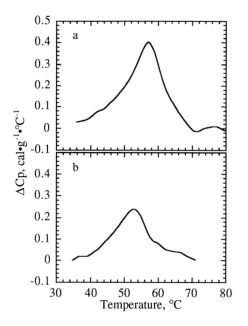

**Figure 1. Thermal Denaturation of
Vitronectin by DSC
a. no heparin; b. plus saturating heparin**

60.0 Kcal•mol^{-1}. The nonidentity of these values indicates that the transition from folded to denatured form indeed is not a two-state transition.

An interesting and perhaps surprising result that came from the calorimetric analyses is that no endotherm was observed in a subsequent second scan of the same protein sample. Under the conditions of the experiment, the "elongated" or "open" form of the protein that is generated by thermal denaturation does not appear to have an ordered structure leading to cooperative denaturation of the protein. This result implies that the product of thermal denaturation in this case does not form a compact structure stabilized by extensive tertiary interactions. This observation is certainly consistent with a model in which vitronectin assumes an elongated or extended conformation upon denaturation.

In figure 1b is the endotherm observed for vitronectin in the presence of saturating amounts of heparin. Strikingly, the presence of ligand actually *decreases* the t_m, the temperature at which the maximum change in heat capacity is observed during the thermal transition. (It should be noted that heparin alone does not give a discrete endotherm in calorimetric analysis.) According to LeChatelier's principle, ligands which bind preferentially to the native or folded conformation of a protein will stabilize the protein to denaturation. Conversely, ligands which favor binding to the denatured form of a protein would be predicted to facilitate denaturation, with the net result in calorimetric studies observed as a decrease in the midpoint of the thermal transition. Thus, the decrease in t_m observed to result from heparin binding to vitronectin provides independent physicochemical support for the contention that the extended form of the protein exhibits enhanced heparin binding. A net destabilization of folding of a protein resulting from ligand binding is not commonly observed in differential scanning calorimetric experiments. Recently, Makhatadze and Privalov (18) have reported results from studies on several model proteins which examine the effects of urea and guanidinium hydrochloride in calorimetric experiments, with the goal of determining the mechanism by which such reagents exert their effects. The addition of either of the chemical denaturants during a differential scanning calorimetric run is observed to lower the t_m for the thermal transition. However, in contrast to the results on urea and guanidinium hydrochloride, the lowering of the t_m for

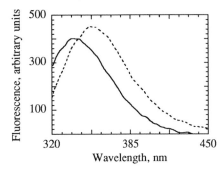

Figure 2. Emission spectra of native (*solid line*) and urea-denatured (*dashed line*) vitronectin.

vitronectin by heparin is of prime interest because the ligand interaction is presumed to be physiologically relevant. For the data acquired in the presence of heparin, the ΔH_{cal} is determined to be 230 Kcal•mol^{-1}, while the ΔH_{vh} is calculated to be 62.6 Kcal•mol^{-1}.

Chemical denaturation of vitronectin -- The scanning calorimetry studies on vitronectin were supplemented by a chemical denaturation approach in which changes in intrinsic protein fluorescence were used as a measure of equilibrium unfolding induced by guanidinium hydrochloride or urea. Denaturation of the protein by either of the reagents is accompanied by only a small change in fluorescence quantum yield; however, a notable red shift in the emission maximum is observed, as expected upon increased exposure of tryptophan residues to solvent (Figure 2). Unfolding curves were generated by following the decrease in fluorescence at 325nm upon addition of denaturant. Equilibrium in the folding/unfolding reaction in the presence of either urea or GdnHCl was observed to be attained only after incubation with denaturant for long times (at least 10 hr). Typical urea denaturation and GdnHCl denaturation curves are shown in Figure 3.

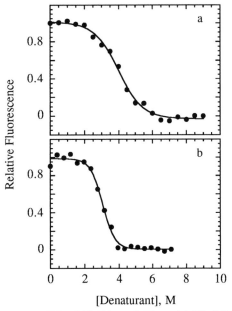

Using native protein, no significant perturbations of the unfolding curves were observed upon varying the protein concentration 10-fold or upon adding saturating amounts of heparin. Since preparation methods which invoke denaturation of vitronectin followed by affinity chromatography on a heparin affinity matrix have been reported to yield protein which is predominantly high molecular weight disulfide-stabilized multimers (10, 11, 19, 20), protein samples were analyzed for multimer formation *during denaturation* by polyacrylamide gel electrophoresis and Western blotting. Electrophoretic analysis demonstrated that multimers are not formed in the presence of

Fig. 3. GdnHCl (b) and Urea (a) Unfolding

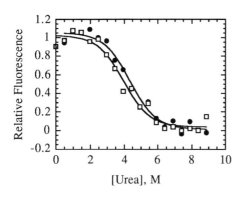

Figure 4. Unfolding of Native (●) and Renatured (□) Vitronectin

denaturant; rather, high molecular weight multimers are observed upon renaturation by removal of either urea or GdnHCl by dialysis. Furthermore, the multi-merization at relatively low protein concentrations (approximately 0.1 mg•ml[-1)] upon renaturation was observed to be minimized by inclusion of Tween-80 in the dialysis buffer. Using this approach, reversibility of the unfolding of vitronectin was observed so that chemical denaturation curves characterizing unfolding of denatured and subsequently renatured protein are superimposable with denaturation curves for native protein (Figure 4).

Iodide quenching of intrinsic fluorescence for native and denatured protein -- Iodide quenching was used to compare native vitronectin, fully denatured vitronectin (in the presence of chemical denaturants), and chemically denatured vitronectin subsequent to removal of the denaturant. As shown in Figure 5, fully denatured vitronectin is more susceptible to iodide quenching, consistent with the red shift observed in fluorescence emission. Multimeric vitronectin exhibits similar quenching behavior to the fully denatured form of the protein.

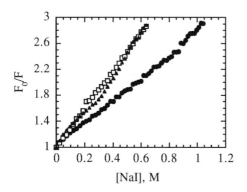

Figure 5. Stern Volmer Plot Native (●), Denatured (□) and Renatured (▲) Vitronectin

DISCUSSION

As a first step in biophysical characterization of various conformational forms of human plasma vitronectin, insight into the thermal and chemical denaturation of the protein have been provided by scanning microcalorimetry and equilibrium chemical denaturation studies. Calorimetric measurements on vitronectin indicate that the protein exhibits a broad endotherm upon temperature denaturation which indicates that a simple two-state approximation cannot be used to interpret thermal unfolding. Vitronectin would not necessarily be expected to adhere to a classic two-state model for reversible protein denaturation since the forms of the protein before and after denaturation exhibit different ligand binding properties. Although the ratio between the ΔH_{cal} and the ΔH_{vh} is often interpreted as the number of overlapping two-state transitions which comprise the total endotherm and correspond to structural units which denature cooperatively, such a conclusion is premature in the absence of detailed three-dimensional structural information describing vitronectin. Strikingly, the presence of saturating levels of heparin actually *decreases* the t_m, the temperature at which the maximum change in heat capacity is observed during the thermal transition, providing yet another independent bit of experimental evidence that the unfolded form of the protein exhibits enhanced heparin binding.

Changes in intrinsic fluorescence of vitronectin upon exposure to denaturants are readily measured by the red shift in the emission maximum resulting from increased exposure of tryptophan residues upon denaturation. From comparison of the unfolding curves generated in GdnHCl and urea, vitronectin denatures with a midpoint the corresponds to a lower denaturation midpoint in GdnHCl than in urea. This behavior is typical of most proteins since GdnHCl is a relatively stronger denaturant than urea. Conditions were used in which the chemical denaturation of vitronectin in urea appeared to be reversible. This was demonstrated by the superimposition of urea unfolding curves for the native protein and protein that was denatured followed by renaturation under appropriate conditions. The reversibility of the unfolding provided an opportunity to evaluate the free energy of unfolding using thermodynamic treatment. The free energy of unfolding, ΔG_{H_2O}, was calculated from the urea denaturation data to be approximately 3.5 Kcal•mol^{-1}. This estimated value appears reasonable in comparison with free energies of unfolding calculated for other monomeric proteins.

Chemical quenching of fluorescence was used to compare tryptophan exposure in native, fully denatured and multimeric vitronectin. Sodium iodide was chosen for the studies since iodide cannot penetrate the interior of a protein by virtue of its negative charge; thus, the ion has been used traditionally to determine the solvent accessibility of tryptophan residues within a protein matrix. The observed difference between the quenching of the native and fully denatured forms of vitronectin is easily rationalized by consideration of the proposed model for "folded" and "extended" forms of the protein. Seven out of 9 of the tryptophans within the protein are located within the C-terminal half of the protein, with several tryptophans found in the local vicinity of the highly positively charged heparin-binding region (tryptophan residues 303, 320, 382 and 405). Indeed, local charge can affect quenching of tryptophan fluorescence within proteins due to the ionic nature of the iodide quencher, and these tryptophans would be predicted to be more susceptible to iodide quenching in the extended form of the protein. The data which demonstrates that the multimeric form of the protein exhibits fluorescence quenching akin to that of fully denatured protein suggests that the susceptible tryptophans are nearly fully exposed in this form of the protein.

Both the thermal denaturation data and the iodide quenching argue for a non-native vitronectin species which is the preferable heparin-binding form. Equilibrium chemical denaturation and fluorescence quenching behavior provide evidence for exposure of tryptophan residues upon denaturation and/or multimerization. Taken together, these data support the proposed model in which treatment of vitronectin with various denaturants disrupts intramolecular contacts and yields an "open" form of the protein which is effective in ligand binding.

REFERENCES

1.   K. T. Preissner, *Ann. Rev. Cell. Biol.* **7**, 275-310 (1991).
2.   B. R. Tomasini, D. F. Mosher, in *Prog. in Hemostasis and Thrombosis* B. S. Coller, Ed. (W. B. Saunders Company, 1991), vol. 10, pp. 269-305.
3.   K. T. Preissner, D. Jenne, *Thromb. Haemostas.* **66**, 189-194 (1991).
4.   K. T. Preissner, D. Jenne, *Thromb. Haemostas.* **66**, 123-132 (1991).
5.   K. T. Preissner, *Blut* **59**, 419-431 (1989).
6.   K. T. Preissner, G. Muller-Berghaus, *J. Biol. Chem.* **262**, 12247-12253 (1987).
7.   M. Hayashi, T. Akama, I. Kono, H. Kashiwagi, *J. Biochem.* **98**, 1135-1138 (1985).
8.   M. Ishikawa, M. Hayashi, *Biochim. Biophys. Acta* **1121**, 173-177 (1992).
9.   P. J. Declerck, *et al.*, *J. Biol. Chem.* **263**, 15454-15461 (1988).
10.   K. Hogasen, T. E. Mollnes, M. Harboe, *J. Biol. Chem.* **267**, 23076-23082 (1992).
11.   M. C. Naski, D. A. Lawrence, D. F. Mosher, T. J. Podor, D. Ginsburg, *J. Biol. Chem.* **268**, 12367-12372 (1993).
12.   T. S. Panetti, P. J. McKeown-Longo, *J. Biol. Chem.* **268**, 11988-11993 (1993).
13.   B. Dahlback, E. R. Podack, *Biochemistry* **24**, 2368-2374 (1985).
14.   U. K. Laemmli, *Nature (Lond.)* **227**, 680-685. (1970).
15.   T. Jovin, A. Chramback, M. A. Naughton, *Anal. Biochem.* **9**, 351 (1964).
16.   M. M. Santoro, D. W. Bolen, *Biochemistry* **27**, 8063-8038 (1988).
17.   J. R. Lakowicz, *Principles of Fluorescence Spectroscopy.* (Plenum Press, New York, 1983).
18.   G. I. Makhatadze, P. L. Privalov, *J. Mol. Biol.* **226**, 491-505 (1992).
19.   B. R. Tomasini, D. F. Mosher, *Blood* **72**, 903-912 (1988).
20.   B. R. Tomasini, M. C. Owen, J. W. F. II, D. F. Mosher, *Biochemistry* **28**, 7617-7623 (1989).

Biology of Vitronectins and their Receptors
K.T. Preissner, S. Rosenblatt, C. Kost, J. Wegerhoff and D.F. Mosher, editors

Distribution of Vitronectin

Dietmar Seiffert[a], Thomas J. Podor[b] and David J. Loskutoff[a]

[a]Department of Vascular Biology, The Scripps Research Institute, 10666 N. Torrey Pines Rd., La Jolla, California, 92037, USA

[b]Hamilton Civic Hospitals Research Centre, 711 Concession Street, Hamilton, Ontario, L8V 1C3, Canada

Introduction

Vitronectin (Vn) belongs to a group of cell adhesion molecules which mediate adhesion through a common RGD-dependent mechanism. It is identical to the S-protein of the complement system, and also appears to serve several regulatory functions in the coagulation and fibrinolytic systems (for reviews, [1,2]). Although initially described as a plasma protein, recent studies provide evidence that Vn also is present in various tissues and body fluids, and that its biosynthesis may be regulated. These studies are reviewed and their implications regarding the potential function(s) of Vn are discussed.

Vitronectin in body fluids

Vn is present in normal *plasma* at concentrations of 200-400µg/ml [3-7] and thus constitutes 0.2 to 0.5 percent of total plasma protein. Unlike fibrinogen and fibronectin, the concentration of Vn in plasma does not significantly differ from that in serum [5]. Reduced plasma levels of Vn have been reported in patients with severe liver failure [6] and cirrhosis of the liver [8,9]. In these patient groups, changes in the plasma Vn levels closely parallel changes in albumin, cholinesterase and prothrombin time, and the plasma Vn levels decrease with the progression of the disease [8,9]. These results suggest that the liver is the major source of plasma Vn. Increased plasma Vn concentrations have been detected in patients undergoing elective orthopedic surgery and in rodents stimulated with acute phase mediators (Seiffert & Podor, unpublished observation), suggesting that Vn may be regulated as an acute phase protein (see below).

Platelets contain approximately 1µg Vn/10^9 platelets [10,11], a value that could account for approximately 0.2 percent of Vn present in serum. Immunohistochemical studies reveal that Vn is present in platelet alpha-granules [12], and it can be released together with platelet factor 4 upon thrombin stimulation [11]. It is unknown whether the Vn in platelets was actually synthesized by megakaryocytes or endocytosed from plasma and incorporated into alpha-granules. The majority of platelet Vn consists of disulfide-bonded multimers [11].

Vn also has been found in *urine, amniotic fluid*, and *bronchoalveolar lavage fluid* [5,13-15]. Although the concentration in amniotic fluid is relatively low, its specific activity (defined as µg Vn/mg total protein) is identical to plasma [5]. The specific activity of Vn in urine is twice that of plasma [5]. Although Vn is present in the lavage fluid from healthy volunteers, the amount is approximately 10-fold higher in patients with interstitial lung disease, including idiopathic pulmonary fibrosis, sarcoidosis and hypersensitivity pneumonitis [13-15]. These results raise the possibility that Vn may serve to regulate inflammatory reactions in the alveoli associated with lung disease. Cultured alveolar macrophages secrete a protein immunological identical to Vn [16] suggesting that the Vn in the lavage fluid may be synthesized locally by alveolar macrophages. On the other hand, the Vn concentration in the lavage fluid correlated with that of albumin [13], raising the possibility that this Vn results from leakage across the alveolocapillary membrane rather than being produced by alveolar cells. In general, there is no detectable age-dependent difference in the Vn concentration in any of these body fluids. However, fetal cord blood contained only 60 to 70 percent of the Vn present in adult blood [5].

Vitronectin in tissues

It has been reported that Vn antigen is present in embryonic tissue, fetal membranes, smooth and skeletal muscle, kidney, supporting stroma of portal triads, and the capsular surface of viscera [17]. However, the monoclonal antibody employed in these studies was later found to cross-react with a 30kD extracellular protein distinct from Vn [18], thus bringing into question the interpretation and significance of these findings [18].

Unlike fibronectin, Vn appears to be absent from the basement membrane of most tissues studied. However, Vn antigen was detected in diseased renal tubular basement membranes [19]. A strong correlation between the deposition of the membrane attack complex of complement and Vn was found in kidneys of patients with glomerulonephritis, arteriosclerosis, and systemic lupus erythematosus [19-22]. There was no evidence for Vn deposits in the absence of membrane attack complex [22]. These observation suggest that Vn may function to regulate immune reactions in the diseased kidney.

Although Vn immunoreactivity seems to be absent in embryonic and fetal spleen, Vn antigen has been detected in the red and white pulp of adult spleen tissue [23]. A similar age-dependent accumulation of Vn has been described in the skin [24]. The reader is referred to the chapter by K. Dahlbaeck for further discussion on the distribution of Vn in the skin.

Vn is also present as an equatorial band on the head of capacitated sperm [25]. In a subpopulation of infertile men, Vn immunoreactivity was reduced, suggesting a yet unidentified role of Vn in fertilization [25]. Vn antigen has also been localized to elastic fibers of the endometrial stroma [26].

Light microscopic studies of the liver revealed vitronectin deposition in the area of focal necrosis and in the portal tracts in patients with acute and chronic viral hepatitis, and in areas of fibrous deposition in the liver of patients with cirrhosis [8]. Vn deposition has also been reported in areas of fibrosis and necrosis for a number of different tissues, including the skin [27], kidney [28], nerve tissue of patients with peripheral neuropathy [29], fibrotic and sclerotic lymphoid tissue [30], germinal center areas in follicular lymphoid hyperplasia [31], fibrillar deposits in the connective tissue matrix around all types of breast cancer [32], myelofibrotic bone marrow [30], and in areas of acute myocardial infarction [33].

Similar staining of sclerotic and necrotic debris was noted in intimal thickenings and fibrous plaques of atherosclerotic arteries in association with collagen bundles, elastic fibers, and cell debris in the vicinity of elastin [34,35]. Moreover, a monoclonal antibody that specifically stained atherosclerotic lesions was shown to detect Vn [36]. These findings point to a possible role of Vn in the pathogenesis of atherosclerotic lesions. However, Vn was also detected in apparently normal vessels in organs including the uterus [26], spleen [23], gut [37], and kidney [22], suggesting that Vn accumulation in the vessel wall is not a specific marker for vascular disease.

Vn staining was also evident in degenerative neurological disease, and co-localized with senile plaques and neurofibrillary tangles in Alzheimer entorhinal cortex [38]. Vn antigen was not detectable in normal brain [38,39]. In agreement with these findings, glial cells in low grade astroglial tumors and in reactive astrogliosis lack Vn staining [39]. However, Vn immunoreactivity was observed in glioblastoma parenchyma, suggesting that it may be a marker of neuroectodermally-derived tumors [39].

Vitronectin biosynthesis

Only a limited number of *cultured cells* have been shown to synthesize and secrete Vn [40]. One problem with some of the published reports is the lack of data to distinguish between the possibility that Vn is actually synthesized by the cells or endocytosed from the medium and released during subsequent incubation steps. Indeed, the later was recently demonstrated for human umbilical vein endothelial cells [41]. Definitive proof for Vn biosynthesis (based for example on metabolic labeling followed by immunoprecipitation with Vn-specific antibodies) has only been provided for two human hepatoma cell lines (Hep G2 and Hep 3B; [40,42,43]). However, data is available to suggest that other cells may also produce Vn, including blood-derived cultured monocytes and macrophages [44], and a teratoma cell line resembling parietal endoderm [45].

Limited information is available on the tissues that *produce Vn in vivo*. Vn cDNAs have been isolated from two independent liver cDNA libraries, indicating that Vn mRNA is present in the human liver [42,46]. In a murine model system, we identified the liver as the major site of Vn biosynthesis in vivo [47]. The steady-state level of Vn mRNA was at least 40-fold higher in the mouse liver than in mouse heart, lung, kidney, spleen, muscle, brain, thymus, uterus, testes, skin, adipose tissue, and aorta [47]. In situ hybridization revealed that the hepatocyte is the primary Vn producing cell type in the liver [47], a finding that was confirmed by liver cell fractionation studies (Seiffert & Podor, unpublished observation). These results suggest that the hepatocytes are the main source of Vn biosynthesis in vivo. It should be noted that Solem et al. [48] also reported the presence of Vn in mouse liver, and in addition could detect low levels of Vn-specific mRNA in mouse brain. We have recently detected Vn mRNA in the brain using more sensitive assays (Seiffert & Loskutoff, unpublished observation).

Preliminary observations raise the possibility that Vn biosynthesis may be regulated in some situations. For example, transforming growth factor beta at picomolar concentrations stimulates Vn biosynthesis in cultured human hepatoma cells [49]. In addition, platelet-derived growth factor and epidermal growth factor increased the steady-state level of Vn mRNA in hepatoma cells, although to a lesser extent [49]. Vn biosynthesis in Hep G2 is also stimulated by interleukin-6, the major acute phase cytokine (unpublished observation).

To examine the possibility that the synthesis of Vn in vivo is regulated as an acute phase reactant, we examined hepatic Vn mRNA expression in rats injected with endotoxin, Freunds complete adjuvant, or turpentine [50]. These agents are all acute inflammatory stimuli which increase serum levels of interleukin-6 and corticosterone. With each stimulus, enhanced levels of hepatic Vn mRNA expression were observed within 3 h, and increased to a maximum between 12 to 18 h. Similar increases were observed in the level of expression of hepatic RNA for cysteine proteinase inhibitor, the major acute phase protein in rats. Immunoblotting analysis also revealed that the plasma Vn concentration is increased during acute inflammation. To determine the effect of interleukin-6 on the expression of hepatic Vn mRNA, rats were injected with recombinant rat interleukin-6. Interleukin-6 stimulated Vn mRNA expression within 1 h, and expression remained elevated when interleukin-6 was administered daily for 9 days. These results indicate that Vn in the rat liver is regulated as an acute phase protein [50].

Vitronectin binding proteins

The demonstration that Vn is produced primarily by hepatocytes raises the possibility that the Vn antigen detected in other tissues by immunohistochemical methods may be plasma derived rather than produced by cells in the vicinity of the deposition [47]. The mechanism(s) by which Vn becomes incorporated into tissues is not known.

Although Vn purified under denaturing conditions (i.e., exposed to 8M urea) binds to type I-VI collagens, plasma Vn does not bind well to any collagen [51]. The importance of the Vn-collagen interaction for the binding of Vn to tissues is further questioned by the finding that Vn binds poorly to collagen under physiological NaCl concentrations [52]. Proteoglycans represent a second group of Vn-binding molecules in tissues, since Vn interacts with heparin, fucoidan and dextran sulfate [2]. This appears to be a specific interaction, since Vn does not bind to dermatan sulfate, chondroitin sulfate, keratan sulfate, or heparan sulfate [2]. These data raise the possibility that glycosaminoglycans coupled to their core protein may bind Vn in vivo. Although several immunohistochemical studies suggested that Vn may be associated with elastin [20,24,30,35], direct binding studies to substantiate these findings have not yet been reported. Also, activated factor XIII and tissue transglutaminases cross-link Vn to itself forming large multimers [53]. In addition, factor XIIIa may also cross-link Vn to yet unidentified extracellular matrix structures [54]. Cross-linking could promote the immobilization of Vn in tissues.

Frequently, Vn has been localized to areas of fibrosis and necrosis [30,33], where intermediate filaments are expected to be exposed. We recently reported that Vn associates with vimentin-type intermediate filaments in vivo, in cell culture, and in a purified protein system [55]. Vn also binds to keratin filament aggregates after incubation of tissue section with purified Vn [56]. Thus, binding of Vn to intermediate filaments in both normal and injured tissues may provide a mechanism for its incorporation into tissues.

In summary, in vivo localization studies have provided clues as to the potential role(s) of Vn in vivo. However, more detailed analysis of its tissue distribution and mechanism of deposition, and the identification of the molecules to which it binds in vivo will be necessary to further increase our knowledge of its function(s) under both normal and pathophysiological conditions.

References

1   Preissner KT. Annu Rev Cell Biol 1991; 7: 275-310.
2   Tomasini BR, Mosher DF. Vitronectin. In: Coller BS, ed. Progress in Hemostasis and Thrombosis. Philadelphia, Sydney: Saunders Company, 1990: 269-305.
3   Barnes DW, Silnutzer J, See C, Shaffer M. Proc Natl Acad Sci USA 1983; 80: 1362-1366.
4   Preissner KT, Wassmuth R, Mueller-Berghaus G. Biochem J 1985; 231: 349-355.
5   Shaffer MC, Foley TP, Barnes DW. J Lab Clin Med 1984; 103: 783-791.
6   Conlan MG, Tomasini BR, Schultz RL, Mosher DF. Blood 1988; 72: 185-190.
7   Hogasen K, Mollnes TE, Tschopp J, Harboe M. J Immunol Meth 1993; 160: 107-115.
8   Inuzuka S, Ueno T, Torimura T, et al. Hepatology 1992; 15: 629-636.
9   Kemkes-Matthes B, Preissner KT, Langenscheidt F, Matthes KJ, Mueller-Berghaus G. Eur J Haematol 1987; 39: 161-165.
10   Parker CJ, Stone OL, White VF, Bernshaw NJ. Br J Haematol 1989; 71: 245-252.
11   Preissner KT, Holzhueter S, Justus C, Mueller-Berghaus G. Blood 1989; 74: 1989-1996.
12   Roger M, Halstensen TS, Hogasen K, Mollnes TE, Solum NO, Hovig T. Nouv Rev Fr Hematol 1992; 34: 47-54.
13   Eklund AG, Sigurdardottir O, Oehrn M. Am Rev Respir Dis 1992; 145: 646-650.
14   Teschler H, Pohl WR, Thompson AB, et al. Am Rev Respir Dis 1993; 147: 332-337.
15   Pohl WR, Conlan MG, Thompson AB, et al. Am Rev Respir Dis 1991; 143: 1369-1375.
16   Pettersen HB, Johnson E, Mollnes TE, Garred P, Hetland G, Osen SS. Scand J Immunol 1990; 31: 15-23.
17   Hayman EG, Pierschbacher MD, Oehgren Y, Ruoslahti E. Proc Natl Acad Sci USA 1983; 80: 4003-4007.
18   Tomasini-Johansson BR, Ruoslahti E, Pierschbacher MD. Matrix 1993; 13: 203-214.
19   Falk RJ, Podack E, Dalmasso A, Jenette JC. Am J Pathol 1987; 127: 182-190.
20   Bariety J, Hinglais N, Bhakdi S, Mandet C, Rouchon M, Kazatchkine MD. Clin Exp Immunol 1989; 75: 76-81.

21   Okada M, Yoshioka K, Takemura T, Akano N, Aya N, Murakami K, Maki S. Virchows Arch A Pathol Anat Histopathol 1993; 422: 367-373.

22   French LE, Tschopp J, Schifferli JA. Clin Exp Immunol 1992; 88: 389-393.

23   Liakka KA, Autio-Harmainen HI. J Histochem Cytochem 1992; 40: 1203-1210.

24   Dahlbaeck K, Loefberg H, Alumets J, Dahlbaeck B. J Invest Dermatol 1989; 92: 727-733.

25   Fusi FM, Lorenzetti I, Vignali M, Bronson RA. J Androl 1992; 13: 488-497.

26   D'Cruz OJ, Wild RA. Fertil Steril 1992; 57: 787-795.

27   Dahlbaeck K, Loefberg H, Dahlbaeck B. Acta Derm Venereol 1988; 68: 107-115.

28   Dahlbaeck K, Loefberg H, Dahlbaeck B. Histochem 1987; 87: 511-515.

29   Zanusso GL, Moretto G, Monaco BB, Rizzuto N. Ital J Neurol Sci 1992; 13: 493-499.

30   Reily JT, Nash JR. J Clin Pathol 1988; 41: 1269-1272.

31   Halstensen TS, Mollnes TE, Brandtzaeg P. Immunology 1988; 65: 193-197.

32   Niculescu F, Rus HG, Retegan M, Vlaicu R. Am J Pathol 1992; 140: 1039-1043.

33   Rus HG, Niculescu F, Vlaicu R. Immunol Lett 1987; 16: 15-20.

34   Niculescu F, Rus HG, Porutiu D, Ghiurca V, Vlaicu R. Atherosclerosis 1989; 78: 197-203.

35   Guettier C, Hinglais N, Bruneval P, Kazatchkine M, Bariety J, Camilleri J-P. Virchows Archiv A Pathol Anat 1989; 414: 309-313.

36   Sato R, Komine Y, Imanaka T, Takano T. J Biol Chem 1990; 265: 21232-21236.

37   Halstensen TS, Mollnes TE, Brandtzaeg P. Gastroenterology 1989; 97: 10-19.

38   Akiyama H, Kawamata T, Dedhar S, McGeer PL. J Neuroimmunol 1991; 32: 19-28.

39   Gladson CL, Cheresh DA. J Clin Invest 1991; 88: 1924-1932.

40   Barnes DW, Reing J. J Cell Physiol 1985; 125: 207-214.

41   Hess S, Poetsch B, Voelker W, Preissner KT. Thromb Haemost 1993; 69: 2373.

42   Suzuki S, Oldberg A, Hayman EG, Pierschbacher MD, Ruoslahti E. EMBO J 1985; 4: 2519-2524.

43   Seiffert D, Wagner NN, Loskutoff DJ. J Cell Biol 1990; 111: 1283-1291.

44   Hetland G, Pettersen HB, Mollnes TE, Johnson E. Scand J Immunol 1989; 29: 15-21.

45   Cooper S, Pera MF. Development 1988; 104: 565-574.

46   Jenne D, Stanley KK. EMBO J 1985; 4: 3153-3157.

47   Seiffert D, Keeton M, Eguchi Y, Sawdey M, Loskutoff DJ. Proc Natl Acad Sci USA 1991; 88: 9402-9406.

48   Solem M, Helmrich A, Collodi P, Barnes D. Mol Cell Biochem 1991; 100: 141-149.

49   Koli K, Lohi J, Hautanen A, Keski-Oja J. Eur J Biochem 1991; 199: 337-345.

50   Podor TJ, Geisterfer M, Gauldie J, Seiffert D. Thromb Haemost 1993; 69: 1719.

51   Gebb C, Hayman EG, Engvall E, Ruoslahti E. J Biol Chem 1986; 261: 16698-16703.

52   Izumi M, Shimo-Oka T, Morishita N, Il I, Hayashi M. Cell Struct Funct 1988; 13: 217-225.

53   Sane DC, Moser TL, Parker CJ, Seiffert D, Loskutoff DJ, Greenberg CS. J Biol Chem 1990; 265: 3543-3548.

54   Sane DC, Moser TL, Greenberg CS. Biochem Biophys Res Comm 1991; 174: 465-469.

55   Podor TP, Joshua P, Butcher M, Seiffert D, Loskutoff DJ, Gauldie J. Ann NY Acad Sci 1992; 667: 173-177.

56   Hintner H, Stanzl U, Dahlbaeck K, Dahlbaeck B, Breathnach SM. J Invest Dermatol 1989; 93: 656-661.

Biology of Vitronectins and their Receptors
K.T. Preissner, S. Rosenblatt, C. Kost, J. Wegerhoff and D.F. Mosher, editors

Vitronectin Expression in Human Germ Cell Tumours and Normal Mouse Development

Cooper, S., Bennett, W., Roach, S. and Pera, M.F.
CRC Growth Factors, Department of Zoology, Oxford University, South Parks Road, Oxford, OX1 3PS, United Kingdom.

Introduction

Distribution of the vitronectin antigen using monoclonal and polyclonal antibodies has been studied in many tissues. We have initiated a study to investigate the expression of vitronectin in the tissues of the developing mouse.conceptus.

Vitronectin has been identified in a fibrillar pattern in the loose connective tissue of embryonic lung, kidney, smooth and skeletal muscle and the capsular surface of all viscera [1]. It has been found associated with dermal elastic fibres [2]) and with serum amyloid protein [3] in the skin. It is apparently absent from the basement membrane of most tissues, unlike the more more extensively studied fibronectin.

In human breast and colon carcinomas [4] vitronectin staining was found in the connective tissue in close association with elastin fibres. However, it has been localised within the tumour mass of glioblastoma and melanoma [5].

The distribution of vitronectin in the embryonic chick [6], rat [7] and human [8] heart suggests that it may be involved in the migration of endocardial cells into the truncal swellings of the atrioventricular cushions during development. The fibrillar network and myocardium facing the atrioventricular cushions immunostain positively for vitronectin.

A 30kD extracellular matrix protein that cross reacts with a monoclonal antibody to vitronectin has recently been identified [9]. This suggests that histochemical studies done using only this antibody warrant re-examination.

To date, the major site for *in vivo* biosynthesis of vitronectin is the liver. Studies of tissues using Northern [10] and *in situ* [11] hybridisation have only demonstrated vitronectin mRNA in the liver and, to a lesser extent, in the brain.

The only cells that have been shown to synthesise vitronectin *in vitro* are hepatic cell lines [12], glioblastoma [5] and human yolk sac carcinoma [13].

Vitronectin expression in the yolk sac

A cell line GCT 44, derived from metastatic yolk sac carcinoma [14] resembles rodent parietal endoderm. Unlike other teratoma cell lines the cells can attach and grow in the absence of serum.

The autocrine production of vitronectin by these cells may account in part for their serum independence *in vitro* and their malignant behaviour *in vivo*.

Northern analysis (Fig 1) of total mRNA from a 17.5 day foetal mouse liver and poly(A)+ mRNA from placenta and yolk sac using the full length cDNA from vitronectin as a probe [15], revealed an abundant mRNA transcript at 1.6kb in the liver as found by other workers [10]. Low levels of transcript were found in the yolk sac, and no detectable transcript in the placenta.

Vitronectin transcript was also detected in cell lines representative of yolk sac. GCT 44 produces vitronectin mRNA and secretes the protein. GCT 72, produces abundant vitronectin mRNA but no detectable levels of secreted protein. No transcripts were detected in any of the embryonal carcinoma cell lines.

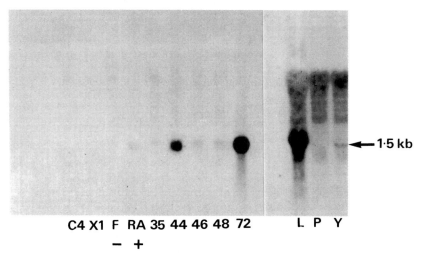

Figure 1
Northern blot analysis using human vitronectin as a probe.
L, mouse liver; P, mouse placenta; Y, mouse yolk sac.
Human cell lines are as follows: C4, GCT 27-C4 (nullipotent embryonal carcinoma); X1, GCT 27-X1 (multipotent embryonal carcinoma); -F, GCT 27-X1 allowed to spontaneously differentiate for 14 days; RA+, GCT 27-X1 treated for 14 days with 1×10^{-6}M retinoic acid; 35, GCT 35 (multipotent embryonal carcinoma); 44, GCT 44 (parietal endoderm-type yolk sac carcinoma); 46, GCT 46 (parietal endoderm-type yolk sac carcinoma);48, GCT 48 (nullipotent embryonal carcinoma); 72, GCT 72 (visceral endoderm-type yolk sac carcinoma).

Yolk sacs were dissected from mouse conceptuses at day 12.5 of gestation, and metabolically labelled for 12 hours in ^{35}S-methionine *in vitro*. Cell lysates and conditioned medium were incubated with anti-vitronectin Ig and then imunoprecipitated using protein A sepharose. The precipitates were analysed by SDS-polyacrylamide electrophoresis (Fig 2).

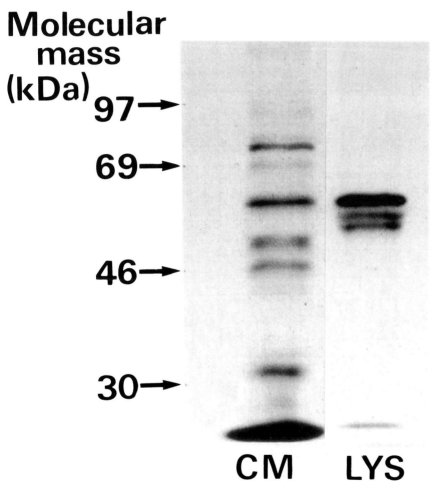

Figure 2
Immunoprecipitation of labelled polypeptides from short term cultures of day 12.5 yolk sac. SDS-polyacrylamide gel electrophoresis was carried out under reducing conditions. CM, conditioned medium; LYS, cell lysate. Molecular weight markers as shown.

A major radiolabelled band of 65kDa was precipitated from both the conditioned medium and cell lysate. This was identical in size to that seen on immunoblots of purified mouse vitronectin and therefore it appears that vitronectin is actually synthesised in the yolk sac.

Vitronectin expression in normal mouse development

   Mouse vitronectin was purified using heparin affinity chromatography [16] and used to prepare specific antisera in rabbit. The antisera immunoblotted a 65kDa band in preparations of purified mouse vitronectin, and was used to examine serial sections from FVB/N mice at days 6.5, 10.5, 12.5 and 16.5. The tissues were fixed in absolute ethanol, embedded in paraffin and 6μm sections made.

Results

The immunohistochemistry results are summarised in Tables 1 -4

Table 1 Day 6.5

Positive tissue	Negative tissue
Decidual cells around the implantation site	Entire conceptus

Table 2 Day 10.5

Positive tissue	Negative tissue
Mid gut	Forebrain
Hindgut	Midbrain
Myocardial cells	Optic cup
Lung buds	Otic vesicle
Hepatic primordium ◊	Neural tube
Placenta (giant cells and blood)	Branchial arches
Visceral yolk sac	Limb buds
Parietal yolk sac	Somites
	Pharynx
	Oesophagus
	Bronchi
	Mesonephros
	Amnion
	Chorionic plate

Table 3 Day 12.5

Positive tissue	Negative tissue
Floor plate of neural tube	Neural tube
Zones of cephalic mesenchyme	Brain
Notochord and surrounding mesenchyme	Ganglia
Sclerotome	Branchial arches
Lung buds	Limb buds
Oesophagus	Surface ectoderm
Myocardium	Endocardial cushion
Pancreatic primordium*	Liver
Visceral yolk sac	Metanephros
Parietal yolk sac	Mesonephros
Placenta (giant cells and blood)	Genital ridge
	Urogenital ridge
	Hindgut
	Amnion
	Chorionic plate

Table 4 Day 16.5

Positive tissues	Negative tissues
Cornea	Neural tissue
Pia-arachnoid	Lens
Zones of cephalic mesenchyme	Retina
Dermis	Spinal chord
Connective tissue	Maturing cartilage
Perichondrial mesenchyme	Bone
Visceral yolk sac	Liver
Parietal yolk sac	Oesophagus
Placental giant cells	Gut (all levels)
	Lungs
	Kidney
	Muscle

◊ weak staining.
* strong staining.

Discussion

During most stages of development the parietal and visceral yolk sac and the placental giant cells are stained positive for vitronectin. This correlates with our finding that human cell lines representative of mouse yolk sac produce vitronectin mRNA (Fig 1) and, that dissected mouse yolk sacs biosynthesise vitronectin protein (Fig 2).Variant forms of vitronectin have been isolated from chick egg yolk [17] and it has been suggested that yolk is the source of vitronectin in the early avian embryo [18].

At midgestation developing epithelia stain positive for vitronectin. The immature lung buds and myocardial cells of the developing heart are positive, but more mature heart and lung are negative later on during development. This is in agreement with the findings of Sumida et al [6,7,8] who described the staining of myocardial cells in embryonic chick, rat and human heart.

Only later in development does the skin stain positive for vitronectin, (see Table 4). Later or full term embryos may reveal a more detailed pattern of staining in mouse skin such as that described by Dahlback et al [19].

A 17.5 day embryo revealed strong staining in the developing tooth. The positive region corresponds to the region of developing dentin, rich in extracellular matrix components, between the ameloblast and the odontoblast.

Despite clear and strong staining for vitronectin antigen in many developing mouse tissues, the liver failed to stain apart from very faint staining in the hepatic primordium at day 10.5. This result may have a number of explanations. It may be that the liver was not fully functional and secreting undetectable amounts of vitronectin at the stages examined. Alternatively, any protein produced by the liver may not have remained in association with the cells that secrete it. Further studies using in situ hybridisation may resolve this enigma and establish whether the staining for antigen observed in various tissues actually represents de novo synthesis if vitronectin.

Vitronectin in Human Embryonal Carcinoma and Seminoma

Vitronectin immunoreactivity in germ cell tumours of the testis is shown in Table 5.

Table 5

Histological classification	Number positive/total
Seminoma	6/6
Mixed germ cell tumour	2/2[a]
Embryonal carcinoma	1/13[b]
Teratocarcinoma	1/4[c]
Testis adjacent to lesion	2/4[d]

Table 5

a	Areas associated with seminoma were positive
b	Areas of yolk sac morphology were positive; most lesions contained hemorrhagic areas which were reactive.
c	Differentiated areas of primitive gut were positive.
d	Thickened peritubular membranes were positive in some areas.

There was strong staining for vitronectin in the interstitial stromal regions of seminoma, but the seminoma cells were negative. Embryonal carcinoma and mixed germ cell tumours stained negatively, except for areas associated with seminoma or yolk sac. The only staining seen with teratocarcinoma corresponded to differentiated regions of primitive gut.

Recently (Looijenga *pers comm*) described the staining of the amorphous eosinophilic material within seminoma, known as "tiger stripe pattern" with vitronectin antibodies. No staining was seen in embryonal carcinoma, however vitronectin staining within the seminiferous tubules of carcinoma *in situ* was observed. The source and function of vitronectin in these tissues is under investigation.

Conclusions

We have described the temporal distribution of vitronectin antigen in the developing mouse. Our results, if confirmed by *in situ* hybridisation, suggest that vitronectin is widely expressed in epithelial and mesenchymal tissue at midgestation. Perhaps this reflects some role in morphogenesis.

Yolk sac synthesis of vitronectin suggests that it may be an important molecule in embryonic development, since these tissues are believed to play supportive roles.

Human yolk sac carcinoma are often aggressive, malignant cells which have minimal growth requirements *in vitro*. Vitronectin secretion could be important for the attachment, and consequently the growth of, embryonal carcinoma cells in mixed tumours *in vivo*.

References

1. Hayman EG, Pierschbacher MD, Ohgren Y and Ruoslahti E. Proc. Natl. Acad Sci. USA. 1983; 80: 4003-4007.
2   Dahlback K, Lofberg H and Dahlback B. Acta Derm. Venereol. 1986; 66: 461-467.
3   Dahlback K, Wulf HC and Dahlback B. J. Invest. dermatol. 1993; 100: 166-170.
4   Loridon-Rosa B, Vielh P, Cuadrado C and Burtin P. Am J. Clin Pathol. 1988; 90: 7-16.
5   Gladson CL and Cheresh DA. J. Clin. Invest. 1991; 88: 1924-1932.
6   Sumida H, Nakamura H. and Satow Y. Arch. Histol. Cytol. 1990; 53: 81-8.

7   Sumida H, Nakamura H and Yasuda M. Cell Tissue Res. 1992; 268: 41-49.
8   Sumida H, Nakamura H and Satow Y. Ann. N. Y. Acad. Sci. 1990; 588: 444-445.
9   Tomasini-Johansson BR, Ruoslahti E and Pierschbacher MD. Matrix 1993; 13: 203-214.
10   Solem M, Helmrich A, Collodi P. and Barnes D. Mol. and Cell. Biochem. 1991; 100: 141-149.
11   Seiffert D, Keeton M., Eguchi Y., Sawdey M. and Loskutoff DL. Proc. Natl. Acad. Sci. USA. 1991; 88: 9402-9406.
12   Barnes D.W. and Reing J. J. Cell. Physiol. 1985; 125: 207-214.
13   Cooper S. and Pera MF. Development. 1988; 7.104: 565-574.
14   Pera MF., Blasco Lafita MJ. and Mills J. Int. J. Cancer.1987; 40: 334-343.
15   Suzuki S., Oldberg A., Hayman EG, Pierschbacher MD. and Ruoslahti E. EMBO J. 1985; 4: 2519-2524.
16   Yatohgo T., Izumi M., Kashiwagi H. and Hayashi M. Cell. Struct. Function. 1988; 13: 281-292.
17   Nagano Y. Hamano T., Nakashima N., Ishikawa M., Miyazaki K. and Hayashi M. J. Biol. Chem. 1992; 5: 24863-24870.
18   Drake C. and Little C. FASEB J. 1993; 7: 1306.
19   Dahlback K, Wulf HC and Dahlback B. J. Invest. Dermatol. 1991; 96: 1016.

Work in this laboratory is funded by the Cancer Research Campaign.

Biology of Vitronectins and their Receptors
K.T. Preissner, S. Rosenblatt, C. Kost, J. Wegerhoff and D.F. Mosher, editors

Vitronectin association with skin

Karin Dahlbäck

Department of Dermatology, University of Lund, 22185 Lund, Sweden

Vitronectin associated with Normal tissue

The skin is an easily accessible organ and therefore well suited to use when studying extracellular matrix components. In the dermis, an abundance of rope-like collagen fibers in a hydrated gel of proteoglycans give strength and substance to the skin and there is also an extensive network of elastic fibers which have elastic properties and give the skin its resilience.

Immunohistochemical staining of normal skin from adults using monoclonal and polyclonal anti-vitronectin antibodies, resulted in an immunoreactivity associated with dermal fibrillar structures (Fig 1) (1).

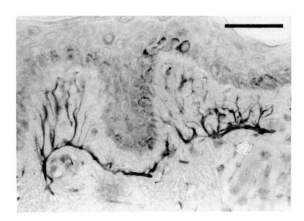

Fig 1. Immunoreactivity of vitronectin associated with elastic fibres in the papillary dermis of normal human skin from an adult indivudal (bar=50um).

Using sequential double staining of the tissue sections, these fibers were shown to correspond to those stained with standard elastin staining procedure using orcein. Vitronectin immunoreactivity was invariably detected in conjunction with elastic fibers in specimens from individuals over 13 years of age (2). In specimens from patients between the ages of 6 and 13, it was faint or absent.

In specimens from individuals younger than 6 years of age, no immunoreactivity of vitronectin was detected in conjunction with elastic fibers, either with the monoclonal or the polyclonal anti-vitronectin antibodies, even though newborns were found to have a plasma vitronectin concentration approximately 67% of the adult level   suggesting active biosynthesis already in the fetus (2).

No vitronectin immunostaining of the epidermis or of the dermal epidermal junction area was found, and no distinct vitronectin immunoreactivity was detected in conjunction with dermal collagen fibers. The dermal tissue distribution of vitronectin appeared to be different from that previously reported for fibronectin.

Fibronectin immunoreactivity was earlier reported to be distributed in a reticulate pattern in the area under the dermal-epidermal junction (3). The anti-vitronectin staining pattern corresponded with that of anti-serum amyloid P component (SAP), suggesting that vitronectin might be associated with the same dermal structures as SAP, a normal plasma and tissue protein of unknown biological function (4).

For comparison, renal specimens from adults also were studied. Vitronectin immunoreactivity was found in conjunction with elastic tissue in renal vessel walls in adults (5). The distribution of vitronectin in kidney did not correspond with that reported for fibronectin, in contrast to an earlier report by Hayman et al. (6). Throughout these studies we consistently found colocalization of vitronectin and SAP immunoreactivity, when the two were compared, with one interesting exception. Unlike that of SAP, vitronectin immunoreactivity was not detected in the basement membrane of morphologically normal glomeruli (5).

Our results may be interpreted as suggesting that vitronectin is deposited on elastic fibers during late childhood and adolescence. Extraction studies indicate that the dermal vitronectin is non-covalently associated with the elastic fibers (7). The absence of vitronectin immunoreactivity in elastic fibers in the very young indicates that vitronectin is not a constituent of elastic fibers before the age of six years.

Our findings of vitronectin in association with elastic fibers have been confirmed by others (8-10). Vitronectin immunoreactivity has been found in association with elastic fibers in lamina muscularis mucosae of the human colon, in specimens of aorta with or without signs of arteriosclerosis, and in conjunction with elastic tissue surrounding human breast and colon carcinoma. In a study of renal biopsies from normal and diseased kidney Falck et al. found vitronectin to be colocalized with C5b-9 in all injured tissue, and to be also present in arterioles and arteries (11).

To determine whether tissue vitronectin is deposited as part of SC5b-9 complex or as uncomplexed protein, vitronectin immunoreactivity was compared with that of C9, using antisera specific for C9 neoantigen (2, 12). As no C9 neoantigen reactivity could be found in association with elastic fibers in skin from people under 30 years of age, it would appear that the vitronectin initially deposited on the elastic fibers is uncomplexed; and as C9 neoantigen reactivity was present in specimens from some individuals over 30, particularly in the elderly and in sun-exposed skin, it would seem that C9 neoantigen reactivity, alone or in complex with vitronectin, may be deposited on elastic fibers later in life.

In the transmission electron microscope, two components of the elastic fibers can be observed, amorphous elastin and a peripheral mantle of microfibrils, 8-12 nm in diameter (13). Similar microfibrils, without concomitant elastin, are present in several tissues besides skin - e.g., in ocular ciliary zonules, periodontal ligament and glomerular mesangium (14). Because of the difficulties involved in the purification of the 8-12 nm microfibrils, their chemical composition and molecular structure is not yet fully understood and the number of constituent molecules is unknown. In 1986 the major constituent, fibrillin was identified, a 350 kDa glycoprotein (15). Later it was found that mutations in the fibrillin gene causing molecular defects in fibrillin is related with the Marfan syndrom (16).

In our first study we were led to hypothesize that vitronectin might be associated with the elastin-associated microfibrils, by the colocalization of the immunostaining obtained by anti-vitronectin and anti-SAP, since SAP earlier was claimed to be associated with 8-12 nm microfibrils. With immuno electronmicroscopy, we were able to demonstrate immunoreactivity both of vitronectin and of SAP at the periphery of elastic fibers but could not with certainty elucidate whether the immunostaining was associated with the microfibrils (2). However, when comparing the patterns of immunostaining produced by anti- vitronectin and anti-SAP antibodies with that produced by anti-fibrillin antibodies we found that the fibrillin reactive dermal fibers formed an extensive network which only partly corresponded with the networks stained by anti-vitronectin, anti-SAP and orcein (standard histochemical elastin stain) (17). The differences in the immunostaining patterns of anti-vitronectin and anti-SAP as compared to that of anti-fibrillin raise doubt as to whether vitronectin and SAP associate with the 8-12 nm microfibrils. The results rather suggest that the epitopes of the anti-vitronectin and anti-SAP antibodies are associated with amorphous elastin at the periphery of the elastic fiber. The consistent finding that fibers stainable by anti-vitronectin and anti-SAP are also stained by orcein supports this hypothesis. The fact that both anti-vitronectin and anti- SAP immunoreactivity is absent from the elastic fibers in youth when microfibrils are most abundant, but present in adult skin when peripheral microfibrils are more scanty, would also seem to contradict the assumption that vitronectin and SAP is associated with microfibrils (13).

In contrast to the findings in human skin, no vitronectin immunoreactivity could be demonstrated in conjunction with the dermal elastic fibres in hairless lightly pigmented mice. A possible explanation to this discrepancy is that the normal lifespan of mice is relatively short (2 years) (18).

Changes related to aging and UV-light

Degeneration of elastic fibers is of vast clinical importance in the development of atherosclerosis, where the arterial elastic fibers are deranged and altered. Changes also occur in the dermal elastic fibers in association with aging and sunexposure. In sun-protected skin in the elderly the elastic fibers are thin and irregular. In sunexposed skin (so-called actinic elastosis), the elastic fibres are thick, numerous and appear curled and tangled; at a later stage of change amorphous material is to be seen in the upper dermis.

No vitronectin immunoreactivity was found on the elastic fibers in specimens from young children (2). The intensity of the immunoreactivity increased with age during childhood and adolescence while being fairly constant in adults. Elastotic material was distinctly immunostained with anti-vitronectin in all the specimens (19,2).

C9 neoantigen immunoreactivity resembled that of vitronectin found in conjunction with dermal elastic fibers, but only in the middle aged and elderly (2). C9 neoantigen immunoreactivity tended to increase with age and was more intensive in sun-exposed than in sun-protected skin. Elastotic material was consistently immunostained with anti C9 neoantigen, though the staining was generally fainter than that obtained with anti-vitronectin (2). Although the significance of C9 neoantigen immunoreactivity found in middle age and later in life remains unclear, hypothetically terminal complement complexes formed in the vicinity of elastic fibers may bind to the elastin-associated vitronectin. An alternative explanation to the elastin associated C9 immunoreactivity may be that soluble SC5b-9 complexes are deposited on elastin fibers. The findings of C9 neoantigen immunoreactivity predominantly in sun-exposed skin, suggests that exposure to the sun may affect its deposition.

In lesions of solar elastosis, the intensity of fibrillin immunoreactivity tended to be fainter than that obtained with anti-vitronectin or anti-SAP (17). Although non-stainable by anti-vitronectin or anti-SAP, a thin zone of fibrillin reactivity was often seen close to the dermal- epidermal junction.

Vitronectin associated with Apoptotic keratin bodies

Keratin bodies are amorphous, eosinophilic globules, chiefly composed chiefly of keratin intermediate filament aggregates, located in the region of the dermo-epidermal junction (20). They are formed by filamentous degeneration of epidermal cells in a process of apoptosis and their presence in normal skin indicates that such apoptosis is a physiological event in skin as in other tissues (20-23). During apoptosis, individual keratinocytes undergo cell-death with condensation and disintegration of the nucleus and aggregation of keratin intermediate filaments. Membrane-bound aggregates of keratin intermediate filaments, if not rapidly phagocytosed, loose their enclosing membrane and drop into the dermis (21). Increased formation of keratin bodies occurs in a number of lymphocyte-mediated skin diseases, including lichen ruber planus, and in response to UV-light (21).

Immunoreactivity of immunoglobulins, mainly IgM and C3, has been demonstrated in keratin bodies. Normal sera contain autoantibodies against keratin intermediate filaments (23,24). This led to speculations as to the function of such auto-antibodies and for complement in facilitating the elimination of cellular debris during apoptosis (23-25).

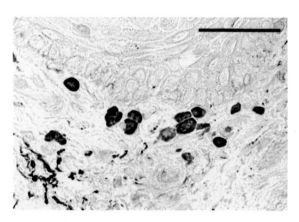

Fig 2. Immunoreactivity of vitronectin associated with apoptotic keratin bodies in a skin lesion of ruber planus (bar=50um)

In skin lesion specimens from patients with lichen ruber planus we found keratin bodies reacting with anti-IgM under the dermal-epidermal junction inclose proximity with the junction, and a band-like cellular infiltrate composed of lymphocytes and some histiocytes under this area

(19). There was no anti-vitronectin or anti-SAP labelling of the dermal inflammatory cell infiltrate. However, the keratin bodies exhibited distinct vitronectin immunoreactivity (Fig 2). They could also be immunostained with anti-SAP (19).

The in vivo immunohistochemical finding of the occurrence of vitronectin in association with keratin bodies derives support from another study, in which in vitro binding of vitronectin to keratin intermediate filaments was demonstrated (25).

The significance of the binding of vitronectin in association with keratin bodies is unclear. Recently macrophage recognition and phagocytosis of apoptotic cells was shown to be dependant on the vitronectin receptor, supporting the hypothesised function in phagocytosis of cell debris from apoptotic cells (26). It might also be speculated that deposition of vitronectin and possibly other complement regulatory proteins, may be involved in the prevention of an inflammatory response during apoptosis.

Vitronectin associated with Amyloidosis

Amyloidosis is a pathological extracellular fibrillar protein deposition that yields a 'cross- ß' pleated sheet on X-ray crystallograph analysis (27). The physiochemical aggregation of the microfibrils gives amyloid its characteristic polariscopic birefringence when stained with Congo red. The microfibrils of the different types of amyloidosis consist of different amyloid-fibril proteins - e.g., immunoglobulin light chains in amyloid light chain (AL) type amyloidosis, and amyloid A in amyloid A (AA) type amyloidosis (27). SAP is known to bind to amyloid microfibrils in vitro and SAP is present in the deposits of most types of amyloidosis (4).

We therefore investigated whether, like SAP, vitronectin is associated with amyloidosis (5,19). We chose to study amyloidosis of the AA type and of the AL type, those being common forms of amyloidosis, and primary localized cutaneous amyloidosis (27,28).

Both in macular and lichenoid primary localized cutaneous amyloidosis there are deposits of amyloid in the upper papillary dermis (28). In contrast to several other types of amyloidosis, the amyloid fibril protein has not yet been identified, owing to difficulties involved in purifying skin amyloid. A variety of different cell types have been considered as possible synthesizers of amyloid in the lichenoid or macular type of primary localized cutaneous amyloidosis (28).

We found vitronectin immunoreactivity in conjunction with congophilic amyloid deposits in glomeruli and vessel walls in patients with AL or AA type of amyloidosis (5). It was also present in amyloid deposits in skin lesions from patients with primary localized cutaneous amyloidosis (19). The distribution of vitronectin immunoreactivity was quite similar to that of SAP in all three types of amyloidosis.The importance of SAP and vitronectin in the pathogenetic events in amyloidosis is unknown. Ultrastructural studies are needed, and it remains to be investigated whether, like SAP, vitronectin binds to amyloid fibrils in vitro. Recently vitronectin was reported to be associated with cerebral amyloid (29)

It has also been found in conjunction with cutaneous amyloid in mice (18).

Vitronectin associated with Nephrosclerosis

When studying renal specimens from elderly patients, an accidental finding was the association both of vitronectin and SAP immunoreactivity with the sclerotic lesions in cases with benign nephrosclerosis (5). Niculescu et al (9) reported on the immunoreactivity of vitronectin in sclerotic lesions in the aorta and Falck et al (11) mentioned its presence in glomerulosclerotic lesions. Future studies are needed to elucidate the significance of these results. The finding of SAP immunoreactivity in sclerotic lesion spresumably may limit the value of anti-SAP as a specific marker of amyloidosis.

Vitronectin in Porphyria

In porphyria cutanea tarda and erythropoietic protoporphyria skin vessel walls are damaged when exposed with UVA due to excessive amounts of porphyrins in the blood. The patients have characteristic PAS (periodic acid Schiff)-positive thickened vessel walls in the upper dermis (30). Similar pathological PAS-positive vessel walls in the upper demis have been observed in sun-exposed skin in normal persons (31). The PAS-positive material has been described as being ultrastructurally composed of reduplicated basal lamina and a finely fibrillar material (30). It has been suggested that the abundant production of basement membrane-like material is a reparative process in response to lasting or repeated injury.

We found that the thickened, PAS-positive vessel walls in the upper part of dermis present in all the specimens investigated from patients with erythropoietic protoporphyria or porphyria cutanea tarda were distinctly immunostained by the anti-vitronectin antibodies in contrast to the vessels in normal sun-protected skin (19). The staining pattern obtained with anti-vitronectin was quite similar to that obtained using anti-SAP.

Vitronectin in Discoid lupus erythematosus, Bullous pemphigoid and Dermatitis herpetiformis

In bullous pemphigoid a linear deposition of IgG and C3 is found at the dermal-epidermal junction zone, and activation of complement is believed to be of importance in the pathogenesis of this disease (32). Patients´ autoantibodies that are directed against the bullous pemphigoid antigen, a glycoprotein found to be associated with hemidesmosomes of basal keratinocytes.

To study the potential deposition of terminal complement complexes in tissue, C9 neoantigen and vitronectin immunoreactivity was examined in parallel in skin biopsies from patients with bullous pemphigoid and also in patients with dermatitis herpetiformis or discoid lupus erythematosus (33). In bullous pemphigoid, we found deposition of C9 neoantigen, but no vitronectin along the dermal epidermal junction in ten out of eleven specimens. This suggests the presence of MAC rather than of SC5b-9 complexes along the dermal epidermal junction, and supports the hypothesis of a direct lytic action of MAC on basal keratinocytes, possibly contributing to the formation of bullae in bullous pemphigoid.

In dermatitis herpetiformis, a granular or fibrillar pattern of IgA and C3 immunoreactivity in the tips of the dermal papillae is characteristic, both in normal and in lesional skin (34). Immunostaining of such granules by C9 neoantigen has also been shown both in lesional and non-lesional skin. The role of IgA and complement in dermatitis herpetiformis is not understood. In contrast to the findings in bullous pemphigoid, colocalization of C9 neoantigen and vitronectin immunoreactivity was seen in the specimens from patients with dermatitis herpetiformis, suggesting that the non-lytic complex SC5b-9 is deposited rather than MAC (33).

In lesional skin in <u>discoid lupus erythematosus</u>, a broad band (referred to as the 'lupus-band') of IgG and/or IgM and C3 immunoreactivity, is usually found along the dermal-epidermal junction (35). Similar to the findings in dermatitis herpetiformis, and in contrast to the findings in bullous pemphigoid, we found colocalization of C9 neoantigen and vitronectin immunoreactivity in the 'lupus band' in lesions of discoid lupus erythematosus (33). French <u>et al</u> recently reported on their findings of immunoreactivity of vitronectin (S protein) in association with the lupus band in some, but not all of the investigated specimens (12 out of 32 skin specimens) (36). They found immunoreactivity of C9 neoantigen in conjunction with the lupus band in 30 of 32 specimens and clusterin (also called complement lysis inhibitor or SP 40-40), another inhibitor of MAC, in conjunction with the lupus band in 20 of the 32 specimens. These finding suggest that both lytic MAC and non-lytic clusterin-SC5b-9 complexes may be present in the 'lupus band'.

Similar findings of colocalized C9 neoantigen and vitronectin in pathological lesions have been reported in cases of several different types of nephritis, including lupus nephritis and IgA nephritis, and in vessel walls in the inflamed colon of patients with ulcerative colitis (11,8). The occurrence of C9 neoantigens together with vitronectin and/or clusterin in the tissue may reflect in situ formation of SC5b-9 complexes and/or clusterin-SC5b-9 complexes or deposition of such fluid phase complexes.

References

1. Dahlbäck K, Löfberg H, Dahlbäck B. Localization of vitronectin (S-protein of complement) in normal human skin. Acta Derm Venerol (Stockh) 1986; 66:461-467
2. Dahlbäck K, Löfberg H, Dahlbäck B. Immunohistochemical demonstration of age-related deposition of vitronectin (S-protein of complement) and terminal complement complex on dermal elastic fibres. J Invest Dermatol 92:727-733, 1989
3. Fyrand O. Studies on fibronectin in the skin. Br J Dermatol 101:263-270, 1979
4. Pepys MB, Baltz ML, de Beer FC, Dyck RF, Holford S, Breathnach SM, Black MM, Tribe CR, Evans DJ, Feinstein A. Biology of serum amyloid P component. Ann NY Ac Sci, 286-298, 1982
5. Dahlbäck K, Löfberg H, Dahlbäck B. Immunohistochemical demonstration of vitronectin in association with elastin and amyloid deposits in human kidney. Histochemistry 1987; 87: 511-515.
6. Hayman EG, Pierschbacher MD, Öhrgren Y, Ruoslahti E. Serum spreading factor (vitronectin) is present at the cell surface and in tissues. Proc.Natl.Acad.Sci.USA 80:4003-4007, 1983
7. Hintner H, Dahlbäck K, Dahlbäck B, Pepys MB, Breathnach SM. Tissue vitronectin in normal adult human dermis is non-covalently bound to elastic tissue. J Invest Dermatol96: 747-753, 1991
8. Halstensen TS, Mollnes TE, Garred P, Brandtzaeg P. Distribution of Terminal complement Complex and S-protein (Vitronectin) in Normal and Inflamed Human Colon. Complement 4: 164, 1987
9. Niculescu F, Rus HG, Vlaicu R. Immunohistochemical localization of C5b-9, S-protein, C3d and apolipoprotein B in human arterial tissues with atherosclerosis. Atherosclerosis 65: 1-11, 1987
10. Loridan-Rosa B, Vielh P, Cuadrado C, Burtin P. Comparative distribution of fibronectin and vitronectin in human breast and colon carcinomas. Am J Clin Pathol; 90: 7-1, 1988
11. Falk RJ, Podack E, Dalmasso AP, Jennette JC: Localization of S protein and its relationship to the membrane attack complex of complement in renal tissue. Am J Pathol 127: 182-190, 1987
12. Mollnes TE, Harboe M: Immunohistochemical detection of the membrane and fluid-phase terminal

complement complexes C5b-9 (m) and SC5b-9. Scand J Immunol 26, 381-386, 1987

13. Franzblau C, Faris B. Elastin. In Cell Biology of Extracellular matrix. ed Hay E. Plenum Press NY, London. 65-89, 1985

14. EG Cleary, MA Gibson. Elastin-associated microfibrils and microfibrillar proteins. Int Rev Connect Tissue Res. 10: 97-209, 1983

15. Sakai LY, Keene DR, Engvall E. Fibrillin a new 350 kDa glycoprotein, is a component of extracellular microfibrils. J Cell Biol 103:2499-2509.1986

16. Dietz HC, Cutting GR, Pyeritz RE, Maslen CL, Sakai LY, Corson HGM, Puffenberger EG, Harnosh A, Nanthakumar EJ, Curristin SM, Stetten G, Meyers DA, Francomano CA. Marfan syndrome caused by a recurrent de novo missense mutation in the fibrillin gene. Nature 352:337-339, 1991

17. Dahlbäck K, Ljungquist A, Löfberg H, Dahlbäck B, Engvall E and Sakai L. Fibrillin immunoreactive fibres constitute a unique network in the human dermis. J Invest Dermatol 93:284-291,1990

18. Dahlbäck K, Wulf HC, Dahlbäck B. Vitronectin in mouse skin: Immuno- histochemical demonstration of its association with cutaneous amyloid. J Invest Dermatol 100:166-170

19. Löfberg H, Dahlbäck B. Immunohistochemical studies on vitronectin in elastic tissue disorders, cutaneous amyloidosis, lichen ruber planus and porphyria. Acta Derm Venerol (Stockh) 68:107-115,1988

20. Hashimoto K. Apoptosis in lichen planus and several other dermatoses. Intra-epidermal cell death with filamentous degenerations. Acta Derm Venerol (Stockh) l976; 56: 187-210.

21. Weeden D, Searle J, Kerr F R. Apoptosis: Its nature and implications for dermatopathology. Am J Dermatopathol 1: 133-144,1979

22. Wyllie AH, Kerr FR, Currie AR. Cell death: the significance of apoptosi. Int Rev Cytol 68: 251-306, 1980

23. Grubauer G, Romani N, Kofler H, Stanzl, Fritsch P, Hintner H. Apoptotic Keratin Bodies as Autoantigen Causing the Production of IgM-Anti-Keratin Intermediate Filament Autoantibodies. J Invest Dermatol 87, 466-471, 1986

24. Hintner H, Romani N, Stanzl U, Grubauer G, Fritsch P, Lawlwy TJ. Phagocytosis of keratin filament aggregates following oponization with IgG-anti-keratin filament antibodies. J invest Dermatol 88: 176-182, 1987

25. Hintner H, Stanzl U, Dahlbäck K, Dahlbäck B, Breathnach SM. Vitronectin shows complement-independent binding to isolated keratin intermediate filaments. J Invest Dermatol. 93:656-661, 1989

26. Savill J, Dransfield I, Hogg N, Haslett C. Vitronectin receptor-mediated phagocytosis of cells undergoing apoptosis. Nature 343: 170-173, 1990

27. Glenner G Amyloid deposits and amyloidosis. The ß-fibrilloses. New Engl. J. Med. 302:1283-1292, 1333-1343, 1982

28. Wong C-K Review Cutaneous Amyloidoses Int J Dermatology 26:273-277, 1987

29. Akiyama H, Kawamata T, Dedhar S, McGeer PL. Immunohistochemical localization of vitronectin, its receptor and beta-3 integrin in Alzheimer brain tissue. J Neuroimmunol 32:19-28, 1991

30. Epstein JH, Tuffanelli DL, Epstein WL. Cutaneous Changes in the Porphyrias. A Microscopic Study. Arch Dermatol 107: 689-698, 1973

31. Braverman IM. Elastic fiber and microvascular abnormalities in aging skin. In Cutaneous Aging. Eds Kligman AM, Takase Y. Tokoy press. Univ Tokyo Press Tokyo. 369-389, 1988

32. Korman N. Bullous pemphigoid. J Am Acad Derm 16: 907-924, 1987

33. Dahlbäck K, Löfberg H, Dahlbäck B. Vitronectin (S-protein of complement) colocalizes with Ig deposits and C9 neoantigen in discoid lupus erythematosus and dermatitis herpetiformis but not in bullous pemphigoid. Br J Dermatol 120: 725-733, 1989

34. Hall RP. The pathogenesis of dermatitis herpetiformis. Recent advances. J Am Acad Derm 16:1129-1144, 1987

35. Provost TT, Reichlin M. Immunopathologic studies of cutaneous lupus erythematosus. J Clin Immunol 8: 223-233, 1988

36. French LE, Polla LL, Tschopp J, Schifferli JA. Membrane attack complex (MAC) deposits in skin are not always accompanied by S-protein and clusterin. J Invest Dermatol 98: 758-763, 1992

Biology of Vitronectins and their Receptors
K.T. Preissner, S. Rosenblatt, C. Kost, J. Wegerhoff and D.F. Mosher, editors

The distribution of vitronectin in resting and thrombin stimulated platelets. An ultra-immunocytochemical study.

E. Morgenstern[a] and K.T. Preissner[b]

[a]Medical Biology of the Saarland University, Homburg/Saar

[b]Haemostasis Research Unit, Max-Planck-Institute Bad Nauheim, Germany.

INTRODUCTION

Vitronectin (VN) is a constituent of the platelets [1] and can bind to the platelet surface via the integrin α_{IIb}-β_3 (fibrinogen-receptor). On ultrathin frozen sections we investigated the distribution of VN in resting platelets using washed or gel-filtered cells, platelets stimulated by ADP or washed platelets under addition of blood plasma (10 Vol %) stimulated for 1 min with thrombin. Both the plasma and the multimeric form of the VN-molecule were applied in order to get information about VN-platelet interaction during various functional states of the molecule, respectively of the cells.

MATERIALS AND METHODS

Suspensions of washed or gelfiltered platelets (GFP) were fixed with 3% paraformaldehyde, 0.1% glutaraldehyde, washed and resuspended in a solution containing 7.5% (w/v) polyvinyl-pyrrolidone K25 (Fluka AG, Buchs, Switzerland) and 2 M sucrose in PBS. For inclusion, droplets of platelet-rich suspensions were placed on 5% gelatin at 37°C. Blocks were plunge-frozen in liquid propane at -180°C (KF80; Reichert, Vienna, Austria). Ultrathin frozen sections (FC4 D; Reichert-Jung, Nußloch/ FRG) were mounted with 2.3 M sucrose droplets on pioloform-coated Ni-grids. After washing in PBS containing 0.05 M glycine, the grids were incubated with primary antibodies S66/67 and S6/7 (rabbit anti-human VN IgG`s [2]) for 60 min at 20°C. Detection of PAI-1 was carried out with K1 (rabbit-anti human PAI IgG, donated by Dr. H. Pannekoeg, CLB/Amsterdam) as a primary antibody.After washing secondary labeling with goat antibody (IgG goat-anti rabbit GAR EM10; Aurion/Biotrend; 20 µl/ml PBS) was carried out for 60 min at 20°C. Finally, the grids were stained with ammonium molybdate or uranyl acetate.

RESULTS

In resting platelets VN was found within the matrix of the α-granules and at a minor quantity within the surface connected system (SCS). VN was present in a high amount on the surface of washed platelets (Fig. 1). This surface labeling was drastically reduced in GFP (Fig. 2). GFP bound much more the multimeric form of the VN (Fig. 3) on the surface than plasma VN (Fig. 4).ADP stimulation increased the number of surface-bound plasma VN (not shown).

PAI-1 was found to be colocated with VN within the α-granules and within the SCS but was sparsely observed on the surface of both washed platelets and GFP (Fig. 5).

In thrombin-stimulated aggregating platelets, VN was observed only to a minor degree on the free surfaces of the platelets and rarely in focal contacts between the aggregating cells. The surface of fibrin fibers was decorated with VN as well (Fig. 6a and b).

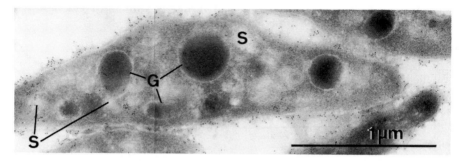

Fig. 1:Washed platelets: VN indicating labels are located at the surface, within the α-granules
(G) and in the SCS (S)

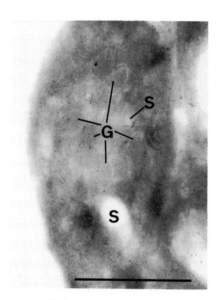

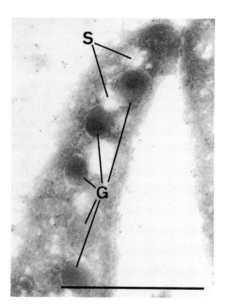

Fig. 2: GFP: VN indicating labels at the surface
are missing (cf Fig. 1). VN is seen within
the α-granules (G) and in the SCS (S).

Fig. 3: GFP + multimeric VN: VN is
bound to the surface, in the α-
granules (G) and in the SCS (S).

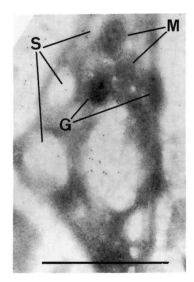

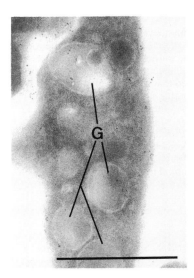

Fig. 4: GFP + plasma VN: VN is not bound on
the surface and seen within he α-granules
(G) and in the SCS (S).M =Mitochondria

Fig. 5: Washed platelet: PAI indicating
labels are located at the surface and
within the α-granules (G).

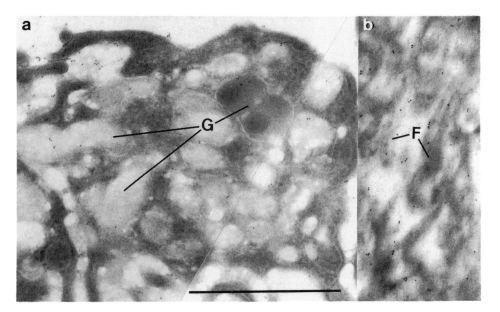

Fig. 6: Platelet in a clot: a) VN indicating labels are located within degranulating α-granules
(G) but infrequently on the platelet surface. b) Labels are seen on the surface of fibrin
fibres (F).

Discussion

The immunocytochemical findings confirm the presence of releasable VN in platelet storage pools [3,4] and clearly demonstrate the localization of VN in α-granules as described by Roger et al.[5]. In contrast to this study VN was found in the SCS and to a high degree on the surface of resting platelets. The clearing of VN from the surface by gel filtration supports this result. It is suggested that a resting platelet - which binds negligible amounts of fibrinogen - is covered with VN in this state. This may be caused to a great extent by VN binding to the fibrinogen-receptor or to other yet unidentified binding sites. When the receptor can bind fibrinogen and forms clusters in focal contacts then the VN binding capacity of the platelet surface decreases - as shown after thrombin addition and platelet aggregation. Furthermore, multimeric molecule binds appreciably to the platelets and ADP stimulation increases the binding of VN when fibrinogen is absent. The ultrastructural data support our hypothesis that reactive (multimeric) forms of VN can bind to non-activated platelets and thereby protect them against interaction with fibrin(ogen).

References:

1 Preissner K T. Structure and biological role of vitronectin. Ann Rev Cell Biol 1991; 7:275-310

2 Preissner K T, Wassmuth R, Müller-Berghaus G. Physico-chemical characterization of hu man S-protein and its function in the blood coagulation system. Biochem J 1985; 231: 349-355

3 Preissner K T, Holzhütter S, Justus C, Müller-Berghaus G. Identification and partial characterization of platelet vitronectin: Evidence for complex formation with platelet-derived plasminogen activator inhibitor-1. Blood 1985; 74: 1989-1996

4 Parker C J, Stone O L, White V F, Bernshaw N J. Vitronectin (S-protein) is associated with platelets. Br J Haematol 1989; 71:245-252

5 Roger T S, Halstensen K, Hogåsen T E, Mollnes, N O, Solum T H. Platelets and vitro nectin: immunocytochemical localization and platelet interaction with exogenously added vitronectin. Nouv Rev Fr Hematol 1992; 34:47-54

Biology of Vitronectins and their Receptors
K.T. Preissner, S. Rosenblatt, C. Kost, J. Wegerhoff and D.F. Mosher, editors

Metabolism of vitronectin complexes.

H.C. de Boer[a], Ph.G. de Groot[a] and K.T. Preissner[b]

[a]Department of Haematology, University Hospital, 3508 GA Utrecht, The Netherlands

[b]Haemostasis Research Unit, Kerckhoff-Klinik, Max-Planck Institute, D-61231 Bad Nauheim, Germany

INTRODUCTION

In human plasma the activity of thrombin (T) is controlled predominantly by antithrombin (AT). This is illustrated when radiolabeled thrombin is added to human fibrinogen-deficient plasma, subjected to SDS-PAGE and visualized on autoradiogram (Figure 1). Within 2 min 80% of the administered thrombin has entered a binary complex with AT, whereas about 10% is inhibited by a_2-macroglobulin (a_2M). The binary T-AT complex is not the final product of thrombin inhibition, since after 2 min about 10% of the T-AT complex has taken part in a ternary complex with a third plasma protein, vitronectin (VN). In time the association of T-AT complex with VN increases up to 40% at 5 min and 80% at 30 min. The initial interaction between T-AT and VN is mediated by electrostatic bonds, which are formed almost immediately (1). This electrostatic interaction causes disruption of the intra-molecular bonds existing in the VN-molecule. As a result the normally folded VN-molecule extends and becomes "activated". The extension of the VN-molecule brings certain cysteines in the amino-terminal portion of VN and the thrombin-moiety of the T-AT complex into close proximity (1), leading to covalent stabilization of the ternary complex.

Complex formation between AT, T and VN has consequences for the individual components. For example, T-AT complex formation neutralizes the heparin affinity of AT (2) and abolishes the proteolytic activity and the RGD-mediated cell-spreading activity of thrombin (3). In contrast, upon T-AT complex formation a receptor-recognition site is generated. This site is responsible for *in vivo* clearance of several serpin-protease complexes in mice via the serpin-receptor-1 (SR-1)(4). The same receptor-recognition site mediates binding to the serpin-enzyme-complex receptor (SEC-receptor) present on cultured hepatocytes and monocytes (5). Furthermore, association of T-AT with VN "activates" the VN-molecule and consequently endows the T-AT complex with specific properties of the VN-molecule (Figure 2). Thus, in the VN-T-AT complex heparin affinity is restored and an RGD-sequence replaced. The consequences of generation of de novo properties with respect to metabolism of the ternary VN-T-AT complex were investigated.

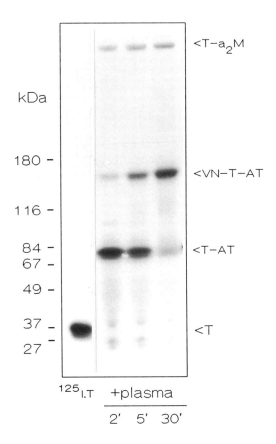

Figure 1. Formation of complexes. Radiolabeled thrombin (0.25 μg) was incubated with 25 μl fibrinogen-deficient citrated human plasma and incubated for the time intervals indicated. Reactions were stopped by adding SDS-containing sample-buffer, and samples were subjected to gradient (4-15%) SDS-polyacrylamide electrophoresis. Complex formation was visualized by autoradiography.
Molecular weight markers are shown along the left margin.

BINDING OF TERNARY COMPLEXES TO HUMAN ENDOTHELIAL CELLS

Covalently stabilized VN-T-AT complexes bind to human umbilical vein endothelial cells (HUVEC) in a time- and concentration dependent manner with an apparent dissociation constant of 16 nM (6). Competition experiments showed that the binding determinant of the VN-T-AT complex was not located on the AT- or thrombin-moiety (Table 1). Also the receptor-recognition site represented by peptide Y105 (5) did not interfere with binding of VN-T-AT complex to the endothelial cells, suggesting that either HUVEC do not express the corresponding receptor or that the receptor-recognition site is lost upon ternary complex formation. The binding determinant could be located on the VN-moiety of the ternary complex (Table 1). The VN-receptor ($\alpha_v\beta_3$) was not involved, since addition of an RGD-containing peptide and blockage of the VN-receptor with a monoclonal antibody (anti-β_3) did not affect the binding.

Table 1. Specificity of VN-T-AT binding to HUVEC

Unlabeled competitor (100x molar excess or as indicated)	% Inhibition
No competitor	0 ± 2
VN-T-AT	100 ± 3
native AT	0 ± 7
thrombin-modified AT	0 ± 1
active-site blocked thrombin	6 ± 6
peptide Y105 (100 μg/ml)	0 ± 5
urea-treated VN	92 ± 1
RGD-containing peptide (10 μM)	4 ± 7
Heparin (100 U/ml)	95 ± 5
antibody 8E6	100 ± 4
VN-peptide 1 (1 μg/ml)	47 ± 1
VN-peptide 2 (1 μg/ml)	74 ± 7
VN-peptide 3 (1 μg/ml)	66 ± 5
VN-peptide 4 (1 μg/ml)	0 ± 5
VN-peptide 5 (1 μg/ml)	0 ± 9

Preincubation of the endothelial cells	
anti-β3 (VN-receptor) antibody	18 ± 7
heparinase (5 U/ml)	42 ± 1
heparitinase (5 U/ml)	14 ± 5
chondroitinase ABC (0.5 U/ml)	11 ± 7
PI-specific phospholipase C (5 U/ml)	0 ± 5

The heparin binding site of VN was responsible for the binding of VN-T-AT to the endothelial cells as was shown by the decreased binding of VN-T-AT to HUVEC in the presence of heparin or an anti-VN monoclonal antibody (8E6) which recognizes the heparin binding form of VN (7). Furthermore, utilizing synthetic peptides 1-5 (Figure 2), the primary binding site could be localized between amino acid residues 348 and 361, which contain two consensus sequences for glycosaminoglycan recognition (8). The involvement of heparan sulfate containing proteoglycans present on the surface of endothelial cells was evidenced by the decreased binding of the complex after treatment of the cells with heparinase and heparitinase or in the presence of purified heparan sulfate (Figure 3). Addition of increasing amounts of purified dermatan sulfate or chondroitin sulfate had no effect.

Until now, at least five different species of heparan sulfate proteoglycans have been characterized.

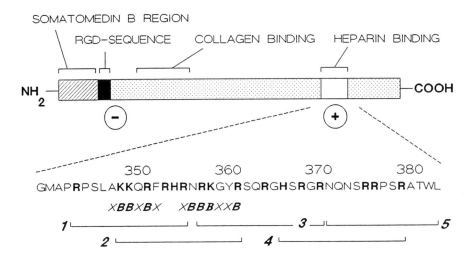

SOMATOMEDIN B REGION

RGD-SEQUENCE   COLLAGEN BINDING   HEPARIN BINDING

NH₂ — ▨ ■ ⬚⬚⬚⬚⬚⬚⬚⬚⬚⬚⬚ ☐ ⬚⬚⬚⬚ — COOH

(−)   (+)

350   360   370   380

GMAPRPSLAKKQRFRHRNRKGYRSQRGHSRGRNQNSRRPSRATWL

XBBXBX   XBBBXXB

1 ⌞_____⌟ ⌞_____3 ⌟⌞_____⌟5

2   ⌞_____⌟   4 ⌞_____⌟

Figure 2. Molecular structure of the extended VN-molecule. The primary structure entails the glycosaminoglycan binding region in which two heparin binding consensus sequences are depicted (B represents the probability of a basic residue and X represents a non-basic residue) (8). The primary structure of 5 overlapping peptides is shown on the bottom.

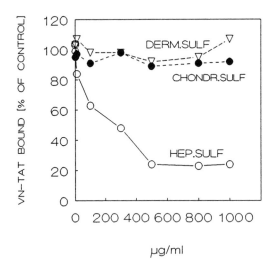

Figure 3. Binding of VN-T-AT to HUVEC expressed as percentage of specific binding in the absence of proteoglycans. Radiolabeled VN-T-AT (2.5 nM) was incubated for 2 h at 4°C in the absence or presence of increasing amounts of purified proteoglycans.

Glypican, a proteoglycan linked to the membrane through a glycosyl phospho-inositol (GPI) anchor (9) is not involved, since disruption of the GPI-anchor with PI-specific phospholipase C did not affect VN-T-AT binding (Table 1). Syndecan is probably not involved either, since this proteoglycan does not bind VN (10). This leaves fibroglycan, amphiglycan and a yet unidentified proteoglycan as possible candidates for binding of VN-T-AT to endothelial cells (11).

The binding experiments described above were performed with covalently stabilized ternary complexes. Also, electrostatically associated VN-T-AT complexes appeared to bind to endothelial cells via the VN-moiety. This was shown with ternary complexes prepared by limited incubation of purified components, which did not lead to complete stabilization of the ternary complex by covalent bonds. These complexes were incubated on endothelial cells in the absence or presence of competitors. Endothelial cells with surface-bound ligands were then subjected to SDS-containing gel electrophoresis and autoradiography (Figure 4). The complex at 83 kDa represents binary T-AT complex bound to the cell-surface. In the presence of competing amounts of urea-treated VN, heparin or covalently stabilized VN-T-AT, the amount of bound T-AT was reduced, indicating that binding of T-AT complex was mediated by VN. Addition of AT or peptide Y105, representing the SEC-receptor recognition site had no effect.

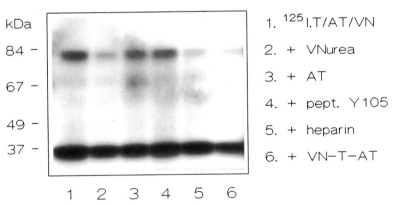

Figure 4. Binding of electrostatically associated VN-T-AT complex to endothelial cells. Radiolabeled thrombin was incubated for 5 min at 37°C with AT and VN in a 1:2:4 ratio (mol:mol). Residual thrombin activity was inhibited with hirudin. Endothelial cells were incubated with radiolabeled ligands for 10 min in the absence or presence of unlabeled competitors (VN$_{urea}$, AT or VN-T-AT: 100x molar excess; peptide Y105 100 μg/ml; heparin 100 U/ml), washed, lysed with SDS-containing sample-buffer and subjected to gel electrophoresis and auto-radiography.

INTERNALIZATION OF TERNARY COMPLEXES

In order to probe the fate of radiolabeled VN-T-AT complex, an internalization assay was developed in which the amount of cell- or matrix-associated complex could be determined. At 37°C, ternary complex was internalized and deposited into the endothelial cell matrix, whereas cells kept on ice and thus metabolically inactive, did not (Table 2). The VN-T-AT complex was deposited into the extracellular matrix as ternary complex and not degraded (data not shown). Furthermore, no trichloroacetic acid soluble breakdown products could be detected in the medium or the cell-compartment. Translocation through the endothelial cells was confirmed by electro-microscopical evaluation of ultra-thin sections (not shown). Addition of cytochalasin B led to a decreased deposition of the complex into the extracellular matrix, suggesting the involvement of the cellular cytoskeleton.

Table 2. Internalization and matrix-deposition of VN-T-AT (cpm per 10^3 cells).

4°C	cells	matrix
[^{125}I].(VN-T-AT)	2511 ± 85	211 ± 21
+ 100x molar excess VN-T-AT	375 ± 35	110 ± 10
+ heparin (100 U/ml)	928 ± 55	85 ± 35
37°C	cells	matrix
[^{125}I].(VN-T-AT)	1054 ± 13	888 ± 19
+ 100x molar excess VN-T-AT	306 ± 10	386 ± 20
+ heparin (100 U/ml)	217 ± 29	342 ± 33

Table 3. Matrix-deposition of VN-T-AT complex	(% of control)
non-treated cells (control)	100 ± 10
ammonium chloride (20 mM)	103 ± 7
primaquin (100 μM)	117 ± 16
chloroquin (100 μM)	100 ± 20
monensin (5 μM)	92 ± 20
cytochalasin B (10 μM)	79 ± 7
cytochalasin B (400 μM)	50
colchicine (0.5 μM)	76 ± 16

The pathway of internalization did not involve lysosomal processing, since lysosomal enzyme inhibitors like ammonium chloride, primaquin, chloroquin or monensin did not interfere with matrix-deposition (Table 3).
These observations suggest that transcytosis rather than endocytosis is involved in the translocation of the complex from the luminal to the baso-lateral side of the cells, although lateral diffusion or transport via cell-junctions can not totally be ruled out.

FUNCTIONAL PROPERTIES OF T-AT ASSOCIATED VN

VN in extracellular matrix has been found to mediate binding and thereby stabilization of plasminogen activator inhibitor-1 (PAI-1) (12) as well as endothelial cell attachment. T-AT associated VN is able to mediate both activities (Figure 5). T-AT associated VN and urea-treated VN could bind PAI-1 with similar dissociation constants (apparant K_d's of 20 and 22 nM respectively), suggesting that PAI-1 binds both VN-forms through the heparin binding domain.

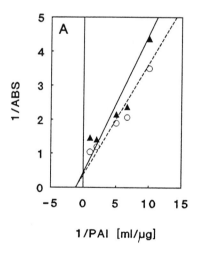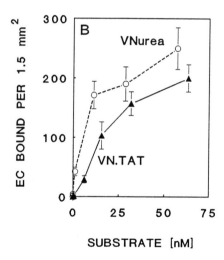

Figure 5. Urea-treated VN (-o-) or VN-T-AT complex (-▲-) was immobilized on 96-well plates via coated antibodies. (A) Concentration dependent binding of PAI-1 (0-1 μg/ml) (35 nM) was determined. Dose-response curves were plotted double-reciprocal and yielded two lines with identical intercepts on the X-axis. (B) An endothelial cell suspension (200.000 cells/ml) was incubated for 1 h at 37°C on immobilized proteins (0-60 nM). Attached cells were counted microscopically.

CONCLUSIONS

When the VN-T-AT complex is formed, some properties of the components are lost, some are generated. The most prominent effect of complex formation is the "activation" of the VN-molecule, which supplies the T-AT complex with a heparin binding domain and an RGD-sequence. The heparin binding domain not only mediates clearance of proteolytically inactivated thrombin, but also leads to enrichment of the extracellular matrix with VN. VN is not taken up in its native (folded) form (13), nor synthesized by endothelial cells. Thus, the proposed mechanism may explain the presence of VN at sites distant from the actual site of synthesis, which is the liver (14). Once deposited into the extracellular matrix, T-AT associated VN can bind PAI-1. Furthermore, the RGD-sequence of VN is accessible to promote endothelial cell attachment, even though this sequence is located directly adjacent to the region to which T-AT complex is covalently linked (1). Both properties are important in maintaining the integrity of the vessel wall.

REFERENCES

1 de Boer HC, de Groot PhG, Bouma BN, Preissner KT. J Biol Chem 1993; 268:1279-1283.
2 Björk I, Ylinenjärvi K, Olson ST, Bock PE. J Biol Chem 1992; 267:1976-1982.
3 Bar-Shavit R, Sabbah V, Lampugnani M.G., Marchisio PC, Fenton II JW, Vlodavsky I, Dejana E. J Cell Biol 1991; 112:335-344.
4 Pizzo SV. Amer J Med 1989; 87 suppl 3B:10S-13S.
5 Perlmutter DH, Glover GI, Rivetna M, Schasteen CS, Fallon CS. Proc Natl Acad Sci USA 1990; 87:3753-3757.
6 de Boer HC, Preissner, KT, Bouma BN, de Groot PhG. J Biol Chem 1992; 267:2264-2268.
7 Tomasini BR, Mosher DF. Blood 1988; 72:903-912.
8 Cardin AD, Weintraub HJR. Arteriosclerosis 1989; 9: 21-32.
9 David G, Lories V, Decock B, Marynen P, Cassiman JJ, Van den Berghe H. J Cell Biol 1990; 111:3165-3176.
10 Elenius K, Salmivirta M, Inki P, Markku M, Jalkanen M. J Biol Chem 1990; 265:17837-17843.
11 Mertens G, Cassiman JJ, Van den Berghe H, Vermylen J, David G. J Biol Chem 1992; 267:20435-20443.
12 Declerck PJ, De Mol M, Alessi MC, Baudner S, Pâques EP, Preissner KT, Müller-Berghaus G, Collen D. J Biol Chem 1988; 263:15454-15461.
13 Hess S, Wijelath E, Völker W, Declerck P, Stockmann A, Pötzsch B, Demoliou-Mason C, Preissner KT. VIIth Int Symp on Biol of Vasc Cells 1992;Abstract 14C.
14 Seiffert D, Keeton M, Eguchi Y, Loskutoff DJ. Proc Natl Acad Sci USA; 1991; 88:9402-9406.

Biology of Vitronectins and their Receptors
K.T. Preissner, S. Rosenblatt, C. Kost, J. Wegerhoff and D.F. Mosher, editors

Receptor mediated endocytosis of vitronectin by fibroblast monolayers

P.J. McKeown–Longo and Tracee S. Panetti

Department of Physiology and Cell Biology, Albany Medical College, Albany, New York 12208

INTRODUCTION

Vitronectin is a conformationally labile plasma protein with at least two identified conformations [1]. These two forms are a native non–heparin binding conformer in the plasma and an altered heparin binding conformer which can be purified from serum [2,3]. The interaction of native vitronectin with components of the coagulation or complement cascade results in exposure of a cryptic heparin binding domain within the vitronectin molecule [1,4]. Thrombin–serpin complexes have been demonstrated to interact with native vitronectin triggering the conformational change [1]. In addition, modified thrombins can induce the conformational change suggesting that thrombin is the important component in the thrombin–serpin complex inducing vitronectin's altered conformation [5]. The ternary complex of vitronectin–thrombin–antithrombin has been purified from serum and from the plasma of patients with disseminated intravascular coagulation [3,6]. The terminal complex of complement, C5b–9, can also bind native vitronectin altering its conformation [4]. The newly exposed heparin binding domain in altered vitronectin mediates binding to heparan sulfate proteoglycans on the surface of endothelial cells [7].

Vitronectin has also been localized to the extracellular matrix of numerous tissues in vivo. Immunofluorescent studies have localized vitronectin to embryonic lung, skin, and kidney basement membrane [8]. Plasma may be the source of vitronectin in non-hepatic tissues, since primarily liver–derived cells have been shown to synthesize vitronectin [9]. In vitro, vitronectin has been identified as a component of the extracellular matrix of both endothelial and fibroblast cell monolayers [7,8]. One function of matrix–associated vitronectin may be to mediate cell adhesion. The specific amino acid sequence, RGD [10], has been demonstrated to be a signal for cell adhesion in many glycoproteins including vitronectin. The RGD sequence in vitronectin has been demonstrated to interact with cell surface integrin receptors to mediate cell adhesion [11,12]. The integrins are a family of heterodimeric cell surface receptors which link the cytoskeleton to the extracellular matrix. Two vitronectin integrin receptors have been well characterized. The $\alpha_v\beta_3$ integrin receptor for vitronectin recognizes vitronectin as well as the other adhesive glycoproteins fibronectin, fibrinogen, and thrombospondin [13,14]. The $\alpha_v\beta_5$ vitronectin receptor mediates cell adhesion in an RGD dependent manner and is thought to interact only with vitronectin [15].

112

TIME COURSE OF BINDING OF NATIVE AND ALTERED VITRONECTIN

To determine whether the native and altered conformations of vitronectin function differently, comparative studies on the metabolism of vitronectin conformers were performed. Confluent fibroblast monolayers were incubated with serum–free nutrient medium supplemented with either ^{125}I–altered vitronectin (aVn) or ^{125}I–native vitronectin (nVn). The binding of both vitronectin conformers to the cell layers increased over time in culture with approximately 3% of altered or native vitronectin bound to the cell layer after 12 hours (Fig 1).

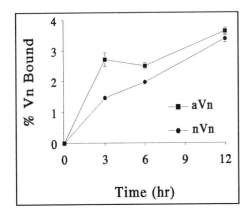

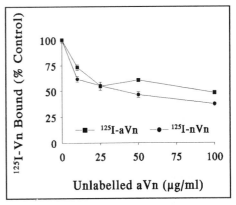

Figure 1   Figure 2

RECIPROCAL INHIBITION OF NATIVE AND ALTERED VITRONECTIN BINDING

To determine if both conformations of vitronectin bound to the same site in the fibroblast monolayer, increasing concentrations of each conformer were compared for their ability to inhibit the binding of either ^{125}I–nVn or ^{125}I–aVn. Altered vitronectin competed equally well for the binding of both ^{125}I–nVn and ^{125}I–aVn to the cell layer (Fig. 2). Similar results were obtained using native vitronectin as competitor (data not shown). These results suggest that both conformers bind to the same site in the cell layer.

LOCALIZATION OF NATIVE AND ALTERED VITRONECTIN IN THE EXTRACELLULAR MATRIX

Earlier studies have shown that the binding of native vitronectin to tissue culture plastic results in the conversion of vitronectin to the altered form [16]. To determine if binding to the extracellular matrix altered the conformation of native vitronectin, we incubated native vitronectin with the confluent fibroblast monolayers for 24 hours.

Localization of vitronectin in the cell layer was performed using either the 8E6 monoclonal antibody (a generous gift of Deane Mosher, Univ. Wisconsin) which recognizes only the altered conformation [1] or a polyclonal antibody prepared against altered vitronectin. The 8E6 monoclonal antibody (Fig. 3A) and the polyclonal antibody (Fig. 3C) detected altered vitronectin both in the extracellular matrix and in intracellular vesicles. As described previously, the intracellular staining represents vitronectin, which has been internalized into endocytic vesicles [17]. The native vitronectin was not detected by the monoclonal antibody (Fig. 3B), indicating that it remained in the native conformation. Native vitronectin was visualized by the polyclonal antibody in the extracellular matrix, but not in the intracellular endosomes (Fig. 3D). These data demonstrate that native vitronectin bound to the matrix remains in the native conformation. As discussed below, the absence of native vitronectin from intracellular endosomes is consistent with the finding that native vitronectin is not degraded by the cells.

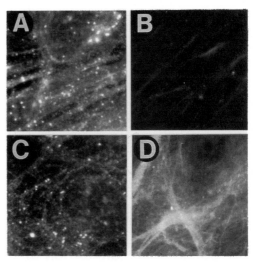

Figure 3

EFFECT OF GLYCOPROTEINS AND HEPARIN ON BINDING OF VITRONECTIN

Altered vitronectin is known to bind to the integrin receptors ($\alpha_v\beta_5$ and $\alpha_v\beta_3$) and to proteoglycans present on the surface of adherent cells. Heparin (Hep) (up to 100 µg/ml) did not block the binding of vitronectin to the cell layers, suggesting that the binding of vitronectin to the cell layer is not dependent on proteoglycans (Fig. 4). To assess the role of integrins in vitronectin binding, ligands known to bind to the $\alpha_v\beta_3$ receptor were tested for the ability to inhibit vitronectin binding. When added at concentrations up to 100 µg/ml, fibrinogen (Fg), fibronectin (Fn) as well as RGD Peptides (500 mM) did not inhibit the binding of vitronectin to cell layers (Fig. 4). Unlabelled vitronectin (100 µg/ml), however, was able to inhibit about 70% of the ^{125}I–Vn to the cell layer when compared to control (C) (Fig. 4). These results indicate that other known ligands for the $\alpha_v\beta_3$ integrin do not compete for the binding of ^{125}I–Vn to the cell layer, suggesting that the $\alpha_v\beta_3$ integrin is not mediating the binding of vitronectin to the cell layer. In addition, antibodies known to block the binding of vitronectin to the $\alpha_v\beta_3$ receptor (LM609; generous gift of Dr. David Cheresh, Scripps), and the $\alpha_v\beta_5$ receptor (P1F6), had no effect on the binding of vitronectin to the cell layer (Fig. 5).

114

These studies indicate that neither proteoglycans nor integrins are involved in the binding of vitronectin to the cell monolayers.

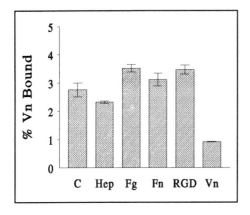

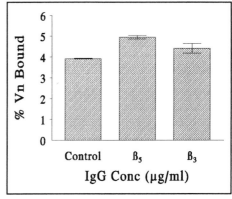

Figure 4   Figure 5

TIME COURSE OF DEGRADATION OF ALTERED AND NATIVE VITRONECTIN CONFORMERS

The degradation of vitronectin conformers was examined by incubating fibroblast monolayers with ^{125}I–aVn or ^{125}I–nVn and measuring the accumulation of trichloro-acetic acid soluble radioactivity in the culture medium (Fig. 6). Approximately 17% of the altered vitronectin was degraded to TCA soluble radioactivity within 12 hours. In contrast, less than 2% of the native vitronectin was degraded during the same time course. Further studies indicated that the degradation of altered vitronectin to TCA soluble radioactivity was saturable, sensitive to chloroquine, and occurred intra-cellularly [17], indicating that altered vitronectin was removed from the matrix by receptor–mediated endocytosis, and degraded in lysosomes.

EFFECT OF GLYCOPROTEINS AND HEPARIN ON DEGRADATION OF ALTERED VITRONECTIN

Although proteoglycans and integrins were not required for the binding of vitronectin to the cell layers, their potential role in mediating the internalization of vitronectin from the matrix was explored. The $\alpha_v\beta_3$ ligands, fibrinogen and fibronectin (100 µg/ml), had no effect on vitronectin degradation, suggesting that the $\alpha_v\beta_3$ integrin is not involved in the endocytosis and degradation of vitronectin. However, heparin inhibited vitronectin degradation by approximately 75% at a concentration of 10 µg/ml (Fig. 7). These results suggest that vitronectin may bind to cell surface proteoglycans prior to being endocytosed. A potential role for proteoglycans in mediating vitronectin internalization is consistent with the previous observation that only the heparin–binding conformer of vitronectin is degraded (Fig. 6).

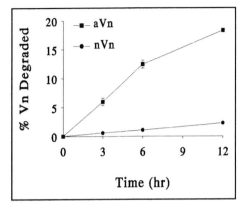

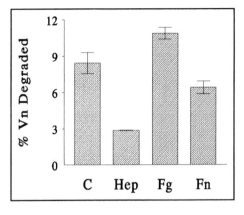

Figure 6   Figure 7

EFFECT OF RGD PEPTIDES AND ANTI–INTEGRIN ANTIBODIES ON VITRONECTIN DEGRADATION

To examine the role of cell surface integrins in mediating the internalization and degradation of vitronectin, radiolabeled vitronectin was incubated with cell mono-layers in the presence of either RGD peptides or monoclonal antibodies against the $\alpha_V\beta_3$ and $\alpha_V\beta_5$ integrin receptors. RGD peptides at a concentration of 0.5 mM in-hibited vitronectin degradation by approximately 75% after a 12 hour incu-bation, suggesting that integrins were involved in vitronectin degradation (Fig. 8). Antibodies (LM609) against a functional domain in the $\alpha_V\beta_3$ vitronectin receptor did not inhibit vitronectin degradation. How-ever, the monoclonal antibody P1F6 against the $\alpha_V\beta_5$ integrin receptor inhibited vitronectin degradation by approximately 80%. The ability of heparin as well as

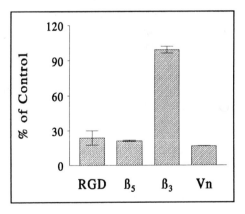

Figure 8

antibodies against the β_5 integrin subunit to block vitronectin degradation, but not binding to the cell layer, suggests that the removal of vitronectin from the matrix occurs through a coordinate interaction of vitronectin with both a cell surface heparan sulfate proteoglycan and the $\alpha_V\beta_5$ integrin receptor.

EFFECT OF THROMBINS ON VITRONECTIN DEGRADATION

Previous studies by Tomasini and Mosher [1,5] have shown that α–thrombin–serpin complexes as well as γ–thrombin could induce the conformational change in vitronectin exposing the cryptic heparin binding domain in vitronectin. To demonstrate that receptor mediated endocytosis of vitronectin was regulated by its conformational state, various thrombin preparations were used to induce the conformational change in native vitronectin. The native vitronectin was pre–incubated with the fibroblast monolayer for 6 hours to allow binding to the extracellular matrix. Various thrombin preparations were added to the monolayer and cell dependent degradation of the radiolabeled vitronectin was determined.

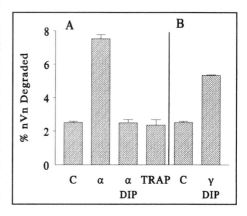

Figure 9

The addition of α–thrombin increased cell–dependent degradation of native vitronectin by 3–fold as compared to control (C) levels (Fig. 9A). α–Thrombin cannot directly alter the conformation of vitronectin, however α–thrombin–serpin complexes can alter the conformation of vitronectin suggesting that the α–thrombin bound to endogenous serpins in the fibroblast matrix, such as protease nexin or PAI–1, triggering the conformational change in native vitronectin. To test this possibility, the active site of thrombin was blocked with diisopropyl fluorophosphate (DIP) prior to incubation with the cell layer. This preparation of thrombin was unable to produce an increase in vitronectin degradation, presumably due to its inability to bind to an endogenous serpin and trigger the conformational change in vitronectin. Since thrombin can also activate intracellular signalling pathways through its binding to the thrombin receptor, the effect of the thrombin receptor agonist (TRAP) on vitronectin degradation was also tested [18]. The TRAP peptide (at 25 µM) had no effect on vitronectin degradation, indicating that thrombin's effect on vitronectin turnover was not due to some effect on cell metabolism triggered by activation of the thrombin receptor. The proteolytic cleavage product of α–thrombin, γ–thrombin, has been shown to bind directly to vitronectin and expose its heparin binding site. DIP–treated γ–thrombin was also able to induce the degradation of native vitronectin (Fig. 9B). These studies suggest that the half–life of vitronectin in the matrix is affected by its conformational state.

SUMMARY

Comparative studies on the metabolism of native and altered vitronectin have shown that both vitronectin conformers bind to the extracellular matrix of cultured fibroblast cell layers. Once bound to the matrix the altered conformer is rapidly removed from the matrix by receptor mediated endocytosis involving the $a_v b_5$ integrin and a cell surface proteoglycan. Native vitronectin remains stable in the matrix but its clearance can be triggered by the addition of thrombin–serpin complexes to the cell layer.

It is becoming clear that vitronectin may play an important role in the regulation of hemostasis. It can affect the enzymatic activity of both thrombin and plasmin, enzymes critical to the regulation of coagulation and fibrinolysis. The relative contribution of the individual conformers to these processes remains to be determined. Altered vitronectin present in the extracellular matrix has been shown to mediate cell adhesion and migration [19–21]. It also binds PAI-1 and promotes the clustering of urokinase–type plasminogen activator receptors on the cell surface. The localization of urokinase–type plasminogen activator into vitronectin–dependent areas of cell-substrate contact promotes the inhibition of urokinase by PAI-1, thereby decreasing pericellular plasmin levels [22,23]. It is not clear whether native vitronectin can be expected to support the regulation of pericellular plasmin activity because definitive studies on whether native vitronectin can support cell adhesion have not been done.

It is clear that native vitronectin can bind thrombin–serpin complexes and that this binding triggers the conformational change in vitronectin to the altered form [1,5]. PAI-1 bound to native vitronectin increases the inhibitory activity of PAI-1 toward thrombin [24]. Therefore, native vitronectin can regulate thrombin activity in the pericellular space. The binding of thrombin to PAI-1–vitronectin complex, may trigger the conformational change, which results in the endocytosis of vitronectin. Therefore, vitronectin may act as an opsonic molecule or molecular bridge mediating the removal and degradation of thrombin–serpin complexes from the tissues. The ability of vitronectin to regulate thrombin and plasminogen–activator defines an important role for vitronectin during processes of wound healing and tissue remodeling.

REFERENCES

1 Tomasini BR, Mosher DF. Blood 1988; 72:903–912.
2 Izumi M, Yamada KM, Hayashi M. Biochim Biophys Acta 1989; 990:101–108.
3 Ill CR, Ruoslahti E. J Biol Chem 1985; 260:15610–15615.
4 Hogasen K, Mollnes TE, Harboe M. J Biol Chem 1992; 267:23076–23082.
5 Tomasini BR, Owen MC, Fenton II JW, Mosher DF. Biochem 1989; 28:7671–7623.
6 de Boer HC, de Groot PG, Bouma BN, Preissner KT. J Biol Chem 1993; 268:1279–1283.
7 de Boer HC, Preissner KT, Bouma BN, de Groot PG. J Biol Chem 1992; 267:2264–2268.

8 Hayman EG, Pierschbacher MD, Ohgren Y, Ruoslahti E. Proc Natl Acad Sci 1983; 80:4003–4007.
9 Barnes DW, Reing J. J Cell Physiol 1985; 125:207–214.
10 Jenne D, Stanley KK. EMBO Jour 1985; 4:3153–3157.
11 Hayman EG, Pierschbacher MD, Ruoslahti E. J Cell Biol 1985; 100:1948–1954.
12 Cherny RC, Honan MA, Thiagarajan P. J Biol Chem 1993; 268:9725–9729.
13 Cheresh DA. Proc Natl Acad Sci 1987; 84:6471–6475.
14 Pytela R, Pierschbacher MD, Ruoslahti E. Proc Natl Acad Sci 1985; 82:5766–5770.
15 Smith JW, Vestal DJ, Irwin SV, Burke TA, et al. J Biol Chem 1990; 265:11008–11013.
16 Tomasini BR, Mosher DF. Prog Hemost Thrombos 1990; 10:269–305.
17 Panetti TS, McKeown–Longo PJ. J Biol Chem 1993; 268:11988–11993.
18 Scarborough RM, Naughton MA, Teng W, Hung DT, et al. J Biol Chem 1992; 267:13146–13149.
19 Charo IF, Nannizzi L, Smith JW, Cheresh DA. J Cell Biol 1990; 111:2795–2800.
20 Schwartz MA, Leavesley D, Cheresh DA. Molec Biol Cell 1992; 3:95a. (Abstract)
21 Leavesley DI, Ferguson GD, Wayner EA, Cheresh DA. J Cell Biol 1992; 117:1101–1107.
22 Ciambrone GJ, McKeown–Longo PJ. J Cell Biol 1990; 111:2183–2195.
23 Ciambrone GJ, McKeown–Longo PJ. J Biol Chem 1992; 267:13617–13622.
24 Keijer J, Linders M, Wegman JJ, Ehrlich HJ, et al. Blood 1991; 78:1254–1261.

Vitronectin receptors and adhesive interactions

Biology of Vitronectins and their Receptors
K.T. Preissner, S. Rosenblatt, C. Kost, J. Wegerhoff and D.F. Mosher, editors

Platelets, Extracellular Matrix and Integrins

J.J. Sixma, H.K. Nieuwenhuis and Ph.G. de Groot

Department of Haematology, University Hospital Utrecht,
P.O. Box 85500, 3508 GA UTRECHT, The Netherlands

INTRODUCTION

Adhesion of cells to substrates and to other cells is a central mechanism in biology. Not only is it essential for the maintenance of the integrity of tissues and organs, but it is also crucial for such diverse processes as differentiation, cell mediated immunity, and oncogenesis. Adhesion is also crucial for haemostasis and its pathological counterpart thrombosis. When the vessel wall is injured, perivascular connective tissue is exposed to flowing blood. Blood platelets from flowing blood adhere to this tissue, spread along it, secrete the contents of their secretion granules, and become adhesive to other platelets. Platelet aggregates or thrombi build up and will form occlusive haemostatic plugs which will prevent further bleeding [1,2]. The platelet plug is then strengthened by clot retraction and eventually, over several hours, replaced by a fibrin network which provides the definite occlusion of the vessel wound [3,4]. When platelet adhesion does not occur, bleeding will continue and no haemostatic plug will form.

Platelet adhesion occurs from flowing blood. Because of this it has developed a number of special characteristics which we will review in this paper. These characteristics are: (1) adhesion has to be fast, (2) it should be able to resist shear stress, and (3) it should lead to platelet activation and thrombus formation.
If further thrombus formation is not needed for the formation of a haemostatic plug, thrombus growth has to be limited. We will also discuss a possible mechanism by which this is attained.

PLATELET ADHESION

Platelet adhesion depends on the interaction of platelet receptors with adhesive proteins in the vessel wall. Major adhesive proteins are collagens, von Willebrand factor, and fibronectin. Laminin and thrombospondin play a minor role. In a wound fibrinogen will bind to perivascular connective tissue and fibrin fibers will form as consequence of the activation of the coagulation system by perivascular tissue factor. Platelets do also adhere to fibrinogen and fibrin [5].

Adhesion to *collagens* has been studied for the major collagens that occur in the vessel wall [6]. Collagen type I, type III and type IV are the most reactive ones among these. Adhesion to these collagens takes place at all shear rates tested up to 1800/sec. Platelets adhere to collagen type V under static but not under flow conditions. Adhesion to collagen type VI and VIII is maximal at a shear rate of 300/sec and drops rapidly at higher shear rates. Platelet adhesion to all these collagens is dependent on the interaction of platelet GPIa-IIa (VLA-2) with the respective collagen. Studies with washed platelets under static conditions have shown that this interaction is dependent on divalent cations, magnesium in particular [7]. Perfusion studies with whole blood over purified collagen coated to a surface showed that adhesion to collagen type IV had a complete dependence on magnesium, whereas adhesion to collagen type III was less dependent and adhesion to collagen type I was almost not dependent at all [8].

Platelet adhesion to *fibronectin* is optimal at shear rates below 500/sec. This adhesion is not calcium dependent. GPIIb/IIIa and GPIc/IIa (VLA-5) are the most important receptors [9]. Fibronectin-dependent adhesion to subendothelium differs from adhesion to isolated fibronectin in that it is also present at higher shear rates [10] and GPIIb-IIIa and GPIc/IIa (VLA-5) are not involved. The platelet receptor that is responsible for fibronectin-dependent adhesion to subendothelium has not been identified yet.

Adhesion to *von Willebrand factor* occurs even at the highest shear rate of 4000/sec tested. It is not dependent on divalent cations. Platelets have two receptors for vWF, GPIb and GPIIb/IIIa. GPIb is the only receptor required for adhesion of platelets to von Willebrand factor under static conditions [11]. GPIb and GPIIb/IIIa are both required for adhesion to purified von Willebrand factor coated to a surface under flow conditions [12]. Von Willebrand factor is essential for adhesion to subendothelium and to collagens. The von Willebrand factor needed for this interaction is either already present in subendothelium or adsorbs from plasma to subendothelium and collagens. When reactive collagens such as thick fibers of collagen type I or III are used, von Willebrand factor can also be provided by secretion from platelet alpha granules [15]. Von Willebrand factor binds specifically to collagen type I and III [13,14]. GPIIb/IIIa-von Willebrand factor interaction is not essential for the support by von Willebrand factor of platelet adhesion to collagen, although it plays a role in platelet spreading [12].

Adhesion to *laminin* occurs at shear rates below 1300/sec [16]. It is strongly dependent on divalent cations. VLA-6 (GPIc'/IIa) is the receptor. The interaction site on laminin is localized on the carboxyterminal heparin binding domain. Studies with a monoclonal antibody against VLA-6 showed that laminin plays only a minor role in adhesion to subendothelium. Laminin is not present in deeper layers of the vessel wall.

Adhesion to *thrombospondin* shows a decrease above shear rates of 1500/sec [17]. Adhesion has an absolute requirement for calcium. Thrombospondin has at least 12 calcium binding sites and it changes in conformation when it is exposed to calcium chelators [18]. Calcium depleted thrombospondin may act as inhibitor of platelet adhesion probably by shielding other adhesive molecules [17].

Platelets do not adhere to *vitronectin* under flow conditions. Under static conditions, platelets have to be activated first to show good adherence. When adhesion to vitronectin under flow conditions was studied after upstream activation by a coverslip coated with collagen, no adhesion to vitronectin was observed. Further studies with upstream activation by local thrombin formation still need to be performed.

The unique features of platelet adhesion

Comparison of the initial rates of platelet adhesion showed that adhesion to von Willebrand factor was faster than to fibronectin. Adhesion to von Willebrand factor was characterized by a very rapid first phase within 3 minutes followed by a slower accumulation after that. Adhesion to fibronectin was linear in time.

Von Willebrand factor is very important as co-factor for platelet adhesion to other proteins at high shear rates. We already mentioned this for adhesion to collagen, but recent studies have shown that von Willebrand factor is also involved in adhesion to fibronectin and fibrinogen/fibrin especially at higher shear rates. This unique role of von Willebrand factor is due to the fact that it is the only adhesive protein able to support adhesion at high shear rates. The GPIb- von Willebrand factor interaction is unique to the platelet. GPIb is not present on other cells.

Among the adhesive proteins there are only two that cause platelet aggregate formation. These two are the collagens and fibrin. Both these two types of surfaces have in common that they concern fibrillar structures on which multiple interaction sites are present. Aggregate formation is decreased on monomeric collagen and absent on fibrinogen. Also cyanogenbromide fragments of collagen support adhesion but not aggregate formation [19]. Comparison of adhesion to fibrinogen on the one hand with 50.000 molecules of GPIIb/IIIa as receptor and collagens with only 2000 receptors shows that it is not the number of receptors that determines aggregate formation. A more likely cause is the clustering of these receptors by a high density of adhesive sites. Studies with the alpha1(I) chain of collagen type I has shown that a single chain has at least three reactive sites [20,21]. A collagen type I monomer is a triple helix consisting of 2 alpha 1 and one alpha 2 chains. If the alpha 2 chain would have a similar number of interaction sites, a single monomer would have 9 closely spaced adhesive sites. Adhesion to monomeric collagen in flow has furthermore a requirement for von Willebrand factor and fibronectin and this will increase the density of adhesive sites further. The observation that von Willebrand factor, and for monomeric collagen also fibronectin, is required as a cofactor [22] indicates that the ability to resist shear is not necessarily a consequence of a high density of adhesive sites.

Thrombus formation will also occur when platelets adhering to the vessel wall are activated by locally generated thrombin. At low shear rates thrombin generation at the vessel wall will lead to formation of fibrin fibers covering the subendothelium on which platelet thrombin may form. At high shear rates platelet thrombi appear on the subendothelium interspersed with fibrin fibers [23,24].

Morphology of platelet thrombi

Immuno-electronmicroscopy of platelet thrombi with antibodies against ligands and receptors using immuno-gold labelling has recently been performed (Heijnen H.G., Lozano,M., Sixma J.J. unpublished). These studies showed that the integrin GPIIb/IIIa was evenly distributed along the platelet plasma membrane. In contrast, the ligands fibrinogen and von Willebrand factor were concentrated on membranes in contact with the adhesive surface or with other platelets with a much lower concentration on the exposed surface of the thrombus. Further study of the GPIIb/IIIa on the exposed surface showed that these GPIIb/IIIa molecules were recognized by anti LIBS-antibodies. These data suggest that thrombus size is limited by absence of ligands on the surface of thrombi and that this absence is explained by reversible binding of ligands that subsequently leave the receptor.

CONCLUSIONS

Adhesion of blood platelets to the perivascular connective tissue involves several ligands and receptors. It has a unique character requiring fast adhesion that is resistant to shear stress. This may be explained by the special system of GPIb and von Willebrand factor. The induction of aggregate formation is seen when platelets interact with fibers in the form of collagens or fibrin or when adhering platelets are activated by concomitantly present thrombin. Limitation of thrombus size in flowing blood may be due to the absence of ligands on the surface of the thrombus due to reversible binding to GPIIb/IIIa.

REFERENCES
1. Hovig T, Dodds WJ, Rowsell HC, Joergensen L, et al. Blood 1967;40:636-668.
2. Sixma JJ, Wester J. Semin Hemat 1977;14:265-299.
3. Hovig T, Dodds WJ, Rowsell HC, Mustard JF. Am J Pathol 1968;53:355-374.
4. Wester J, Sixma JJ, Geuze JJ, Heijnen HFG. Lab Invest 1979;41:182-192.
5. Hantgan RR, Hindriks GA, Taylor R, Sixma JJ, De Groot PhG. Blood 1990;76:345-353.
6. Saelman EUM, Gralnick HR, Hese KM, Sixma JJ, Nieuwenhuis HK. Submitted 1993
7. Staatz WD, Rajpara SM, Wayner EA, Carter WG, Santoro SA. J Cell Biol 1989;108:1917-1924.
8. Saelman EUM, Hese KM, Sixma JJ, De Groot PhG, Nieuwenhuis HK. Submitted 1993
9. Beumer S, Orlando E, De Groot PhG, Sixma JJ. Thromb Haemost 1991;65:811 (abstract).
10. Houdijk WPM, Sixma JJ. Blood 1985;65:598-604.
11. Savage B, Shattil SJ, Ruggeri ZM. J Biol Chem 1992;267:11300-11306.
12. Lankhof H, Vink T, Schiphorst ME, Wu YP, IJsseldijk MJW, De Groot PhG, Sixma JJ. Thromb Haemostasis 1993;abstract New York.
13. Roth GJ, Titani K, Hoyer LW, Hickey MJ. Biochemistry 1986;25:8357-8361.
14. Houdijk WPM, Girma J-P, Van Mourik JA, Sixma JJ, Mayer D. Thromb Haemost 1986;56:391-396.
15. D'Alessio P, Zwaginga JJ, De Boer HC, Federici AB, Rodeghiero F, Castaman G, Mariani G, Mannucci PM, De Groot PhG, Sixma JJ. Thromb Haemost 1990;64:227-231.
16. Hindriks GA, IJsseldijk MJW, Sonnenberg A, Sixma JJ, De Groot PhG. Blood 1992;79:928-935.

17. Agbanyo FR, Sixma JJ, De Groot PhG, Languino LR, Plow EF. J Clin Invest 1993;92:288-296.
18. Lawler J, Simons ER. J Biol Chem 1983;258:12098-12101.
19. Saelman EUM, Nieuwenhuis HK, Hese KM, Barnes MJ, Sixma JJ. Blood (in press).
20. Morton LF, Peachey AR, Barnes MJ. Biochem J 1989;258:157-163.
21. Zijenah LS, Barnes MJ. Thromb Res 1990;59:553-566.
22. Houdijk WPM, Sakariassen KS, Nievelstein PFEM, Sixma JJ. J Clin Invest 1985;75:531-540.
23. Weiss HJ, Turitto VT, Baumgartner HR. J Clin Invest 1986;78:1072-1082.
24. Weiss HJ, Turitto VT, Vicic WJ, Baumgartner HR. Blood 1984;63:1004-1015.

Biology of Vitronectin and their Receptors
K.T. Preissner, S. Rosenblatt, C. Kost, J. Wegerhoff and D.F. Mosher, editors

Megakaryocyte specific expression of the αIIb integrin

Gérard Marguerie and Georges Uzan

Laboratoire d'Hématologie.
Département de Biologie Moléculaire et Structurale.
INSERM unité 217
CENG 85 X, 38041 Grenoble, cedex, France

INTRODUCTION

Historically different strategies have been developed to study the role of platelet in vascular biology. Early studies have focused on agonists and signal transduction mechanisms and on the secretion of soluble messengers. A tremendous amount of work has also been done during the past decade on membrane glycoproteins and adhesive reactions. To day, a new strategy which allow an in vivo analysis of the physiological relevance of specific molecules is emerging. This strategy involves targeted gene expression in the platelet of transgenic mice.

To address this problem a promoter of a gene which is exclusively expressed in megakaryocyte, the progenitor cell of circulating platelet, should be identified. A good candidate is the gene coding for the integrin αIIb subunits.
The molecule is the α chain of the RGD sensitive platelet integrin αIIb β3, which expresses receptor function for fibrinogen, fibronectin and von Willebrand factor (1), and is exclusively implicated in platelet aggregation. While the β3 subunit is expressed in a variety of cells, including megakaryocyte fibroblasts, endothelial cells and macrophages, the αIIb mRNA is only produced in megakaryocyte and at an early stage of megakaryocytopoiesis (2). For these reasons we have focused our attentions on the transcriptional regulation of this gene, to identify megacaryocyte specific transcriptionnaly active sequences

STRUCTURE OF THE αIIb GENE

The gene for αIIb spans approximately 17kb (3-4) and is located on the long arm of chromosome 17 at position q21.1-q21.2 (5-6-7). Though functional domains of the αIIb subunit have not been all, identified, there is no obvious relationship between the position of the 30 exons of the gene and the functional structure of the protein (Heidenreich et al 1990). In addition, the αIIb gene contains seven intragenic complete AluI repeats, the function of which is unknown at the present time.

Since αIIb expression is limited to megakaryocyte the structure of the 5'flanking region of the gene was carefully examined. The promoter contains a single transcriptional start site which has been identified by independent groups using different approaches including RNase mapping, primer extension and S1 nuclease experiments (3-4). The site corresponds to an A, located 33 base pair upstream from the only ATG in correct reading frame (Fig. 1). The region upstream the transcriptional start site does not contain the canonical TATA and CAAT boxes which are frequently present in proximity to the start site of polymerase II transcribed genes. This is not unusual, since a variety of TATA less promoters have now been described. Some of these TATA less promoters contain multiple GC boxes and are mainly involved in the transcription of house keeping genes. Other TATA less promoter are GC poor and mostly involved in the transcription of genes that are implicated in cell proliferation or differentiation (8). The αIIb gene seems to fall into the latter category.

128

```
                                GAATTCCCCGGATCAGAAAATAGAAATCAAAAGGAAAATGTGGCTATGGTTAC  -1201
CCCTAGCGGACCTCTTAAATCTTCCTGAGAACCTGCTTTTTTGGGAAGGCATGAGTGCCAGTAAGACTTGGCACTCCTCC  -1121
TCTTCCGCTTACCGAGAGAAAATGACTTTGCCTTTCTGCTCAAAACTCATCCCTTCACTTTGTCACCCTATGTTTGCATC  -1041
TTCCATCCTTAGTGTGTGTTTCCATCCAGTCTTTCAGCAATACACGTACTACACATTGGACTCTTGGGTAGTCTCTAGGG   -961
CTGGAGCAAGGAGCCTTGCTCCCAAGGGACTCATTTACACAATCCTGTGAACGGACCAAGAGTAAACAGTGTGCTCAATG   -881
                  Bgl 1
                    ↓
CTGTGCCTACGTGTGTTAGCCCACGCGGCCAGCCTGAGGAGTCAGGGAAGGCTCCCCTAGGCAAAGCCCCAACCAGAATC   -801
AAGTCTTAATGGTTAAAGAGCTCCATCACCCAAAAAGGATTGAGGGCCTACCTTCAACTGAACAGCTAATGCATAATCTC   -721
AGAAACTGTGAGTCAAAATTCCCTGGAATAACTCCACTTTATCCCCAATCTCCTTGCCACCTAGACCAAGGTCCATTCAC   -641
                                                                        ets
CACCCTGTCCCCAGCACTGACRGCACTGCTGTGGCCACACTAAAGCTTGGCTCAAGACGGAGGAGGAGTGAGGAAGCTGC   -561
                                 ets                                    IR1
TGCACCAATATGGCTGGTTGAGG|CCGCCG|AAGGTCCTAGAAGGAGGAAGTGGGTAAATGCCATATCCAAAAAGATACAGA   -481
                                                                       Pvu II
                                                                         ↓
AGCCTCAGGTTTTATCGGGGGCAGCAGCTTCCTTCTCCTTCCCCGACCTGTGGCCAAGTCACAAAGCACCACAGCTGTAC   -401
                      IR1
AGCCAGATGGGGGAAGGGAGGAGATTAGAACTGTAGGCTAGAGTAGACAAGTATGGACCAGTTCACAATCACGCTATCCC   -321
AAGCCAGAAAGTGATGGTGGCTTGGACTAGCACGGTGGTAGTAGAGATGGGGTAAAGATTCAAGAGACATCATTGATAGG   -241
|C|AGAACCAAT|AGGACATGGTAATAAATCTATTCTCAGGAAAGGGGAGGAGTCATGGCTTTCAGCCATGAGCATCCACCCT   -161
                              IR2                                     Dde I
                                                                       ↓
CTGGGTGGCCTCACCCACTTCCTGGCAATTCTAGCCACCATGAGTCCAGGGGCTATAGCCCTTTGCTCTGCCCGTTGCTC    -81
    IR2
AGCAAGTTACTTGGGGTTCCAGTTTGATAAGAAAAGACTTCCTGTGGAGGAATCTGAAGGGAAGGAGGAGGAGCTGGCCC     -1
+1
├→
ATTCCTGCCTGGGAGGTTGTGGAAGAAGGAAG ATG GCC AGA GCT TTG TGT CCA CTG CAA GCC CTC TGG    68
                                 Met Ala Arg Ala Leu Cys Pro Leu Gln Ala Leu Trp

CTT CTG GAG TGG GTG CTG CTG CTC TTG GGA CCT TGT GCT GCC CCT CCA GCC TGG GCC TTG    128
Leu Leu Glu Trp Val Leu Leu Leu Leu Gly Pro Cys Ala Ala Pro Pro Ala Trp Ala Leu
                                                    Bgl I
                                                     ↓
AAC CTG GAC CCA GTG CAG CTC ACC TTC TAT GCA GGC CCC AAT GGC AGC CAG TTT GGA TTT    188
Asn Leu Asp Pro Val Gln Leu Thr Phe Tyr Ala Gly Pro Asn Gly Ser Gln Phe Gly Phe
            Ⓐ                    Nco I
TCA CTG GAC TTC CAC AAG GAC AGC CAT GGG AG GTGAGCCGTAAGGGAAGTTGGGGTATTGGGAGAGAGC   258
Ser Leu Asp Phe His Lys Asp Ser His Gly

GGGACCCCTCCCCATCACTGCTTCTGGGGGCTTCGAGTTTCCCATTTGCGATAGCAGTTGAGCAAGGTGACTTGTGGGGG   338
CCTATTCAGGTTGATTTCTTGTCAAGATGTGGGTCCCAGGGACTGGCTCAGGTGAAGGTATAAGGGCAGGGCACATGTGG   418
GCTGATGGGCACTGAAAACTACAGCAAGAACAAAGGGAAGACAATAGTTGATGCTTTATTTTTTCCCAAGGGTCAGTTGT   498
ATGAACCACTCCACCCTCAACACCTTGAAATGCAGAGAGGAGGCCGGGCGCGGTGCTCATGCCTGTAATCCCAGCACTTT   578
GGGAGGCCGAGGCGGGCAGATCACCTGAGGTCGAGAATT                                          616
```

Figure 1. DNA sequence of the first exon and 5'flanking region of the human GPIIb gene. The A designated as + 1 is the transcription start site the arrow below shows the direction of transcription. Potential regulatory elements are boxed. Horizontal arrows show inverted repeats. The 5'splice site is indicated with a dotted line From Prandini et al 1988.

Comparison of the αIIb promoter with the sequence of known promoters of other integrins does not reveal obvious similarities (Fig. 2). The 5' flanking domain of the gene encoding the α subunit of the fibronectin receptor α5ß1 has recently been sequenced (9). This region also lacks the TATA box but contains three SP1 binding sites with the consensus sequence CCGCCC. The gene coding for the α4 subunit which associates with the ß1 subunit to form the α4 ß1 complex which functions as a receptor for fibronectin and VCAM1 has also been characterized (10). Expression of this gene is restricted to the lymphoid and the myeloid lineages (11). In contrast to the αIIb and α5 promoters the α4 promoter contains TATA and CAAT boxes and is poor in GC sequences. Thus it is likely that cell specific expression of the different α subunits of the integrin family is driven by different sets of transcriptional factors. Interestingly enough, the αIIb, α5 and α4 promoters all contain multiple binding sites for the C-ets proteins, with the consensus motif GAAGGA. The translation products of the C-ets proto-oncogenes constitute a family of transcriptional factors involved in cell proliferation and acting in cooperation with AP1 (12-13) which mediates transcriptional stimulation by phorbol esters or TGFß (14-15). In that respect, it is worth noting that the α5 and α4 promoters also contain consensus binding site for the AP1, complex. Since, α5ß1 and α4 ß1 integrins are both regulated by these inducers, it is attractive to speculate that during growth stimulation, their expression is regulated by a combination of AP1 and C-ets protein. In the GPIIb promoter, the C-ets proteins appear to be associated with the transcription factor GATA1, suggesting (15-16) that specific expression of αIIb during megakaryocytopoiesis may be controlled by an entirely different mechanism.

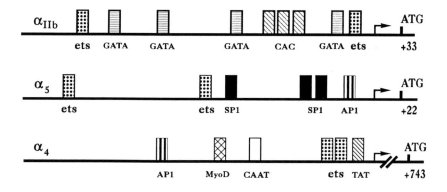

Figure 2.   Comparison of the structure of the αIIb promoter and promoters of other integrins a subunits does not reveal obvious similarities suggesting that they may be controlled by different mechanisms.

Comparison of the nucleotide sequence of the αIIb promoter with erythroid specific promoters reveals interesting features (Fig.3). The αIIb promoter contains multiple binding sites for the erythoid factor GATA 1, with the motif TGATAA. The role of this different GATA 1 binding sites inside the promoter have recently been funcionnaly examined (16). The results of these studies indicated that the -463 and the -54 sites play a significant role in the activity of the promoter. This transcriptional factor is a zinc finger proteins which belongs to the GATA family of which three members, GATA1, GATA2 and GATA3, have been identified. GATA1 is expressed in erythrocytes, megakaryocytes and mast cells. This factor is critical for the development of the erythroid lineage as gene knock out experiments in transgenic mice, resulted in a profound anaemia and was lethal for most of the homozygote animals (17). The presence of multiple binding site for GATA1 in the

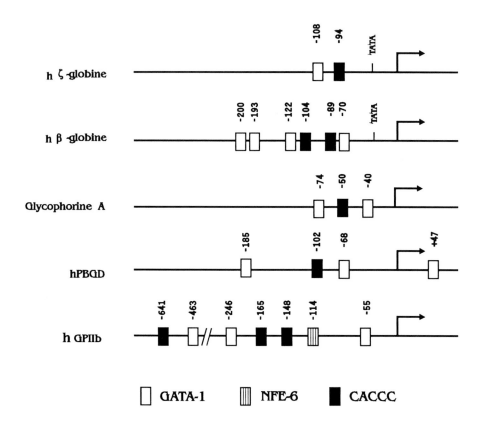

Figure 3.   Comparison of the structure of the promoter of GPIIb gene with erythiroid promoters.

sequence of the αIIb promoter, suggests that this factor may also be important in regulating gene expression in megakaryocyte. In addition to the GATA1 sites, the αIIb promoter contains multiple CACCC motifs present a position -165, -155 and -148 and a

CCAAT motif at position -228. The DNA binding proteins interacting with theses sites have been shown to cooperate with the GATA 1 sites in erythroid genes (18). When analysed in DNasesI protection assays these sites are protected by nuclear factors indicating that they may be functional. In summary, one can conclude from sequence analysis, that the αIIb promoter contains all the elements that are implicated in the expression of erythroid genes. Yet, αIIb is not produced in these cells.

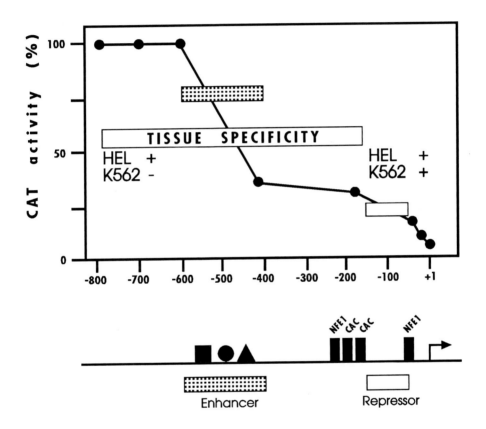

Figure 4.   Deletion analysis of αIIbaIIb gene promoter indicates the presence of three different functional domains.
The uncleotide sequence located between - 400 and - 600 functions as an Erythroid-Megakaryocytic specific enhancer. The domain located between nucleotides - 70 to - 150 exhibits negative regulatory function and in probably responsible for the tissue specific expression of the gene.

This does not seem to be unique to αIIb. Other megakaryocyte gene have recently been examined, and contain elements of erythroid like promoter, including PF4 gene and the GPIb gene. It is therefore attracting to speculate that these erythroid like promoters are not functional in erythrocyte and are turned off by some unknow mechanism.

TISSUE SPECIFIC ACTIVITY OF THE αIIb PROMOTER.

A series of experiments have been performed to verify the tissue specific activity of the αIIb promoter. Using transient CAT assays we have demonstrated that a 676 pb fragment of the promoter contains sufficient informations for cell-specific expression. When fused to the CAT gene and tranfected in cell line this promoter region was able to drive the expression of the enzyme in HEL cells but not in K562 or Hela cells (19. The activity of the αIIb promoter in this case corresponded to 15% of that observed with the ubiquituous RSV promoter taken as reference. From these results, it was concluded that the 5'flanking region of the αIIb gene contains responsive DNA sequences that control its transcription in a tissue specific maner and that these DNA elements are located within the first 600 bp of the promoter domain.

THE αIIb PROMOTER CONTAINS A SPECIFIC ERYTHRO-MEGAKARYOCYTIC ENHANCER.

Nucleotide sequences that are implicated in the transcriptional control of the αIIb gene were then examined, using 5' deletion mutants of the promoter, fused with the CAT reporter gene and transfection experiments in HEL cells (19-3). The 5'flanking region contains at least four distinct functional domains (Fig.4).

The region located in between nucleotides -584 and -414 functions as a positive regulatory domain. Removal of this domain produces a 70% decrease of the promoter activity. Further deletion, to nucleotide -113, had no detectable effect, while deletion of the sequence between nucleotide -113 to nucleotie -29 produced an additional 20% decrease of the activity. Finally, the sequence, adjacent to the transcriptional start site, from -13 to +33 exhibited a significant residual activity of about 6% when compared to a promoter less construct, suggesting that this TATA less domain contains specific sequences that are critical for the positioning of the transcriptional complex.

The promoter sequence between -584 to -414 functions as an erythro-megakaryocytic specific enhancer. This domain alone, produces a 5 to 6 fold increase of the activity of aN ubiquitous enhancer less viral promoter SV40. This enhancer activity is orientation and position independent and was observed in HEL cells, a megakaryocytic cell line which expresses αIIb, in K562 cells, an erythroid cell line which does not express αIIb, but not in the fibroblastic Hela cells. DNA foot printing and mobility shift assays, have shown that this enhancer domain contains multiple binding sites for nuclear factors that are present in K562 and HEL cells but not in Hela cells. Mutagenesis of the contact site of these nuclear proteins abolished the enhancer activity, thus establishing the direct implication of these factors in the activity of the αIIb promoter.

Transcriptional enhancer have been discovered in eukaryotic and prokaryotic organisms. They are binding sites for proteins and may function either as positive or negative regulatory domains. It has been proposed that proteins interacting with enhancers may control the rate limiting step in initiation of transcription by interfering with the TFIID transcriptional complex. It is worth noting that the αIIb enhancer contains binding sites that are homologous to the binding site for the erythroid factor GATA1 and for the C-ets protein. A cooperation of these two factors in the transcriptional activity fo the αIIb promoter cannot be excluded. It remains however, to know how this enhancer communicates with other functional domain of the promoter.

NEGATIVE REGULATION OF THE αIIb GENE IN NON MEGAKARYOCYTIC CELLS.

The presence of a tissue specific enhancer within the structure of the αIIb promoter does not explain why the transcription of the gene is restricted to the megakaryocyte. As indicated by the above summazized results, this enhancer interacts with erythroid and megakaryocytic factors and is active in both lineages. Since the promoter contains informations to direct a magakaryocyte specific expression of the gene, DNA responsive elements that are responsible for this lineage specificity, must be present within the structure of this promoter. To delineate these elements, 5'progressive deletion mutants were generated and analysed for their capacity to express the CAT reporter gene in megakaryocytic and non megakaryocytic cells. These experiments allowed the identification of a DNA core sequence from nucleotide -113 to nucleotide -75 which functions as a potential negative regulatory element in non megakaryocytic cells (Fig.4). Mutation of the corresponding sequence resulted in a complete loss of the cell specific activity of the promoter, but did not affect promoter activity in megakaryocytic cells.

The presence of negative responsive elements with silencing activity have been identified in a variety of genes, some of these elements are promoter specific, others are ubiquitous. We still don't know wether the negative regulatory sequence present within the structure of the αIIb promoter is αIIb specific. A comparison of the sequence with other known sequences for protein binding sites did not reveal obvious similarities. It is tempting to speculate that the silencer is not functional in magakaryocyte because theses cells do not provide functional nuclear factors with binding capacity to this care sequence. This negative element is located precisely within the erythroid like promoter. Thus, one can also hypothesize that this element turns off the erythroid like promoter in non megakaryocytic cells. Further investigations are required to understand the communication between this negative domain and other functional domains of the promoter and its precise implication in the lineage restricted transcription of the αIIb gene.

CONCLUSION

Megakaryocytes play a central role in the production of circulating platelets and consequently in the human hemostatic response. These cells differ from other hematopoietic cells by their capacity to undergo non mitotic developpment prior to the production of platelet. They are directly or indirectly involved in a number of physiological disorders including inflammation, atheriogenesis, thrombosis and immunological diseases. Therefore understanding the mechanisms that are implicated in the establishment or the maintaining of the megakaryocyte phenotype is of real interest. In our efforts to understand these mechanisms the αIIb gene whose expression is limited to this lineage was used. Domains of the promoter of this gene, that are necessary for a tissue specific activity have been identified. It remains to verify if these elements are also functional in vivo. Transgenesis is the method of choice to address this question. Production of transgenic mice, using megakaryocyte specific promoters will indeed allow targeting of genes in the megakaryocytic lineage and in platelet. This open a new exiting approach to understand the role of platelet in vascular biology in vivo.

REFERENCES

1   Plow EF, Marguerie G, and Ginsberg MH. 1987, pp 267-275. Alan R. Liss. Inc. New York.

2   Molla A, Andrieux A, Chapel A, Schweitzer A, Berthier R, and Marguerie G. Brit J Haematol 1992;82:635-639.

3   Prandini MH, Denarier E, Frachet P, Uzan G, and Marguerie G. Biochem Biophys Res Comm 1988;156:595-4601.

4   Heidenreich R, Eisman R, Surrey S, Delgrosso K, Bennett JS, Schwartz E, and Poncz. M.Biochem 1990, 29:1232-1244.

5   Bray PF, Rosa JP, Johnston G, Shin DT, Cook RG, Lan C, Kan YW, McEver RP, and Shuman MA. J Clin Invest 1987;80 1812-1817.

6   Sosnoski DM, Emanuel BS, Hawkins AL, Van Tuinen P, Ledbetter DH, Nussbaum RL, Laos F-T, Schwartz E, Phillips D, Bennett J, Fitzgerald LA, and Poncz M. J Clin Invest 1988;81:1993-1998.

7   Van Cong N, Uzan G, Gross MS, Jegun-Foubert C, Frachet P, Boucheix C, Marguerie G, and Frezal. J Hum Genet 1988;80:389-392.

8   Smale ST, and Baltimore D. Cell 1989;57:103-113.

9   Birkenmeir TM, Quillan JJ, Boedeker ED, Argraves WS, Ruoslahti E and Dean DG. J Biol Chem 1991;266:20544-20549.

10   Rosen GD, Birkenmeier TM, and Dean DG. Proc Natl Acad Sci USA 1991;88:4094-4098.

11   Hemler ME, Elices MJ, Parker C, and Takada Y. Immunol Rev 1990;114:45-65.

12   Watson DK, Mc Williams MJ, Lapis P, Lantenbgerger JA, Schweinfast CW, and Pagas TS. Proc Natl Acad Sci USA 1988;85:7862-7866.

13   Wazylyk B, Wazylyk C, Flores P, Beque A, Deprince D, and Stehelin D. Nature 1990;346:191-193.

14   Angel P, Imagawa M, Chin R, Imbra RJ, Rahmsdorf MJ Jonat C Herrlich P and Karin M. Cell 1987;49:729-739.

14   Lemarchandel V, Ghysdouel J, Mignotte V, Rahuel C and Romeo PM Mol Cell Biol 1993;13:668-676.

15   Kim SJ, Angel P, Kagyatis R, Hattori K, Sporn MB, Karin M, and Roberts AB. Mol Cell Biol 1990;10:1492-1497.

16   Martin F, Prandini MH, Thevenon D, Marguerie G, and Marguerie G. J Biol Chem 1993;268 in press.

17   Pevny L, Simon MC, Robertson E, Klein WH, Tsai SF, D'Agati V, Orkin SM, and Constantini F. Nature 1991;349:257-260.

18   Frampton J, Walker M, Plumb M, and Harrison PH. Mol Cell Biol 1990;10:3838-3842.

19   Uzan G, Prenant M, Prandini MH, Martin F and Marguerie G J Biol Chem 1991;266:8932-8939.

Biology of Vitronectins and their Receptors
K.T. Preissner, S. Rosenblatt, C. Kost, J. Wegerhoff and D.F. Mosher, editors

Vitronectin Receptors in Melanoma Tumor Growth and Metastasis

B. Felding-Habermann[a], B.M. Mueller[b], L.C. Sanders[b] and D.A. Cheresh[b]

The Scripps Research Institute, 10666 N. Torrey Pines Rd., La Jolla, CA 92037 U.S.A.

[a] Department of Molecular and Experimental Medicine, SBR 8
[b] Department of Immunology, IMM 13

Introduction

To date at least four vitronectin receptors have been identified. They belong to the integrin family of adhesion receptors and are composed of non-covalently linked α and ß subunits. These complexes recognize their ligands in a cation dependent manner (1). Integrins are usually classified by shared ß subunits, however, certain α subunits may as well associate with multiple ß subunits, and thus allow to group the corresponding integrins by their α chains. The vitronectin receptors share either the αv subunit or the ß3 subunit and include integrins αvß1, αvß3, αvß5 and αIIbß3 (2). These vitronectin receptors share the RGD sequence within their ligands as a binding motif, and particularly αvß3 and αIIbß3 interact with RGD in a variety of adhesive proteins including vitronectin, fibrinogen, fibronectin, thrombospondin and von Willebrand Factor. As known so far, integrin αvß5 exhibits the most restricted binding specificity, recognizing only vitronectin among the extracellular matrix proteins. Additional, non-RGD containing sequences within their ligands were identified as binding sites for αIIbß3, and more recently for αvß3 and αvß5 and these may help stabilizing receptor ligand interaction (3-7).

The vitronectin receptors exhibit distinct tissue distributions with αvß5 being the most widespread member expressed by a variety of cell types, and αIIbß3 - with the exception of certain tumors (8) - being restricted to megakaryocytes and platelets where it functions as an activation dependent receptor which mediates thrombus formation (9). Integrins αvß1 and αvß3 are limited to certain normal tissues, but may become expressed upon neoplastic transformation. One of the most striking changes in integrin expression upon transformation is the appearance of integrin αvß3 in human melanoma along with the transition from horizontal growth to vertical invasion of dermal tissue. αvβ3 enables the melanoma cells to interact with a variety of adhesive proteins which they may encounter during metastatic dissemination. Moreover, αvβ3 seems to be required for efficient cell migration on certain substrates (10) and can act in concert with other integrins, particularly α5β1, to promote matrix dependent cell locomotion (11). One of the most prominent characteristics of integrin αvβ3 is its ability to bind vitronectin with high affinity. Vitronectin is primarily synthesized in the liver and circulates in its monomeric form in the bloodstream (12). However, it also localizes to certain tissues where it is associated with the extracellular matrix and thereby alters its conformation resulting in the exposure of its RGD-containing cell attachment site (12). Characteristically, vitronectin was found in the stroma of wound tissue (13), lymphnodes (14) and elastin fibers of the skin where vitronectin deposition increases with age and exposure to the sun (15). Both, aging and sun exposure increase the likelihood of contracting melanoma. Therefore, the expression of integrin $\alpha v \beta 3$ on melanoma cells and its interaction with vitronectin in the skin may provide a specific advantage for the progression of this tumor. In order to test this hypothesis, we generated an experimental model system that allows direct analysis of the requirement of αv integrin expression by human melanoma cells for their tumorigenic capacity.

Results and Discussion

M21 human melanoma cells express the αv integrins $\alpha v \beta 3$ and $\alpha v \beta 5$ (16,17). From this cell line a stable spontaneous mutant was selected based on the lack of αv expression using a monoclonal antibody that specifically recognizes the human αv subunit. These M21-L cells do not form either of the αv containing integrins due to a defect in αv gene expression at the transcriptional level (16). M21-L cells, however, display indistinguishable ratios of $\beta 1$ integrins when compared to the parental M21 cell line (18). Since M21-L cells may differ from M21 cells not only in αv integrin expression but potentially in further unknown characteristics, it was necessary to generate a valid control to directly address the role of αv-integrins using these cell lines. We therefore transfected M21-L cells with a vector containing a full length αv cDNA and a gene encoding neomycine resistance (18). From drug resistant subpopulations, M21-L4 cells were selected for αv integrin expression at levels comparable to the parental M21 cell line. M21-L12 cells were selected as non-αv-expressing control transfectants. In M21-L4 cells αv transfection restored cell surface expression of both, $\alpha v \beta 3$ and $\alpha v \beta 5$, as was demonstrated by FACS analysis and immunoprecipitation of surface labeled cells using αv specific as well as complex specific anti $\alpha v \beta 3$ or anti $\alpha v \beta 5$ monoclonal antibodies (18). Interestingly, the ratio of $\alpha v \beta 3$ to $\alpha v \beta 5$ expression in M21-L4 cells was indistinguishable from the ratio found in the parental M21 cells with $\alpha v \beta 3 : \alpha v \beta 5$ being approximately 20:1. This indicates that the expression levels of these integrins are controlled by β subunit synthesis when αv is present. Both, $\alpha v \beta 3$ and $\alpha v \beta 5$, expressed by M21-L4 cells upon αv transfection were fully functional in *in vitro* binding studies. In all four cell lines M21 ($\alpha v+$), M21-L ($\alpha v-$) and the transfectants, M21-L4 ($\alpha v+$) and M21-L12 ($\alpha v-$), the levels of $\beta 1$ integrin expression were unchanged as was demonstrated by using a $\beta 1$ specific monoclonal antibody (18).

In order to test the requirement of αv integrins, $\alpha v \beta 3$ and/or $\alpha v \beta 5$, for malignant phenotype expression the human melanoma cell variants M21, M21-L, M21-L4 or M21-L12 cells were injected subcutaneously into the flanks of athymic nude mice and were analyzed for their tumorigenicity (18). Inoculation of 1×10^6 M21 cells resulted in rapidly growing tumors in 7/8 animals while animals injected with this number of M21-L cells developed relatively small tumors in only 2/8 animals and these were considerably delayed in their appearance. M21-L cells failed to produce tumors when smaller cell numbers were injected (1×10^5). The impaired tumorigenicity of αv lacking M21-L cells was restored upon αv transfection in M21-L4 cells. 1×10^6 M21-L4 cells produced fast growing tumors in 6/8 animals whereas the same number of αv-negative control transfectants, M21-L12, caused only a small tumor in 1/8 animals. Similar results were obtained in four independent experiments, thus showing that both tumor take rate as well as the median volume of the developing tumors depended to a large extent on the expression of integrin(s) αvβ3 and/or αvβ5 on the injected melanoma cells (18). It is important to point out that cells cultured from tumors induced by αv-negative M21-L or M21-L12 cells did not express αv integrins indicating that these tumors did not result from an *in vivo* selection of reverted αv-bearing cells. To rule out a general growth defect in the αv-lacking M21-L or M21-L12 melanoma cells their *in vitro* proliferation rates were analyzed and compared to those of the αv-positive M21 and M21-L4 cells. In the presence of fetal bovine serum all four cell lines displayed indistinguishable growth behaviors (18). Therefore the impaired tumorigenicity of αv-negative M21-L and M21-L12 cells must be due to a lack in their *in vivo* growth capacities.

Tumor cells are generally characterized by their altered dependence on growth factors (19). An increase in malignancy correlates with a loss of growth factor requirement from the surrounding cells or tissue and may ultimately lead to complete autonomy of the tumor cells (20). Since we found that the expression of αvβ3 and/or αvβ5 was crucial for M21 melanoma cell growth *in vivo*, this system provides an experimental basis to study the consequences of αv-integrin occupancy on cell proliferation characteristics, such as responsiveness to growth factors or their production, and possibly the loss of exogenous growth factor dependence. Indeed, when αv-positive and αv-negative M21 cells were compared in *in vitro* proliferation tests in the absence of serum or other exogenous growth factors - but provided with immobilized extracellular matrix proteins we demonstrated that αv-expressing M21 and M21-L4 cells but not αv-negative M21-L or M21-L12 cells proliferated in response to immobilized vitronectin (21). All four cell types showed increased proliferation when seeded onto fibronectin- or collagen type I- matrices and this is likely due to their equivalent expression of β1 integrins (18). These results suggest that vitronectin can promote a proliferative signal to human melanoma cells by interacting with their αv integrin(s). Further evidence supports a role of αv integrin ligation not only for *in vitro* but also for *in vivo* tumor cell proliferation. When vitronectin was coinjected with M21 cell variants subcutaneously into nude mice, αv expressing M21 and M21-L4 cells induced tumors with a fourfold increase in proliferation over cells injected without vitronectin (21). This increased proliferation was not observed when αv-negative M21-L or M21-L12 cells were coinjected with vitronectin. As a control, all four cell types showed increased tumor growth upon coinjection with collagen. Together, these results indicate that the presence of an αv integrin ligand, specifically vitronectin, in an *in vivo* microenvironment favors proliferation of cells that express αv integrins. Therefore, the presence of vitronectin and the expression of αv-containing vitronectin receptors on melanoma cells may stimulate their proliferation and thus favor melanoma tumorigenicity, which may subsequently potentiate metastatic disease.

The metastatic stage of melanoma progression involves the escape of melanoma cells from the primary tumor site, their invasion into surrounding tissues and lymphatic or blood vessels, transport to distant sites of the body where the tumor cells then extravasate and home to preferred organs. Hypothetically, αv-integrin ligand interaction may be involved in several of these steps and thereby promote melanoma cell metastasis. Fibrin is known to be one of the major constituents of primary melanoma tumor stroma (22). Adhesion of melanoma cells to fibrin within the tumor stroma may be mediated by integrin αvβ3, since fibrinogen is specifically recognized by this receptor (23,24). Remodeling and degradation of fibrin rich stroma is induced by tumor cell factors which stimulate fibrinogen release from surrounding

138

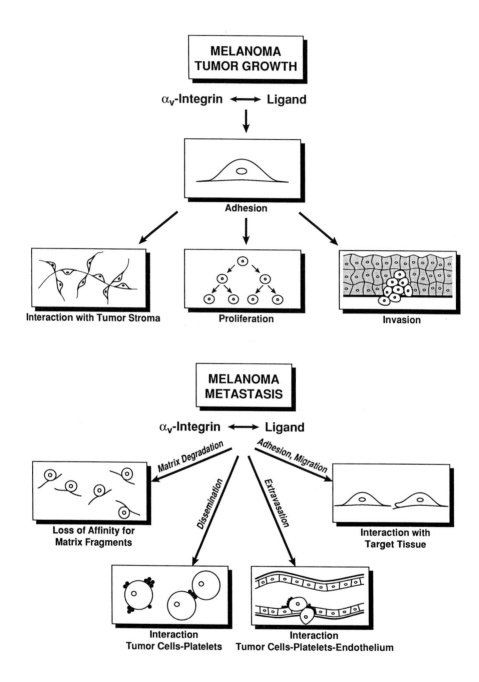

Figure 1.   Potential involvement of αv-integrin ligand interaction in various stages of primary melanoma growth and metastasis.

blood vessels and degradation of established stromal fibrin fibers. During these processes melanoma cells may interact either with the intact fibrin or fibrinogen molecules or with their fragments and this can promote distinct cellular responses (6). The interaction of αvβ3 with fibrinogen or its fragments is characterized by a decrease in binding affinity during plasmic fragmentation due to the sequential loss of recognition sites within this ligand including the C-terminal RGD sequence of the fibrinogen α chain. This is associated with loss of cell spreading in response to fibrinogen fragments as compared to an intact fibrinogen matrix (6). It therefore appears likely that this mechanism promotes the release of metastatic melanoma cells from a fibrin containing tumor stroma.

At later stages of metastatic dissemination melanoma cells may again utilize integrin αvβ3 to specifically interact with cells and tissues as required for their distribution. Vitronectin localized to lymphnode tissue supports adhesion of αv-integrin expressing melanoma cells, but is not recognized by αv-lacking melanoma cells as was shown in an experimental system (14). During hematogenous dissemination of metastasizing melanoma cells they may associate with blood cells for their protection from host immune defense. Tumor cell interaction with platelets has been proposed for this mechanism (25). Platelets express integrin αIIbβ3 which mediates thrombus formation by binding of fibrinogen upon platelet activation (9). Fibrinogen serves as a divalent crosslinking molecule in the latter process and could theoretically play a similar role in the interaction between αvβ3-containing melanoma cell and platelets during metastatic dissemination. For tumor cell extravasation circulating micro thrombi composed of tumor cells and platelets may then bind to the vessel wall at micro lesions of the endothelium where platelets and/or tumor cells can attach to von Willebrand Factor within the subendothelial matrix. Platelet adhesion to this molecule is mediated by platelet protein GPIbα and this interaction is particularly favored under high shear stress conditions as found in capillaries (26) which are likely the preferred sites for tumor cell extravasation. Direct attachment of melanoma cells to von Willebrand Factor and other constituents of the subendothelium, such as vitronectin and fibronectin, could be mediated by integrin αvβ3. Alternatively, circulating melanoma cells or tumor cell containing micro thrombi could attach to intact endothelium using divalent adhesive molecules, such as fibrinogen, or cell surface immobilized integrin ligands as mediators for their interaction. Endothelial cells have the potential to express integrin αvβ3 and this is regulated by their state of activation. Particularly during inflammatory stages factors like IL-1 promote upregulation of αvβ3 in endothelial cells (27) which may, in concert with other adhesion receptors, facilitate contact formation between metastasizing tumor cells or their micro thrombi and the vessel wall. A schematic summary on the potential involvement of αv-integrin ligand interaction in various stages of primary melanoma growth and metastasis is shown in Figure 1.

References

1   Hynes RO (1987) Cell 48: 549-554.
2   Felding-Habermann B, Cheresh DA (1993) Current Oppinion Cell Biol. 5: 864-868.
3   Kloczewiak M, Timmons S, Lukas TJ, Hawiger J (1984) Biochemistry 23: 1767-1774.
4   Santoro SA, Lawing WJ (1987) Cell 48: 867-873.
5   D'Souza SE, Ginsberg MH, Burke TA, Plow EF (1990) J. Biol. Chem. 265: 3440-3446.
6   Felding-Habermann B, Ruggeri ZM, Cheresh DA (1992) J. Biol. Chem. 267: 5070-5077.
7   Vogel BE, Lee SJ, Hildebrand A, Craig W, Pierschbacher MD, Wong-Staal F, Ruoslahti E (1993) J. Cell Biol. 121: 461-468.
8   Chen YQ, Gao X, Timar J, Tang D, et al. (1992) J. Biol. Chem. 267: 17314-17320.
9   Phillips DR, Charo IF, Scarborough RM (1991) Cell 65: 359-362.
10   Leavesley DI, Ferguson GD, Wayner EA, Cheresh DA (1992) J. Cell Biol. 117: 1101-1107.
11   Bauer JS, Schreiner CL, Giancotti FG, Rouslahti E, Juliano RL (1992) J. Cell Biol. 116: 477-487.
12   Stockmann A, Hess S, Declerck P, Timpl R, Preissner KT (1993) J. Biol. Chem. (in press)
13   Grinnell F, Ho CH, Wysocki A (1992) J. Invest. Dermatol. 98: 410-416.
14   Nip J, Shibata H, Loskutoff DJ, Cheresh DA, Brodt P (1992) J. Clin. Invest. 90: 1406-1413.
15   Dahlback K, Lofberg H, Alumets J, Dahlback B (1989) J. Invest. Dermatol. 92: 727-733.
16   Cheresh DA, Spiro RC (1987) J. Biol. Chem. 262: 17703-17711.
17   Wayner EA, Orlando RA, Cheresh DA (1991) J. Cell Biol. 113: 919-929.
18   Felding-Habermann B, Mueller BM, Romerdahl CA, Cheresh DA (1992) J. Clin. Invest. 89: 2018-2022.
19   Dickson C, Peters G (1987) Nature 326: 833.
20   Cross M, Dexter TM (1991) Cell 64: 271-280.
21   Sanders LC, Felding-Habermann B, Mueller BM, Cheresh DA (1992) Cold Spring Harbor Symposia on Quantitative Biology, Vol LVII. Cold Spring Harbor Laboratory Press, pp. 233-240.
22   Dvorak HF (1986) N. Engl. J. Med. 315: 1650-1659.
23   Cheresh DA, Berliner S, Vincente V, Ruggeri ZM (1989) Cell 58: 945-953.
24   Smith JW, Ruggeri ZM, Kunicki TJ, Cheresh DA (1990) J. Biol. Chem. 265: 12267-12271.
25   Nierodzik ML, Plotkin A, Karpatkin S (1991) J. Clin. Invest. 87: 229-236.
26   Ruggeri ZM, Ware J (1992) Thrombosis Haemostasis 67: 594-599.
27   Dinarello CA (1992) Immunol. Rev. 127: 119-146.

Biology of Vitronectins and their Receptors
K.T. Preissner, S. Rosenblatt, C. Kost, J. Wegerhoff and D.F. Mosher, editors

Peptide and Non-peptide Antagonists of β3 Integrins

Beat Steiner, William C. Kouns and Thomas Weller

From the Pharma Division, Preclinical Research, F. Hoffmann-La Roche Ltd,
CH-4002 Basel, Switzerland

INTRODUCTION

Adhesion of platelets to extracellular matrices, platelet aggregation and thrombus formation are crucial events in hemostasis and thrombosis. The membrane glycoprotein IIb-IIIa complex (GP IIb-IIIa) on the platelet surface is essential for platelet aggregation and is also important for normal adhesion of platelets to matrix proteins exposed in the subendothelium of a damaged vessel [1,2]. This Ca^{2+} dependent heterodimeric complex is a member of the integrin family of cell adhesion receptors [3-5]. Several of these structurally related receptors, including GP IIb-IIIa (integrin $\alpha_{IIb}\beta_3$) and the closely related vitronectin receptor $\alpha_v\beta_3$, recognize the Arg-Gly-Asp (RGD) sequence that is present in many adhesive proteins [6]. The β_3 integrins $\alpha_{IIb}\beta_3$ and $\alpha_v\beta_3$ demonstrate rather low ligand selectivity since they both interact with several RGD-containing ligands. $\alpha_{IIb}\beta_3$ binds the adhesive proteins fibrinogen, fibronectin, von Willebrand factor and vitronectin, while $\alpha_v\beta_3$ binds vitronectin, fibrinogen, von Willebrand factor, osteopontin and thrombospondin [6].

Short synthetic RGD-containing peptides inhibit the binding of all ligands to both integrins [7,8], supporting the hypothesis that these adhesive proteins interact with the receptor via their RGD domains. Since $\alpha_{IIb}\beta_3$ contains only a single binding site for RGD-containing peptides [9], all the ligands appear to bind to the same domain on $\alpha_{IIb}\beta_3$. The synthetic dodecapeptide HHLGGAKQAGDV, corresponding to the C-terminus of the fibrinogen γ-chain, also inhibits ligand binding to $\alpha_{IIb}\beta_3$ [10], although this sequence is present neither in fibronectin, von Willebrand factor nor in vitronectin. On the other hand, the dodecapeptide is quite selective for $\alpha_{IIb}\beta_3$ in that it only weakly interacts with $\alpha_v\beta_3$ [11]. These data suggested that it should be possible to synthesize antagonists that are selective for platelet $\alpha_{IIb}\beta_3$ and inhibit the binding of all adhesive ligands.

Since the dodecapeptide appeared to be rather complex in structure and has a low affinity for $\alpha_{IIb}\beta_3$, RGDS served as a lead compound for the chemical synthesis of potent and selective $\alpha_{IIb}\beta_3$ antagonists. Solid-phase receptor assays were used to determine the inhibitory potency and the receptor selectivity ($\alpha_{IIb}\beta_3$ versus $\alpha_v\beta_3$) of newly synthesized compounds. Structural elements were identified which increase the potency and/or the selectivity of the antagonists.

EXPERIMENTAL PROCEDURES

Synthesis of peptides, pseudopeptides and non-peptide antagonists. The peptides were synthesized by the classical technique using various coupling procedures and a combination of acid-labile protecting groups. The pseudopeptides and the non-peptide antagonists were synthesized as described previously [12].

Purification of integrin $\alpha_{IIb}\beta_3$ from human platelets. Outdated, washed human platelets were lysed at 4 °C for 15 h with 1% Triton X-100, 150 mM NaCl, 1 mM $CaCl_2$, 1 mM $MgCl_2$, 0.02% NaN_3, 10 μM leupeptin, 0.5 mM PMSF (phenylmethylsulfonyl fluoride), 2 mM N-ethylmaleimide and 20 mM Tris/HCl, pH 7.3. The glycoproteins were isolated using a concanavalin A (Con A)-Sepharose 4B column and applied on an aminoethylglycine (Aeg)-RGDS affinity column as described previously [9]. The $\alpha_{IIb}\beta_3$ retained on the Aeg-RGDS column (referred to as active $\alpha_{IIb}\beta_3$ [13]) was specifically eluted with buffer A (0.1% Triton X-100, 150 mM NaCl, 1 mM $CaCl_2$, 1 mM $MgCl_2$, 0.05% NaN_3 and 20 mM Tris/HCl, pH 7.0) containing 3 mM RGDV. This preparation of the receptor was immobilized on microtiter plates.

Purification of the vitronectin receptor $\alpha_V\beta_3$ from human placenta. This integrin was purified as described previously [14]. Briefly, the placental proteins were extracted by incubating small sliced pieces at room temperature for 1 h in a lysis buffer containing 1% Triton X-100, 150 mM NaCl, 1 mM $CaCl_2$, 1 mM $MgCl_2$, 0.02% NaN_3, 0.5 mM PMSF, 1 mM leupeptin, 2 mM N-ethylmaleimide and 20 mM Tris/HCl, pH 7.4. The vitronectin receptor $\alpha_V\beta_3$ was then purified according to the procedure described above for platelet $\alpha_{IIb}\beta_3$ and immobilized on microtiter plates.

In an additional experiment, the purified $\alpha_V\beta_3$ was extensively dialyzed vs. buffer A to remove RGDS and then applied on an RNRDAPEGC-affinity column. The synthetic peptide RNRDAPEG, corresponding to amino acids 214-221 of the β_3 integrin subunit, was previously shown to bind to $\alpha_{IIb}\beta_3$ [15]. Following washing of the affinity column with buffer A, the bound proteins were eluted by including 3 mM RGDV in buffer A.

Inhibition of fibrinogen binding to immobilized $\alpha_{IIb}\beta_3$ or $\alpha_V\beta_3$. The wells of plastic microtiter plates (Nunc-Immunoplate MaxiSorp) were coated overnight at 4 °C with purified active $\alpha_{IIb}\beta_3$ or $\alpha_V\beta_3$ at 0.5 μg/ml (100 μl/well) in a buffer containing 150 mM NaCl, 1 mM $CaCl_2$, 1 mM $MgCl_2$, 0.0005% Triton X-100 and 20 mM Tris/HCl, pH 7.4. Blocking of nonspecific binding sites was achieved by incubating the wells with 3.5% bovine serum albumin (BSA, Fluka) for at least 1 h at 20 °C. Prior to initiation of the binding assay, the plates were washed once with 150 mM NaCl, 1 mM $CaCl_2$, 1 mM $MgCl_2$ and 20 mM Tris/HCl, pH 7.4 (buffer B). Fibrinogen (fibronectin free, IMCO, Sweden) was diluted in buffer B containing 1% BSA to 0.5 μg/ml for the binding to $\alpha_{IIb}\beta_3$ and to 1.5 μg/ml for the binding to $\alpha_V\beta_3$. The receptor-coated wells were incubated with fibrinogen in the absence or presence of various concentrations of antagonists (100 μl/well) for 4 h or overnight at room temperature. Nonbound fibrinogen was removed by three washes with buffer B and bound fibrinogen was detected by an enzyme-linked immunosorbent assay (ELISA) as described previously [13].

RESULTS AND DISCUSSION

The goal of this program was to synthesize potent antagonists of the platelet integrin $\alpha_{IIb}\beta_3$ (GP IIb-IIIa) in order to test their usefulness for the treatment and prevention of arterial thrombosis. Initially, the tetrapeptide RGDS served as a chemical lead compound. However, the target molecules had eventually to fulfill the following properties: (i) they should inhibit platelet aggregation induced by all physiological platelet agonists with a much higher potency than RGDS; (ii), they should, like RGDS, inhibit the binding of fibrinogen as well as von Willebrand factor to $\alpha_{IIb}\beta_3$, since both these adhesive proteins are involved in linking the platelets together in vivo and (iii), the antagonists should, unlike RGDS, only bind to the $\alpha_{IIb}\beta_3$ integrin.

Solid-phase receptor assays were established to measure the potency and selectivity of newly synthesized molecules. Inhibition of fibrinogen binding to purified platelet $\alpha_{IIb}\beta_3$ was used to determine their inhibitory potency relative to RGDS, while inhibition of fibrinogen binding to the purified vitronectin receptor $\alpha_v\beta_3$ was used to determine the receptor selectivity of these compounds. The antagonist concentration required for 50 % inhibition of ligand binding to a given receptor is generally expressed as IC_{50}. This value, however, is highly dependent on the ligand concentration used in a given assay. For instance, the IC_{50} for RGDS to inhibit platelet aggregation in platelet rich plasma (PRP) is about 100 μM, while in the solid-phase assay the IC_{50} to inhibit fibrinogen binding to purified $\alpha_{IIb}\beta_3$ is 3.6 $\pm$ 0.4 μM (mean $\pm$ SEM, n=165). Similarly, the IC_{50} for RGDS to inhibit endothelial cell adhesion to immobilized vitronectin or fibrinogen is 20-30 μM [14], while the IC_{50} for RGDS to inhibit fibrinogen binding to purifed $\alpha_v\beta_3$ is 9.4 $\pm$ 1.8 nM (mean $\pm$ SEM, n=9). For these reasons the potency and selectivity of the compounds shown in Tables I and II were indicated relative to RGDS. These factors are largely independent of the assay system used.

Potency and selectivity of peptides. Table I shows that the dodecapeptide HHLGGAKQAGDV inhibits fibrinogen binding to $\alpha_{IIb}\beta_3$ with a 3.1-fold lower potency than RGDS and is 640 times less potent than RGDS in inhibiting fibrinogen binding to $\alpha_v\beta_3$. Overall, these results indicated that the selectivity of the dodecapeptide for $\alpha_{IIb}\beta_3$ vs. $\alpha_v\beta_3$ was 210-fold better relative to RGDS. The peptide RNRDAPEG, corresponding to amino acids 214-221 of the β_3 subunit, was also tested for its ability to inhibit fibrinogen binding to these two closely related integrins. Interestingly, this peptide inhibited fibrinogen binding to both integrins but it was more than 1,000 times more selective for $\alpha_{IIb}\beta_3$ than RGDS. As previously demonstrated, this β_3 derived peptide binds to $\alpha_{IIb}\beta_3$ and not to the ligand fibrinogen [15]. Therefore, we investigated whether the vitronectin receptor $\alpha_v\beta_3$ could be purified on a β_3-peptide affinity column. $\alpha_v\beta_3$

bound to the RNRDAPEGC-affinity column and could be specifically eluted by including 3 mM RGDV into buffer A. This indicated that the peptides RNRDAPEG and RGDV compete for the same binding site on $\alpha_v\beta_3$. It is possible that this region of β_3 plays an important role in the regulation of the ligand binding site on the β_3 integrins [15].

Table I. Relative Inhibitory Potency and Selectivity of Peptides and Pseudopeptides.

No.	Compound	$\alpha_{IIb}\beta_3$-FG $\frac{IC_{50}\ comp}{IC_{50}\ RGDS}$	$\alpha_v\beta_3$-FG $\frac{IC_{50}\ comp}{IC_{50}\ RGDS}$	Selectivity[a] rel. to RGDS
1	(RGDS)	1	1	1
2	HHLGGAKQAGDV	3.1	640	210
3	RNRDAPEG	3.1	3600	1160
4		0.01	650	65000
5		1	570	570
6	(Ro 43-5054)	0.0003	4500	$1.5 \cdot 10^7$

[a]Selectivity is expressed as rel. potency in $\alpha_v\beta_3$ assay/rel. potency in $\alpha_{IIb}\beta_3$ assay.

Potency and selectivity of pseudopeptides. From studies in the field of inhibitors of serine proteases like trypsin and thrombin it has been known for quite some time that the *p*-amidinophenyl moiety can mimic the arginine side chain [16]. As indicated in Table I, compound **4** in which the N-terminal arginine of the lead peptide RGDS has been replaced by racemic *p*-amidino-phenylalanine, exhibited a 100-fold higher affinity for $\alpha_{IIb}\beta_3$ than RGDS, which can not be attributed to the presence of the C-terminal valine as shown previously [12]. In contrast, this pseudopeptide was a significantly less potent inhibitor of fibrinogen binding to $\alpha_v\beta_3$, resulting already in a high selectivity for platelet $\alpha_{IIb}\beta_3$. The reasons for the increased selectivity are not fully understood. However, compared to the arginine side chain the *p*-amidinophenyl group is considerably less flexible. This might lead to a preference of a conformation which is more suitable for the interaction with $\alpha_{IIb}\beta_3$. Although a significant loss of affinity for $\alpha_{IIb}\beta_3$ was observed for the amidinobenzoyl derivative **5**, its affinity for $\alpha_v\beta_3$ was comparable to that of the *p*-amidinophenyl derivative **4**. Insertion of a single methylene group by replacement of the glycine moiety in **5** by a β-alanine portion resulted in compound **6** (Ro 43-5054) which exhibited an affinity and selectivity for $\alpha_{IIb}\beta_3$ far superior to those found for **4**. This points to the importance of the distance between the basic amidino group and the carboxylic acid function of the aspartate side chain. Ro 43-5054 inhibits ADP-induced platelet aggregation in PRP with an IC_{50} of 50 nM and has no effect on endothelial cell adhesion to various adhesive proteins [14]. Like RGDS, this pseudopeptide potently inhibits the binding of all adhesive proteins to $\alpha_{IIb}\beta_3$.

Potency and selectivity of non-peptide antagonists. As indicated in Table II the replacement of glycine by the rigid *m*-aminobenzoic acid building block in combination with an N-terminal arginine as in **7** was accompanied by an increase of inhibitory potency compared to the lead peptide RGDS, while the selectivity for $\alpha_{IIb}\beta_3$ warranted further improving. This result is remarkable because it has been reported that the glycine residue cannot be replaced by other amino acids without significant loss of activity [17]. Truncation of the C-terminal end and re-placement of the arginine by *p*-amidinophenylalanine as realized in **8** was accompanied by a significant loss of affinity for $\alpha_{IIb}\beta_3$ as compared to **7**. From model considerations it might be concluded that the guanidino group of **7** can easily be matched to the guanidino group of RGDS due to the flexibility of the arginine side chain present in both molecules. Such a match is hardly possible with **8,** because here the amidino group is attached to a rigid aromatic ring. Therefore, the distance to the carboxylate of β-alanine lies outside the optimal range for high affinity binding to $\alpha_{IIb}\beta_3$. The situation looks quite different when using the shorter *p*-amidinobenzoyl group, as demonstrated with the derivative **9** for which not only high affinity but also high selectivity for $\alpha_{IIb}\beta_3$ versus $\alpha_v\beta_3$ was observed.

The *p*-amidinophenylsulfonamide **10** (Table II) was identified by random screening for platelet aggregation inhibitors. Inspection of molecular models showed that upon superimposing the side chain carboxylate function of aspartic acid in RGDV with the carboxylate group of **10**, the guanidino group of the former

Table II. Relative Inhibitory Potency and Selectivity of Non-peptides.

No.	Compound	$\alpha_{IIb}\beta_3$-FG $\dfrac{IC_{50}\ comp}{IC_{50}\ RGDS}$	$\alpha_v\beta_3$-FG $\dfrac{IC_{50}\ comp}{IC_{50}\ RGDS}$	Selectivity[a] rel. to RGDS
7	Boc-Arg—...—Asp-Val-OH	0.06	250	4200
8	β–Ala-OH	10	2500	250
9	β–Ala-OH	0.0005	8000	$1.6 \cdot 10^7$
10	COOH	0.34	1700	5000
11	COOH	0.004	8400	$2.1 \cdot 10^6$
12	COOH	0.008	> 8000	$> 1 \cdot 10^6$
13	COOH (Ro 44-9883)	0.0002	17500	$8.8 \cdot 10^7$

[a] Selectivity is expressed as rel. potency in $\alpha_v\beta_3$ assay/rel. potency in $\alpha_{IIb}\beta_3$ assay.

can be perfectly matched to the amidino group of the latter if both molecules are allowed to adopt an extended conformation. Two structural variations proved to be beneficial to further improvements: (i) replacement of the sulfonamide function by a carboxamide and (ii), introduction of a carbonyl group in the benzyl position as seen in compound **11**. Model considerations suggested that this ketone function could play the role of the amide carbonyl group of arginine in RGDV. The high selectivity of **11** for platelet $\alpha_{IIb}\beta_3$ vs. $\alpha_v\beta_3$ is remarkable.

In order to arrive at a more flexible scheme to introduce different substituents at the α-position to the ketone carbonyl group in **11**, the possibility of using any α-amino acid from the chiral pool was assessed. One of the easiest ways to make use of α-amino acids in organic synthesis is the preparation of amides. As a consequence, a prototype compound structurally closely related to **11** was the glycine derivative **12**, where the aromatic carbocyclic phenoxyacetic acid portion has been replaced by a saturated heterocyclic piperidinyloxyacetic acid building block. Indeed, inspection of molecular models demonstrated that the two important functional groups in **11** and **12**, i.e. the amidino group and the carboxylate, can be nicely matched if the piperidine ring of **12** adopts a chair conformation with the acetic acid side chain in an equatorial position. As indicated in Table II, **12** already exhibited a promising potency as well as an excellent selectivity. It has to be stressed that the glycine in **12** does not play the same role as the glycine in the lead peptide RGDS. With the tyrosine derivative **13** (Ro 44-9883), the most potent example which also exhibited remarkable selectivity for $\alpha_{IIb}\beta_3$ vs. $\alpha_v\beta_3$ was identified. The non-peptide antagonist Ro 44-9883 inhibits platelet aggregation in PRP induced by all agonists with an IC_{50} of 25 nM, i.e. the inhibitory potency is 4,000 times higher than that of the initial lead peptide RGDS. Like RGD-containing peptides Ro 44-9883 inhibits the binding of all ligands to $\alpha_{IIb}\beta_3$ with a similar potency.

Thus, potent and selective antagonists of platelet $\alpha_{IIb}\beta_3$ can be synthesized which may become useful anti-thrombotic agents. Using a similar approach it might also be possible to develop antagonists that are selective for $\alpha_v\beta_3$.

REFERENCES

1. George, J.N., Nurden, A.T. and Phillips, D.R. (1984) N. Engl. J. Med. **311**, 1084-1098
2. Sakariassen, K.S., Nievelstein, P.F.E.M., Coller, B.S. and Sixma, J.J. (1986) Br. J. Haematol. **63**, 681-691
3. Hynes, R.O. (1987) Cell **48**, 549-554
4. Ruoslahti, E. and Pierschbacher, M.D. (1986) Cell **44**, 517-518
5. Ruoslahti, E. and Pierschbacher, M.D. (1987) Science **238**, 491-497
6. Steiner, B. and Kirchhofer, D. (1991) Schweiz. med. Wschr. **121**, Suppl. 43, p. 206
7. Plow, E.F., Pierschbacher, M.D., Ruoslahti, E., Marguerie, G.A. and Ginsberg, M.H. (1985) Proc. Natl. Acad. Sci. U.S.A. **82**, 8057-8061
8. Pytela, R., Pierschbacher, M.D., Ginsberg, M.H., Plow, E.F. and Ruoslahti, E. (1986) Science **231**, 1559-1562

148

9. Steiner, B., Cousot, D., Trzeciak, A., Gillessen, D. and Hadvary, P. (1989) J. Biol. Chem. **264**, 13102-13108
10. Kloczewiak, M., Timmons, S. and Hawiger, J. (1984) Biochemistry **23**, 1767-1774
11. Lam, S.C-T., Plow, E.F., Cheresh, D.A., Frelinger, A.L.III, Ginsberg, M.H. and D'Souza, S.E. (1989) J. Biol. Chem. **264**, 3742-3749
12. Alig, L., Edenhofer, A., Hadvary, P., Hürzeler, M., Knopp, D., Müller, M., Steiner, B., Trzeciak, A. and Weller, T. (1992) J. Med. Chem. **35**, 4393-4407
13. Kouns, W.C., Hadvary, P., Häring, P., and Steiner, B. (1992) J. Biol. Chem. **267**, 18844-18851
14. Kouns, W.C., Kirchhofer, D., Hadvary, P., Edenhofer, A., Weller, T., Pfenninger, G., Baumgartner, H.R., Jennings, L.K. and Steiner, B (1992) Blood **80**, 2539-2547
15. Steiner, B., Trzeciak, A., Pfenninger, G. and Kouns, W.C. (1993) J. Biol. Chem. **268**, 6870-6873
16. Wagner, G., Horn, H., Richter, P., Vieweg, H., Lischke, I. and Kazmirowski, H.-G. (1981) Pharmazie **36**, 597-603
17. Samanen, J., Ali, F., Romoff, T., Calvo, R., Sorensen, E., Vasko, J., Storer, B., Berry, D., Bennett, D., Strohsacker, M., Powers, D., Stadel, J. and Nichols, A. (1991) J. Med. Chem. **34**, 3114-3125

Biology of Vitronectins and their Receptors
K.T. Preissner, S. Rosenblatt, C. Kost, J. Wegerhoff and D.F. Mosher, editors

Peptides from the second calcium binding domain of integrin chain α_{IIb} inhibit fibrinogen binding to $\alpha_{IIb}\beta_3$ by direct interaction.

Beate Diefenbach[a], Brunhilde Felding-Habermann[b], Alfred Jonczyk[a], Friedrich Rippmann[a]

[a]Preclinical Research, E. MERCK, 64271 Darmstadt, Germany
[b]Program of Vascular Biology, Scripps Research Institute, La Jolla, California, USA

Introduction

Platelet adhesion and aggregation play a major role in the blood clotting cascade, thrombus formation and wound healing. One initial step of the cascade is mediated by the adhesive ligand fibrinogen binding to platelet glycoprotein $\alpha_{IIb}\beta_3$ (1,2). $\alpha_{IIb}\beta_3$ is an integrin receptor found only on platelets (3). Activated $\alpha_{IIb}\beta_3$ on platelets interacts with soluble fibrinogen via the RGD peptide sequence, a common motif conserved in a variety of matrix proteins (4). It is present in positions 95-97 and 572-574 on the fibrinogen α-chain (5,6,7). $\alpha_{IIb}\beta_3$ recognizes RGD with an affinity almost identical to its recognition of the non-RGD containing dodecapeptide sequence (amino acid 400-411: HHLGGAKQAGDV) at the carboxy-terminus of the fibrinogen γ-chain (8,9).
Synthetic RGD and dodecapeptide inhibit fibrinogen binding to $\alpha_{IIb}\beta_3$ by direct interaction with the receptor (6,10). Previous studies identified on β_3 109-171 and/or 211-222 as the RGD binding site (11,12) and the region 294-314 on α_{IIb} as a binding site for the dodecapeptide (13). This study focusses on the interaction of a series of synthetic peptides derived from the second calcium binding domain of α_{IIb} 294-314 which is a binding site for the dodecapeptide sequence of the fibrinogen γ-chain (13) on $\alpha_{IIb}\beta_3$.

Experimental Procedures

Protein purifications

Human fibrinogen was purified from fresh human blood according to Kazal et al.(14). Vitronectin and fibronectin were purified from human plasma by heparin affinity chromatography (15,16). Vitronectin receptor $\alpha_v\beta_3$ was isolated from human placenta as described by Smith and Cheresh (17,18). Platelet fibrinogen receptor $\alpha_{IIb}\beta_3$ was purified using a modification of the method of Pytela and Pierschbacher (10).

Protein labelling

Integrins and ligands were biotinylated with biotin N-hydroxysuccinimidoester and biotinylation was verified by ligand or receptor binding in the solid phase assay described below.

Solid phase receptor assay

96-well microtiter plates were coated overnight with integrin or integrin ligand. After blocking non specific protein binding sites, biotinylated ligand or receptor and competitor were added. After 3 hours at 30°C, unbound protein was washed away and goat anti-biotin antibody conjugated to alkaline phosphatase was added and incubated for one hour. After three washes, substrate solution (p-nitrophenylphosphate) was added and color was developed in the dark. The reaction was stopped by addition of NaOH and read in the ELISA reader at 405nm.

Peptide plate assay

Freshly diluted peptides containing a terminal cysteine residue were coupled to BSA coated 96-well plates using SPDP. After blocking the plates with BSA, biotinylated ligand or receptor was added, incubated, washed and detected as described above. In competition assays, increasing concentrations of competitor were added at the same time as biotinylated integrin $\alpha_{IIb}\beta_3$.

Results

Effect of α_{IIb} peptides on fibrinogen binding to $\alpha_{IIb}\beta_3$

The second calcium binding domain of α_{IIb} has been identified as a binding region for the fibrinogen γ-chain dodecapeptide (13). To examine the fibrinogen-$\alpha_{IIb}\beta_3$ interactions, this region was used as a basis for designing peptides. Peptides encompassing region 296-313 (DVNNGDGRHDLLVGAPLY) amino acids from the N-terminus of the α_{IIb} chain were prepared lacking increasing numbers of terminal residues. The activity of the peptides was compared in the solid phase receptor assay. As shown in Figure 1A only peptide α_{IIb} 300-306 showed significant inhibition of fibrinogen binding to $\alpha_{IIb}\beta_3$. The removal of amino acid 306 resulted in a partial loss of activity and of amino acid 305 and 306 in a complete loss of inhibitory activity. Addition of one amino acid at the C-terminal end also resulted in a complete loss of activity. Peptides containing C-terminal residues from 308 to 313 were also inactive in the fibrinogen binding assay (data not shown). Figure 1B shows the activity of N-terminal truncated peptides. Only the loss up to position 303 resulted in a decrease of inhibitory activity. Addition of amino acids on the N-terminus up to position 296 had no effect on inhibitory capacity (data not shown). Thus, α_{IIb} 302-306 was identified as the smallest effective unit. By contrast with other active α_{IIb} peptides α_{IIb} 302-306 does not contain a reversed RGD sequence.

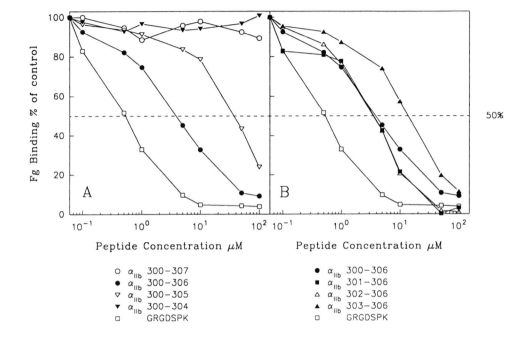

Figure 1: Biotin labelled fibrinogen was allowed to bind to immobilized $\alpha_{IIb}\beta_3$ in the absence (control) or presence of peptide. Labelled fibrinogen was detected by an anti-biotin antibody ELISA. OD_{405nm} data were converted to % of control (no inhibitor = 100% binding). Linear GRGDSPK peptide was always used as reference. All points were run in triplicate with standard deviation less than ±10%.

Relevance of charge pattern and configuration of α_{IIb} 300-306 peptide

To investigate whether the charge pattern of the active peptides was important for interaction, peptides with a different charge pattern were designed. The aspartic acids at position 301 and/or 305 were converted to uncharged asparagine. As shown in Table 1A the negative charge in position 305 is necessary to maintain inhibitory activity. Converting position 301 to an asparagine had no effect on inhibitory activity. This is not surprising because, as shown earlier, the amino acids 300 and 301 are not required for inhibition.

It appears that the configuration of single amino acids plays a major role. Shifting D-amino acids through the sequence α_{IIb} 300-306 caused a complete loss of activity when a D-amino acid was in position 303, 304, 305 or 306 (Table 1B).

Table 1: IC_{50} values of derivatized α_{IIb} 300-306 peptides

Table 1A: Effect of derivatized α_{IIb} peptides on fibrinogen binding to $\alpha_{IIb}\beta_3$

Table 1B: Effect of D-amino acid containing α_{IIb} peptides on fibrinogen binding to $\alpha_{IIb}\beta_3$

IC_{50} (µM)

Origin	Sequence	$\alpha_{IIb}\beta_3$/Fg
α_{IIb} 300-306	GDGRHDL	7
α_{IIb} 300-306 D 301 to N	GNGRHDL	2
α_{IIb} 300-306 D 305 to N	GDGRHNL	>100
α_{IIb} 300-306 D 301 and 305 to N	GNGRHNL	>100
	GRGDSPK	0.15

IC_{50} (µM)

Sequence	$\alpha_{IIb}\beta_3$/ Fg
GDGRHDL	7
aDGRHDL	2
GdGRHDL	3
GDaRHDL	2
GDGrHDL	>100
GDGRhDL	>100
GDGRHdL	>100
GDGRHDl	>100

Binding site of α_{IIb} 300-306

small letters: D-amino acids

After demonstrating the inhibitory effect of α_{IIb} 300-306, the question of whether the peptide was interacting with the ligand or the receptor was asked. From the sequence the prediction was that this region would rather interact with a ligand of $\alpha_{IIb}\beta_3$. The binding of biotinylated ligands and integrins $\alpha_{IIb}\beta_3$ and $\alpha_v\beta_3$ was tested in a solid phase receptor assay (Figure 2A). In a second set of experiments the peptide was cross-linked to plates precoated with BSA using SPDP and the labelled set of proteins was assayed for binding. As shown in Figure 2B (I,II), only $\alpha_{IIb}\beta_3$ bound to immobilized α_{IIb} 300-306. Neither fibrinogen, vitronectin or fibronectin nor $\alpha_v\beta_3$ were able to interact with the peptide. In contrast both receptors were able to bind to immobilized CGG-GRGDSPK peptide consistent with previously published data (4,8).

Interaction of α_{IIb} 300-306 with other peptide inhibitors of ligand binding

The interaction of $\alpha_{IIb}\beta_3$ with such a small portion of its α_{IIb} subunit raised the question about the nature of this interaction. Therefore $\alpha_{IIb}\beta_3$ was incubated with different peptides on immobilized α_{IIb} 300-306. As expected the soluble α_{IIb} 300-306 competes with its immobilized form for binding to the receptor (Figure 3). The substituted form, how-ever, of the α_{IIb} peptide containing a D-leucine in position 306 had no effect on receptor binding. The fibrinogen γ-chain peptide (amino acid 405-411) (13) was also able to interfere with receptor binding to immobilized α_{IIb} 300-306 peptide while soluble GRGDSPK was the strongest inhibitor for receptor binding. This suggests that the α_{IIb} 300-306/$\alpha_{IIb}\beta_3$ interaction is specific.

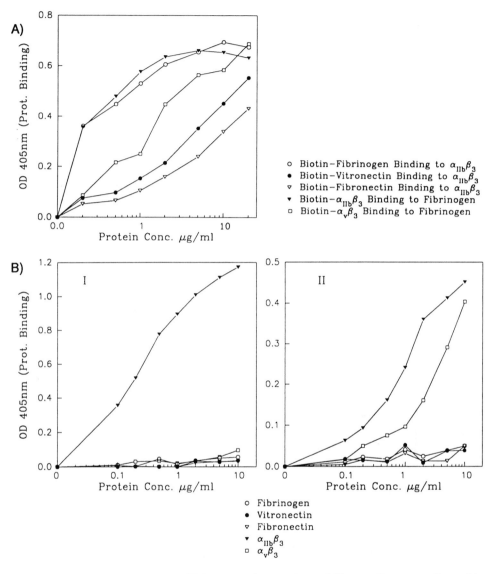

Figure 2:   A) Biotin labelled fibrinogen, vitronectin and fibronectin were allowed to bind to immobilized $\alpha_{IIb}\beta_3$; biotin labelled $\alpha_{IIb}\beta_3$ and $\alpha_v\beta_3$ were allowed to bind to immobilized fibrinogen. Labelled proteins were detected by an anti-biotin antibody ELISA as described under Figure 1. All points were run in triplicate with standard deviation less than ±10%.

   B) Peptides containing a terminal cysteine residue (I:CGG-α_{IIb} 300-306, II: CGG-GRGDSPK) were coupled to BSA-coated plates using SPDP. Biotin labelled ligands or receptors were allowed to bind to the conjugated plates and the bound protein was detected as described under Figure 1. All points were run in triplicate with standard deviation less than ±10%.

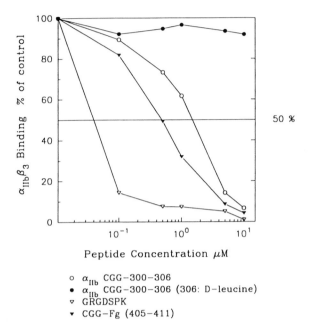

Figure 3: The α_{IIb} 300-306 peptide containing a terminal cysteine residue was SPDP-coupled to BSA coated plates. Biotin-labelled $\alpha_{IIb}\beta_3$ (5µg/ml) was incubated in the absence or presence of peptide. Bound $\alpha_{IIb}\beta_3$ was detected as described under Figure 1. OD_{405nm} data were converted to % of control (no inhibitor = 100% binding). All points were run in triplicate with standard deviation less than ±10%.

Discussion

Glycoprotein $\alpha_{IIb}\beta_3$ is found on platelets and reacts with soluble fibrinogen as a major ligand after platelet stimulation. This is the basis for thrombus formation in blood coagulation and wound healing. Dysregulation of the fibrinogen-$\alpha_{IIb}\beta_3$ interaction can lead to cardiovascular disorders such as myocardial infarction or stroke. This medical relevance of controlled interaction of fibrinogen $\alpha_{IIb}\beta_3$ led us to study the $\alpha_{IIb}\beta_3$ sites which participate in ligand binding. Previous studies of the ligand binding site on $\alpha_{IIb}\beta_3$ identified the region 294-314 on α_{IIb} as a binding site for the dodecapeptide from the carboxyterminus of the fibrinogen γ-chain (13). A synthetic peptide encompassing this region of α_{IIb} (296-306) was able to inhibit platelet aggregation and binding of fibrinogen to platelets and to purified $\alpha_{IIb}\beta_3$ (19).
In this study peptides from the second calcium binding region of α_{IIb} (amino acids 296-313) were designed and tested in vitro for their ability to interfere with the interaction of intact fibrinogen and isolated $\alpha_{IIb}\beta_3$.
The experiment identified α_{IIb} 302-306 as the smallest effective unit. Truncation of either the N- or C-terminal end of α_{IIb} 302-306 resulted in partial or complete loss of

inhibitory activity. Adding residues on the C-terminus (up to 307 to 313) abolished inhibition whereas adding residues on the N-terminus had no effect. It seems that only the short form can assume the correct conformation.

The smallest effective unit contains one negative and two positive charges and is therefore highly hydrophilic. Changing the charge pattern of the molecule by converting the aspartic acid in position α_{IIb} 305 to an asparagine abolished inhibition. Exchanging the natural L-amino acids in position 303, 304, 305 or 306 for D-isomers resulted in a complete loss of activity. These results show that the length, the charge pattern and the conformation of the peptide are important for interaction.

After demonstrating the inhibitory effect of α_{IIb} 302-306, localization of the binding site was the next goal. As shown only $\alpha_{IIb}\beta_3$ but neither ligand nor $\alpha_v\beta_3$ were able to bind to the immobilized peptide. The direct interaction of α_{IIb} 300-306 with $\alpha_{IIb}\beta_3$ inhibits fibrinogen binding to $\alpha_{IIb}\beta_3$. A possible explanation for this finding could be the binding of the peptide to a complementary sequence on the β_3 chain which results in a partial dissociation of the subunits so that the receptor is no longer functional.

Looking more closely on the nature of this peptide-receptor interaction, it became obvious that not only the soluble form of the α_{IIb} 300-306 peptide but also RGD-peptide and fibrinogen γ-chain peptide (Fg γ-chain 405-411) were able to inhibit receptor binding to immobilized α_{IIb} 300-306 peptide. It could be that the Fg γ-chain peptide and α_{IIb} 300-306 interact directly occupying all binding sites for the receptor. Another possibility could be that there are similar or overlapping binding sites of all three peptides on the receptor.

In summary, we show that the sequence GDGRHDL from the α_{IIb} integrin chain 300-306 is involved in modulating $\alpha_{IIb}\beta_3$ interaction with its ligand. This data should allow the design of reagents which can modify thrombolytic cascades.

Acknowledgment

We are grateful to Dr SL Goodman and DJ Sander for their critical reviews and valuable suggestions in the preparation of this manuscript.

References

1.   Phillips DR, Charo IF, Parise LV, Fitzgerald LA. 1988; Blood 71-4:831-843
2.   Plow EF, Ginsberg MH. 1989; Prog. Hemost. Thromb. 9:117-156
3.   Jennings LK, Phillips DR. 1982; J. Biol. Chem. 257:10458-10466
4.   Ruoslahti E, Pierschbacher MD. 1987; Science 238:491-497
5.   Pierschbacher MD, Ruoslahti E. 1984; Nature 309:30-33
6.   Plow EF, Pierschbacher MD, Ruoslahti E, Marguerie GA, Ginsberg MH. 1985; Natl. Acad. Sci. U.S.A. 82:3931-3936
7.   Plow EF, Pierschbacher MD, Ruoslahti E, Marguerie GA, Ginsberg MH. 1987; Blood 70:110-115

156

8.   Smith JW, Ruggeri ZM, Kunicki TJ, Cheresh DA. 1990; J. Biol. Chem.
     265:12267-12271
9.   Plow EF, Srouji AH, Meyer D, Marguerie GA, Ginsberg MH. 1984;
     J. Biol. Chem. 259:5388-5391
10. Pytela R, Pierschbacher MD, Ginsberg MH, Plow EF, Ruoslahti E. 1986;
     Science 231:1559-1562
11. D'Souza SE, Ginsberg MH, Burke TA, Lam SCT, Plow EF. 1988; Science
     242:91-93
12. Charo IF, Nannizzi L, Phillips DR, Hsu MA, Scarborough RM. 1991;
     J. Biol. Chem. 266:1415-1421
13. D'Souza SE, Ginsberg MH, Burke TA, Plow EF. 1990; J. Biol. Chem.
     265:3440-3446
14. Kazal LA, Amsel S, Miller OP, Tocantins LM. 1963; Pro. Soc. Exp. Biol. Med.
     113:989-994
15. Ytohgo T, Izumi M, Kashiwagi H, Hayashi M. 1988; Cell. Struct. Funct.
     13:281-292
16. Richter H. 1988; personal communication
17. Smith JW, Cheresh DA. 1988; J. Biol. Chem. 263:18726-18731
18. Cheresh DA, Spiro, RC. 1987; J. Biol. Chem. 262:17703-17711
19. D'Souza SE, Ginsberg MH, Matsueda GR, Plow EF. 1991; Nature 350:66-68

Biology of Vitronectins and their Receptors
K.T. Preissner, S. Rosenblatt, C. Kost, J. Wegerhoff and D.F. Mosher, editors   157

Selectivity of the non-peptidic GPIIb/IIIa-receptor-antagonist BIBU 52 in respect to other RGD-dependent integrin receptors using an *in vitro* cell adhesion assay

E. Seewaldt-Becker, F. Himmelsbach and Th. H. Müller,

Department of Pharmacology, Dr. Karl Thomae GmbH, Birkendorferstr. 65, 88397 Biberach, Germany

INTRODUCTION

The synthetic glycoprotein IIb/IIIa receptor antagonist BIBU 52 inhibits collagen, ADP- and thrombin-induced aggregation in human platelet rich plasma with an IC_{50} of 80 nM suggesting a therapeutic use for this compound. The fibrinogen receptor gpIIb/IIIa (or $\alpha_{IIb}\beta_3$) is a member of the integrin family of adhesion molecules.

Because of the relative high homology between gpIIb/IIIa and the vitronectin-receptor (VNR), which share a common β_3 subunit, the first question to be addressed by this study was whether BIBU 52 also interacts with the VNR.

A characteristic of some integrins (including gp IIb/IIIa and VNR) is their binding to a very specific sequence on the ligand consisting of the peptide RGD (Arg-Gly-Asp).

RGD-binding dependency has been documented for the fibronectin receptor $\alpha_5\beta_1$, the receptor $\alpha_2\beta_1$ (responsible for the binding of collagen I) as well as for the binding of laminin to its receptor $\alpha_6\beta_1$.

The second aim of this study was to evaluate if BIBU 52 interacts with the binding of these RGD dependent adhesion receptors to their ligands.

The VNR as well as the receptors for fibronectin ($\alpha_5\beta_1$), collagen ($\alpha_2\beta_1$) and laminin ($\alpha_6\beta_1$) have been described to be present on the surface of endothelial cells (ref.1). Therefore in most of the experiments we used endothelial cells as an intact cell system to evaluate their adhesion to the ligands: vitronectin, fibronectin, collagen I and laminin in the presence of the gp IIb/IIIa receptor-antagonist BIBU 52.

We also evaluated the effect of BIBU 52 on the attachment of human smooth muscle cells, human fibroblasts and human carcinoma cells (B16) to vitronectin.

MATERIALS AND METHODS

Culture of human umbilical vein endothelial cells (HUVEC)
Human endothelial cells were isolated from human umbilical veins and cultured according to the method of Jaffe (ref.2). HUVECs were used in the first or second passage for the adhesion assay.

Culture of human vascular smooth muscle cells (HVSMC) and human umbilical cord arterial smooth muscle cells (HUASMC)
Segments of saphenous veins from patients undergoing a vein stripping procedure or human umbilical cords were collected into sterile HEPES buffered PBS solution (supplemented with 100 U/ml penicillin; 100 µg/ml streptomycin).
The respective tissue was then minced to be stirred for 16 hours at 37°C in a mixture containing elastase (0.5 mg/ml), collagenase (2 mg/ml) and trypsin inhibitor (1.0 mg/ml).
The cell suspension was then passed through a 100 µm pore Nylon mesh to remove undigested tissue fragments. The pellet was resuspended in medium M199 containing 10% (v:v) FCS and 5% (v:v) human serum and plated into 25 cm^2 culture flasks. For the assay human vascular smooth muscle cells were used in the fourth passage, human umbilical cord smooth muscle cells were used as primary culture.
The cells were characterized as smooth muscle cells by fluorescent staining with α-actin antibodies.

Human umbilical cord fibroblasts (HUC-Fm) were purchased from European Collection of Animal Cell Cultures and used in the third passage.

Human melanoma cells B16 (a gift from Deutsches Krebsforschungszentrum) were used in the third passage.

Adhesion Assay
96 well plates were coated with 10 µg/ml (1µg/well) of vitronectin, fibronectin or collagen I (diluted in PBS solution). Laminin was used at a concentration of 30 µg/ml (3 µg/well).
Plates were incubated for 120 min at 37°C, the protein solutions were removed and the plates were washed with PBS solution once. To eliminate unspecific binding the plates were incubated with 250 µl/well of bovine serum albumin (2% in PBS solution) for 2 hours at room temperature and washed with PBS solution twice.
Cells were allowed to attach to the respective matrix protein in 96-multiwell polystyrene plates under static conditions. Cells and the respective matrix protein were preincubated with BIBU 52 (or GRGDS; antibody) for 30 min at 37°C (preincubation period). Then cells together with the compound (or GRGDS; antibody) were transferred to the precoated multiwells and incubated for 60 minutes with the matrix protein at 37°C (attachment period). Nonadherent cells were removed by gently washing the plates twice with PBS solution.
The number of adherent cells was quantified using MUH (Methylumbelliferylheptanoate) substrate. For each experiment HUVECs from a different donor were used (HUVEC preparation). As control the attachment of cells to the matrix in the presence of solvent but without BIBU 52 was defined to be 100% (=control). The results of table 1 and 2 are expressed as % of control (mean value ± SE). The result of one experiment is based on three parallel measurements. In each experiment the peptide GRGDS was used as a positive control.
In the binding experiments of HUVECs (preparation 1-4) to vitronectin, BIBU 52 was dissolved in DMSO/HCl (9/1). In all other experiments BIBU 52 was diluted in H$_2$O / hydroxypropyl-ß cyclodextrin / 0.1 N HCl (1:1:1).

RESULTS:

1. Characterisation of the adhesion assay:

1.1 Effect of the peptide GRGDS on the adhesion of endothelial cells to vitronectin, fibronectin, laminin and collagen I.
GRGDS inhibited the adhesion of endothelial cells to vitronectin dose dependently with IC_{50} values of: 0.032 ±0.007 mM; 0.044±0.008 mM; 0.083±0.008 mM (depending on the endothelial cell preparation). The adhesion of endothelial cells to fibronectin was also inhibited dose dependently by GRGDS with IC_{50} values (depending on the endothelial cell preparation) of 0.47 ±0.13; 0.56 ± 0.2 and 0.36± 0.08 mM. As negative control, antibodies to vitronectin were used under the same experimental conditions and did not show any effect on adhesion.
GRGDS has a much lower potency inhibiting the adhesion of EC to laminin and collagen I. At a dose of 2mM only 50% reduction of EC adhesion to laminin and to collagen I was found.
1.2 Effect of antibodies on the adhesion of EC to vitronectin, fibronectin, laminin and collagen I
Antibodies to either human vitronectin or to the VNR effectively inhibited the adhesion of EC to vitronectin (figure 1).

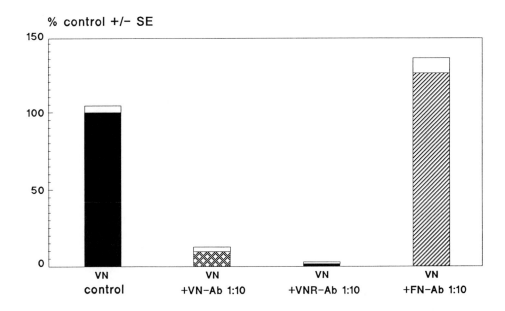

Figure 1:
Inhibition of vitronectin (VN) mediated attachment of endothelial cells by an antibody (Ab) to vitronectin and an antibody to the vitronectin-receptor (VNR).
Data are mean values ± SE of triplicate determinations

Under the same experimental conditions antibodies to fibronectin (as control-antibody) had no effect on the adhesion of EC to vitronectin.

Antibodies to human fibronectin reduced the adhesion of EC to fibronectin to 57 ± 8% of control. Under the same experimental conditions antibodies to vitronectin (as negative control antibody) had no effect on adhesion.

Antibodies to human laminin reduced the adhesion of EC to laminin to 34 ± 0.5% of control. Under these conditions vitronectin-antibodies had no effect on adhesion.

Antibodies to human collagen reduced the adhesion of EC to collagen I to 28 ± 0.9 % of control. In the same experimental setup antibodies to fibronectin had no effect on the adhesion of endothelial cells to collagen.

2. Effects of BIBU 52 on the adhesion of human umbilical vein endothelial cells to vitronectin, fibronectin, collagen I and laminin

As shown in table 1 and figure 2 , BIBU 52 had no significant effect on the adhesion of endothelial cells to the glycoproteins vitronectin, fibronectin, collagen I and laminin at a dose of 100 μM.

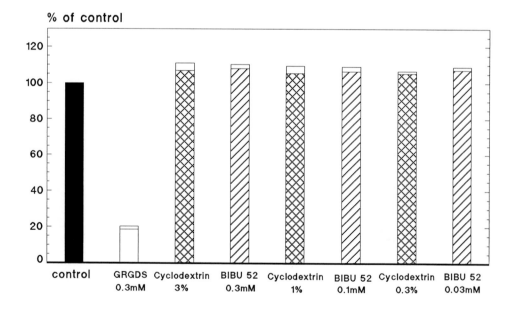

Figure 2:
One typical experiment showing the effect of BIBU 52 on the adhesion of human umbilical vein endothelial cells (HUVEC) to vitronectin
Results are expressed as mean values ± SE.

Table 1:
Effect of BIBU 52 (100 µM) on the adhesion of human umbilical vein endothelial cells
(HUVEC) to vitronectin, fibronectin, collagen I and laminin.
Results are expressed as mean values ± SE.

cells	matrix	% of control	±SE
HUVEC 1.preparation	Vitronectin	84.23	2.74
HUVEC 2.preparation	Vitronectin	91.36	3.42
HUVEC 3.preparation	Vitronectin	104.20	0.81
HUVEC 4.preparation	Vitronectin	102.54	2.77
HUVEC.5.preparation	Vitronectin	93.44	2.30
HUVEC 1.preparation	Fibronectin	115.24	2.44
HUVEC 2.preparation	Fibronectin	93.93	2.49
HUVEC 3.preparation	Fibronectin	108.85	0.74
HUVEC 1.preparation	Collagen I	112.09	2.19
HUVEC 2.preparation	Collagen I	92.54	1.24
HUVEC 3.preparation	Collagen I	110.70	1.37
HUVEC 1.preparation	Laminin	100.41	0.41
HUVEC 2.preparation	Laminin	100.21	1.53
HUVEC 3.preparation	Laminin	96.53	3.64

3. Effects of BIBU 52 on the adhesion of human vascular smooth muscle cells , human umbilical fibroblasts and melanoma cells (B16) to vitronectin

Table 2
Effect of BIBU 52 (100 µM) on the adhesion of human umbilical cord arterial smooth muscle
cells , human venous smooth muscle cells , human umbilical fibroblasts and B-16 human
melanoma cells to vitronectin. Results are expressed as mean values ± SE.

cells	matrix	% of control	± SE
HUASMC	Vitronectin	95.25	9.69
HVSSMC	Vitronectin	96.04	2.70
HUC-Fm	Vitronectin	131.99	7.97
HUC-Fm	Vitronectin	95.62	2.73
B-16	Vitronectin	87.39	2.23
B-16	Vitronectin	124.44	3.76

In assays using human smooth muscle cell, human fibroblasts or human melanoma cells (B16), BIBU 52 (100µM) had no inhibitory effect on adhesion to vitronectin.

DISCUSSION

The non-peptidic compound BIBU 52 is highly selective for the fibrinogen receptor gpIIb/IIIa in comparison to other receptors of the integrin family. Using an intact cell system BIBU 52 had no effect on even the closely related vitronectin-receptor. Other RGD dependent receptors like the fibronectin receptor $\alpha_5\beta_1$, the receptor for collagen I $\alpha_2\beta_1$ and the receptor for laminin $\alpha_6\beta_1$ were also uneffected by BIBU 52. Endothelial cells used for the present experiments have on their surface the vitronectin receptor, the fibronectin receptor and at least 8 additional integrins. Therefore, a given ligand may not bind exclusively to only one integrin, but to several receptors on the cell surface. Consequently the assay has been characterized by using antibodies specific for a matrix protein and/or antibodies binding to the receptor on the endothelial cell. The fact that these antibodies could inhibit the adhesion of endothelial cells to the respective matrix protein (at least in part) supports the conclusion that these receptors serve an adhesive function. Using antibodies to vitronectin or antibodies to the VNR cell adhesion was essentially eliminated. These results suggest that vitronectin promoted endothelial cell attachment to vitronectin is indeed mediated by the VNR. The attachment promoted by fibronectin, laminin and collagen I could not be inhibited by 100% by antibodies and seems to be mediated by more than one type of receptor. Another explanation is that the antibodies used for these proteins were not as specific as those used for vitronectin.

Binding specificity may be affected by cell-type-specific factors as well as by environmental factors. Using intact cells as a test system includes, at least in part, the environmental factors. In the intact cell the receptor is located in its natural environment, the receptor-phospholipid interactions as well as conformation changes and receptor-cytoskeleton interactions remain undisturbed. Isolated receptors could have been used as a test-system with the advantage of having a better defined test system. However, the natural enviroment of the receptor is lost, which might be very important for the function of the receptor.

Additional experiments performed using human smooth muscle cells, human fibroblasts and B16 cells as a source of VNR confirmed the results found with endothelial cells.

We conclude that BIBU 52 inhibits very selectively the binding of fibrinogen to its receptor gpIIb/IIIa but does not interfere with the structurally related vitronectin receptor as well as with other RGD dependent receptors of the integrin family of adhesion molecules like the receptors for fibronectin, collagen I and laminin at a concentration of up to 100 µM.

REFERENCES

1   Mourik van J.A. et al. Biochem. Pharmacol. 1990; 39 (2):233-239.
2   Jaffe E.A., Nachman R.L., Becker C.G., Minick C.R. J. Clin. Invest . 1973; 52:2745-56.

Biology of Vitronectins and their Receptors
K.T. Preissner, S. Rosenblatt, C. Kost, J. Wegerhoff and D.F. Mosher, editors

Tyrosine phosphorylation of a 38 kDa protein upon interaction of urokinase-type plasminogen activator (u-PA) with its cellular receptor

I. Dumler, T. Petri, and W.-D. Schleuning

Research Laboratories of Schering AG, D-13342 Berlin, Germany

INTRODUCTION

Biological processes requiring tissue remodelling and cell migration such as trophoblast invasion, spermatogenesis, organogenesis during embryonic life, wound healing and malignant growth are frequently associated with plasmin generation from plasminogen, catalyzed by plasminogen activators (PAs) (1). The biosynthesis of the components of the plasminogen activating system is highly regulated in space and time (2). Recently a high affinity cell surface receptor for urokinase-type plasminogen activator (u-PAR) was isolated and characterized (3,4). u-PAR, a cysteine rich 313-residue single polypeptide with five potential N-linked glyco-sylation sites, is fixed to the plasma membrane via a COOH-terminal glycosyl phosphoinositol (GPI) anchor (5,6) and believed to provide a lever for the regulation of extracellular proteolysis by temporal and topological restriction of u-PA activity. As other GPI-linked proteins such as members of the Ly-6 family have been implicated in signal transduction (7) we have explored a similar role of u-PAR. Previous work has demonstrated that u-PA stimulates the differentiation of HL 60 cells, the migration of bovine endothelial cells and the chemotaxis of human neutrophils (8).

Our present study shows that u-PA induces a dose dependent phosphorylation of a 38 kDa protein (p38) on tyrosine in U-937 (histiocytic lymphoma) cells.

EXPERIMENTAL

Materials
Chemicals were purchased from Sigma (St. Louis, MO), Pharmacia (Uppsala, Sweden) or Serva (Heidelberg, Germany). ^{32}P-ATP, ^{35}S-methionine and ^{125}I were obtained from Amersham International. Purified rabbit anti-phosphotyrosine polyclonal antibodies were from Dianova (Hamburg, Germany), PI-specific PLC from Sigma, the protein tyrosine kinase inhibitor Herbi-mycin A (Streptomyces sp.) from Calbiochem Biochemical (San Diego, USA), and u-PA from Serono (Freiburg, Germany).

Cell culture
U-937 cells were provided by the American Type Culture Collection and grown at 37 °C in RPMI 1640 medium containing 5 % fetal bovine serum. For ^{35}S-methionine labeling of the cellular proteins 2×10^5 U-937 cells/ml were seeded into 9 cm petridishes in 8 ml RPMI 1640 containing 10 % of the standard methionine concentration, 5 % fetal bovine serum and 3 μCi/ml ^{35}S methionine. Cells were incubated for 3 days and harvested by centrifugation (1000 g, 10 min, 4 °C).

Purification of u-PAR

u-PAR was purified in two steps, consisting of temperature induced phase separation and affinity chromatography with immobilized u-PA, as previously described (9).

Radioiodination and cross-linking assay

DFP-inhibited u-PA was iodinated by the IODO-GEN procedure (10). The iodinated protein was separated from free iodine by gel filtration on a Sephadex G-25 column. The specificity activity of labeled u-PA was approximately 15 μCi/mg protein.

For cross-linking experiments crude cell lysates or purified receptor preparations were incubated with DFP-inactivated radiolabeled u-PA for 1 h at 4 °C, followed by incubation with 1 mM DSS for 15 min at room temperature, and with 10 mM CH_3COONH_4 for a further 10 min. Samples were analyzed by SDS-PAGE, followed by radioautography.

Preparation of rabbit anti-u-PAR serum

The protein band corresponding to purified u-PAR was sliced out of the gel after SDS-PAGE and homogenized in 100 μl of elution buffer (50 mM NH_4HCO_3, 0.1 % SDS). The first injections (subcutaneous) consisted of 10 μg of protein per animal and were followed by two further treatments (5 μg of protein per animal). The serum was prepared 5 days after the last booster injection.

Electrophoresis, Western blotting and radioautography

SDS-PAGE was carried out in slab gels (7.5 or 10 %) as described (11). Samples were reduced immediately before electrophoresis in the presence of 20 mM DTT for 5 min at 95 °C or analysed under nonreducing conditions.

Gels were electroblotted onto nitrocellulose sheets which were subsequently blocked with 1 % BSA or 30 % fetal calf serum. Alkaline phosphatase-conjugated goat anti-rabbit Ig was used as a second antibody and the sheets were developed with nitro blue tetrazolium/5-bromo-4-chloro-3-indolyl phosphate. Radioautography of ^{125}I, ^{35}S, and ^{32}P-labeled proteins was performed with dried polyacrylamid gels using Konica X-ray film. In some experiments gels containing samples from the immune protein tyrosine kinase assay were soaked in 1 N KOH at 55 °C for 2 h to hydrolyze phosphate on serine and threonine (12).

Immunoprecipitation and immune complex kinase assay

Protein samples were incubated for 2 h at 4 °C with 25 μl of protein A sepharose (Sigma), that had been incubated previously with 10 μl of anti-u-PAR antibody, overnight. The resin was subsequently sedimented by centrifugation and washed twice with 300 μl of 25 mM Hepes, pH 7.4, 150 mM NaCl, 1 mM sodium orthovanadate or 0.5 % Nonidet P-40, and twice with buffer without Nonidet P-40. To perform the immune complex kinase assay, the immunoprecipitate was washed further with 25 mM Hepes, pH 7.4, 10 mM $MnCl_2$ containing a protease inhibitor mixture (1 mM phenylmethylsulfonyl fluoride, and 10^{-6} M each of aprotinin, pepstatin and cystatin). Following the addition of 5 mM p-nitrophenylphosphate and 30 nM ^{32}P-ATP the immunoprecipitates were incubated for 10 min at room temperature and subsequently washed for a final time in 20 mM Hepes, pH 7.4. After addition of SDS-sample buffer (with DTT) the immunoprecipitates were boiled and electrophoresed on 10 % SDS-PAGE. After the separation the gels were dried and subjected to radioautography.

Protein determination
Protein was quantified with a bicinchoninic acid (BCA)-reagent from Pierce, using BSA as a standard.

RESULTS AND DISCUSSION

U-PAR purified from U-937 cells was electrophoretically homogeneous (Fig. 1a). Cross-linking with iodinated u-PA revealed a single band corresponding to the complex of u-PA/u-PAR exhibiting a M_r of 110 kDa (Fig. 1b). Antisera raised in rabbits against pure u-PAR recognized the receptor in purified fractions, as well as in crude cell extracts (Fig. 1c). Immunoprecipitation

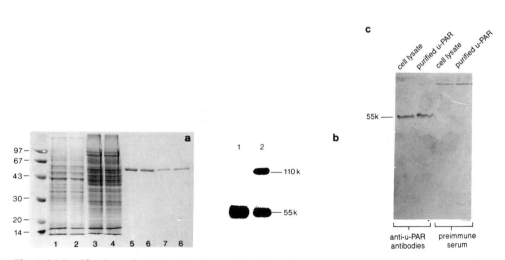

Fig. 1 (a) Purification of u-PAR. The lysis of U-937 cells, temperature-induced phase-separation and affinity chromatography were done as described under "Materials and Methods". Lanes 1,2 - oily phase fractions; lanes 3,4 - water-soluble phase fractions; lanes 5-8 - fractions after affinity chromatography. (b) Cross-linking of radioiodinated u-PA with u-PAR. DFP-treatment of u-PA and chemical cross-linking with DSS were performed as described. Lane 1 - cross-linked control with ^{125}I-u-PA; lane 2 - cross-linked u-PAR purified by affinity chromatography. (c) Characterization of antibodies raised to u-PAR. 20 /ug of cell lysates or 0.5 /ug of purified u-PAR were electrophoresed on 10 % SDS-PAG, transferred to nitrocellulose, and probed with immune or preimmune serum (as indicated). Immunoreactive proteins were identified by incubation with alkaline phosphatase-conjugated second antibody.

166

of protein extracts of metabolically (^{35}S-methionine) labeled native U-937 cells was performed and two main protein bands were visualized on radioautogramms (Fig. 2a). One of them, according to M_r and immunoreactivity corresponded to u-PAR, whereas the identity of the other band (M_r 38 kDa: p38) has as yet not been established. A third minor band comigrated with the front line and contained material, which has not been characterized and is therefore not further considered in the context of this work.

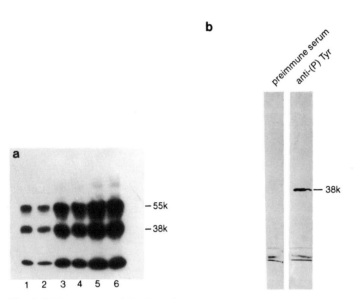

Fig. 2 (a) Immunoprecipitation of in vivo labeled U-937 cells. Cells were biosynthetically labeled with ^{35}S-methionine and immunoprecipitation using anti-u-PAR was done as described. All samples were resolved by SDS-PAGE, after which the gels were dried and analysed by autoradiography. Different amount of cell lysates were used for immunoprecipitation: lanes 1,2 - 25 μg protein; lanes 3,4 - 50 μg protein; lanes 5,6 - 100 μg protein. (b) Western blotting of coimmunoprecipitated proteins with anti-(P)Tyr antibodies. The lysate of 2×10^6 U-937 cells preincubated with 100 nM u-PA for 1 h at 37 °C was used for immunoprecipitation with anti-u-PAR antibodies. Commercial polyclonal anti-(P)Tyr antibody and alkaline-phosphatase-conjugated second antibody were used for Western blot analysis.

With previous data on the association of several GPI-anchored proteins with protein tyrosine kinase activity in mind (7), we used commercial polyclonal anti-(P)Tyr antibody to investigate whether it reacted with p38. The results indicated the presence of phosphotyrosine (Fig. 2b).
In vitro phosphorylation of proteins precipitated with anti-u-PAR antibody resulted in labeling of p38. The band was resistant to alkaline hydrolysis, indicating phosphorylation on tyrosine (data not shown). The phosphorylation was dependent on the dose of u-PA in the cell culture medium (Fig. 3a).

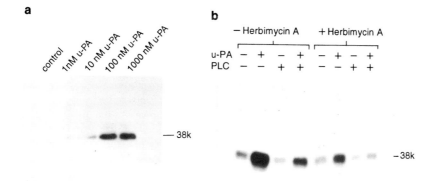

Fig. 3 (a) Dose-dependency of p38 protein phosphorylation on the concentration of u-PA. U-937 cells (2×10^6) were incubated at 37 °C for 1 h in the presence of different concentration of u-PA. After that the immunoprecipitation and phosphorylation were done as described. (b) Effect of treatment of U-937 cells with PI-PLC ($2 \, \mu l / 2 \times 10^5$ cells) and Herbimycin A ($1 \, \mu M / 2 \times 10^5$ cells) on coprecipitation and phosphorylation of p38. The cells were treated with the indicated amount of drugs for 1 h at 37 °C. Immunoprecipitation and in vitro kinase assay were performed as described.

In another experiment we removed GPI-linked proteins from the surface of the cells by treatment with PI-specific PLC before immunoprecipitation. This treatment significantly decreased the phosphorylation of p38 (Fig. 3b). Hence intact GPI-linked u-PAR is necessary for the association with PTK and phosphorylation.

Preincubation of U-937 cells with the PTK inhibitor Herbimycin A also abolished the u-PA-induced phosphorylation of p38 (Fig. 3b).

Taken together our results demonstrate that (1) u-PAR is associated with p38, a protein that is coprecipitated by polyclonal anti-u-PAR antisera and (2) phosphorylated on tyrosine upon the addition of ATP. Several aspects of these findings solicit some additional comment.

U-PAR has so far been purified exclusively from cells stimulated with PMA. This treatment increases the number of receptors but concomitantly leads to a lower affinity for u-PA (13,14). Such an effect could be explained either by the biosynthesis of a modified receptor or the dissociation of a cofactor from the receptor upon addition of PMA. As we wanted to obtain antisera against the constitutive high affinity receptor, we chose to purify it from native U-937 cells and to utilize this material for immunization. It seems surprising that p38 did not copurify with u-PAR on u-PA-sepharose. This may be due to the extraction procedure employed or to conditions chosen during charging and washing of the column. Alternatively p38 could be synthetized at a higher rate than u-PAR, leading to higher percentage of ^{35}S-incorporation, or contain a significantly higher proportion of methionine so that the roughly equal intensity of bands in Fig. 2a could represent similar amounts of radioactivity but not equimolar amounts of protein. If γ^{32}P-ATP is added to the immunoprecipitates, p38 is phosphorylated on tyrosine. This is demonstrated both by the use of phosphotyrosine specific monoclonal antibody (Fig. 2b) and by basic hydrolysis of the modified tyrosine residues (data not shown). The tyrosine phosphorylation is dose dependent (Fig. 3a), abrogated by (a) the tyrosine kinase inhibitor Herbimycin A, and (b) treatment of the cells with phospholipase C. Taken together these data establish an unequivocal link between the occupation of the receptor and tyrosine kinase activity. The phosphorylation may be autocatalytic or catalyzed by an as yet unidentified enzyme. The biological function of p38 tyrosine phosphorylation is presently unknown. As the u-PA receptor has been implicated in signal transduction (5,15), tyrosine phosphorylation may represent an early event in the activation of an u-PAR mediated signalling pathway or - alternatively - play a role in the mechanism of receptor internalization.

REFERENCES

1   Danø, K., Anderson, P.A., Grondahl-Hansen, J., Kristensen, P., Nielsen, L.S. and Skriver, L. (1985) Adv. Cancer Res. 44, 139-266.

2   Saksela, O. (1985) Biochim. Biophys. Acta. 823, 36-65.

3   Kristensen, P., Eriksen, J; Blasi, F. and Danø, K. (1991) J. Cell. Biol. 115, 1763-1771.

4   Roldan, A.L., Cubellis, M.V., Masucci, M.T., Behrendt, N., Lund, L.R., Danø, K., Appella, E. and Blasi, F. (1990) The EMBO J. 9, 467-474.

5   Palfree, R.G.E. (1991) Immunology Today. 12, 170-171.

6   Ploug, M., Ronne, E., Behrendt, N., Jensen, A.L., Blasi, F. and Danø, K. (1991) J. Biol. Chem. 266, 1926-1933.

7   Stefanova, I., Horejsi, V., Ansotegui, I.J., Knapp, W. and Stockinger, H. (1991) Science 254, 1016-1019.

8   Gudewicz, P.W., and Gilboa, N. (1987) Biochem. Biophys. Res. Commun. 147, 1176-1181.

9   Behrendt, N., Ronne, E., Ploug, M., Petri, T., Lober, D., Nielsen, L.S., Schleuning, W.-D., Blasi, F., Appella, E., and Danø, K. (1990) J. Biol. Chem. 265, 6453-6460.

10   Fraker, P.J., and Speck, J.C. (1978) Biochem. Biophys. Res. Commun. 80, 849-857.

11   Laemmli, U.K. (1970) Nature (London) 227, 680-685.

12   Kamps, M.P., and Sefton, B.M. (1989) Anal. Biochem. 175, 22-26.

13   Lund, L.R., Ronne, E., Roldan, A.L., Behrendt, N., Romer, J., Blasi, F., and Danø, K. (1991) J. Biol. Chem. 266, 5177-5181.

14   Nielsen, L.S., Kellerman, G.M., Behrendt, N., Piccone, R., Danø, K., and Blasi, F. (1988). J. Biol. Chem. 263, 2358-2363.

15   Ullrich, A., and Schlessinger, J. (1990). Cell 61, 203-212.

The abbreviations used are:

u-PA - urokinase-type plasminogen activator; u-PAR-u-PA receptor; PMA - phorbol 12-myristate 13-acetate; DFP - diisopropyl fluorophosphate; DSS - disuccinimidyl suberate; PBS - phosphate buffered saline; SDS - sodium dodecyl sulphate; PAGE - polyacrylamide gel electrophoresis; DTT - dithiothreitol; BSA - bovine serum albumine.

Biology of Vitronectins and their Receptors
K.T. Preissner, S. Rosenblatt, C. Kost, J. Wegerhoff and D.F. Mosher, editors

Modulation of the protein tyrosine kinase pp60$^{c\text{-}src}$ upon platelet activation: A protein kinase C inhibitor blocks translocation

Liebenhoff, U. and Presek, P.

Rudolf-Buchheim-Institut für Pharmakologie, Frankfurter Str. 107, 35392 Gießen, Federal Republic of Germany

Summary: Human platelets contain abundant levels of protein tyrosine kinases of the src-family, which correlates with high levels of protein tyrosine phosphorylation in response to stimulation events. We have shown that overall pp60$^{c\text{-}src}$ kinase activity increases on stimulation of platelets with agonists that directly or indirectly activate PKC. Thrombin stimulation was studied in more detail and modulation of pp60$^{c\text{-}src}$ could be attributed to 2- to 3-fold increase in substrate affinity. This was accompanied by phosphorylation of pp60$^{c\text{-}src}$ at Ser-12, a residue known to be phosphorylated by PKC. A specific PKC inhibitor, Ro-31-8220, blocked thrombin-induced translocation of pp60$^{c\text{-}src}$ to the cytoskeleton. We suggest that phosphorylation at Ser-12 in the membrane-binding domain is a signal for dissociation of pp60$^{c\text{-}src}$ from the plasma membrane, which may in turn position the enzyme closer to putative substrates and result in the increased substrate affinity.
Stimulation of platelets with various agonists, such as thrombin, collagen, platelet-activating factor, vasopressin, ADP, thromboxane A_2 analogs, the calcium ionophore A23187 and the phorbolester phorbol 12-myristate, 13-acetate (PMA), causes a dramatic increase in phosphorylation of multiple proteins at tyrosine residues. Since none of the known receptors for platelet agonists possess intrinsic protein tyrosine kinase activity, the non receptor protein tyrosine kinases of the src-family are good candidates that may be involved in these signalling events. Five members of the src-family have been detected in human platelets: pp60$^{c\text{-}src}$, p59^{fyn}, p62$^{c\text{-}yes}$, p59^{hck}, and p54/58^{lyn}. pp60$^{c\text{-}src}$ is the most abundant and represents as much as 80 % of total tyrosine kinase activity, suggesting an important role in platelet physiology (for review see refs. 1, 2). As platelets also contain high phosphotyrosine phosphatase activity, this may likewise contribute to agonist-induced tyrosine phosphorylation (3, 4). Two protein tyrosine kinases, p125^{FAK} and pp60$^{c\text{-}src}$, have recently been shown to be activated upon stimulation events (5, 6, 7), but their specific role in platelet function and their mechanism of activation are not fully understood. As shown in Fig. 1, all agonists tested, including thrombin, collagen, PMA and the calcium ionophore A23187, which directly or indirectly activate the Ca^{2+}-, phospholipid-dependent protein kinase C (PKC), increased overall kinase activity of pp60$^{c\text{-}src}$ about 1.6- to 3-fold. Elevation of cyclic AMP directly by forskolin or indirectly by the platelet antagonist prostaglandin E_1 (PGE$_1$) or elevation of cyclic GMP by sodium nitroprusside did not affect the kinase activity of the enzyme.

To examine the kinetic behaviour of pp60$^{c\text{-src}}$ we determined that the differences in enzyme activity of resting and thrombin-stimulated platelets could be attributed to a 2- to 3-fold increase in the affinity for ATP and exogenous substrates. The V_{max} values were only slightly altered (7).

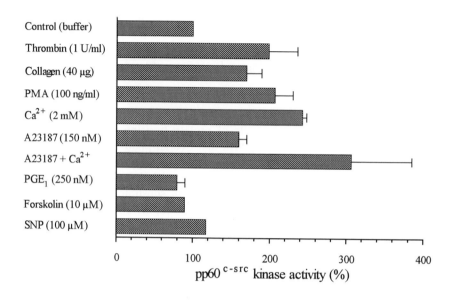

Figure 1. pp60$^{c\text{-src}}$ kinase activity of human platelets stimulated by various biologically active agents. Isolated platelets were stimulated for 30 sec, lysed, and kinase activity of the immobilized enzyme was determined in a solid phase assay as described previously (8). The enzyme activity is given as mean values ± S.D. and the percentage values are determined relative to untreated controls (= 100%).

To examine modifications of pp60$^{c\text{-src}}$ as a probable cause of these events, we analysed tryptic phosphoamino peptides of immunoprecipitated ^{32}P-labeled pp60$^{c\text{-src}}$ from unstimulated and stimulated platelets. The agonists listed above induced an approximately 2-fold increase in pp60$^{c\text{-src}}$ phosphorylation primarily at Ser-12, a residue known to be phosphorylated by PKC, as shown for thrombin in Fig. 2. In contrast, forskolin, PGE$_1$, and sodium nitroprusside did not cause modifications in the phosphorylation pattern of pp60$^{c\text{-src}}$. Surprisingly, phosphate incorporation at Ser-17, the major site of serine phosphorylation in pp60$^{c\text{-src}}$ by the cyclic AMP-dependent protein kinase (A-kinase) (9) was not enhanced after incubation of platelets with PGE$_1$ or forskolin, although less than 60 % of the pp60$^{c\text{-src}}$ molecules are phosphorylated at Ser-17 (9). To further investigate whether Ser-17 is phosphorylated by A-kinase, we analysed tryptic phosphopeptides of *in vitro*-phosphorylated, purified pp60$^{c\text{-src}}$ by purified A-kinase. We found no phosphorylation of the peptide containing Ser-17, but an additional, unidentified peptide was phosphorylated (Fig. 2).The question of whether pp60$^{c\text{-}src}$ is a substrate of A-kinase in intact platelets, as has been concluded from studies with its viral counterpart in transformed fibroblasts (9), must be addressed. The physiological significance of this phosphorylation site is still unknown, because deletion or substitution of Ser-17 does not result in a detectable alteration of pp60$^{c\text{-}src}$ kinase activity (9). On the other hand, these results do not exclude other functional consequences of Ser-17 phosphorylation like translocation or association with other proteins.

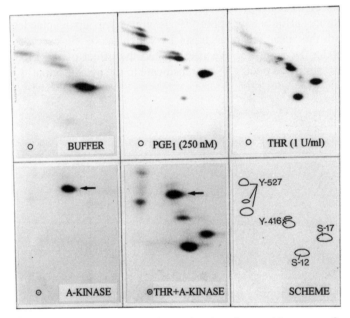

Figure 2. Two-dimensional tryptic phosphopeptide maps of pp60$^{c\text{-}src}$ from resting and stimulated platelets. ^{32}P-labeled platelets were incubated for 30 seconds with buffer, PGE$_1$, or thrombin (THR), lysed and immunoprecipitated with a polyclonal anti-pp60$^{c\text{-}src}$-antiserum as described previously (8). In parallel, purified pp60$^{c\text{-}src}$ was phosphorylated by A-kinase using [γ-^{32}P]ATP. pp60$^{c\text{-}src}$ was eluted from the SDS gel and digested with trypsin. The resulting phosphopeptides were seperated in two dimensions as described (8). The autoradiograms are shown. THR+A-KINASE shows a mixture of tryptic phosphopeptides of pp60$^{c\text{-}src}$ immunoprecipitated from ^{32}P-labeled thrombin-stimulated platelets and from purified pp60$^{c\text{-}src}$ in vitro phosphorylated by A-kinase. The arrows indicate the A-kinase-dependent phosphorylation site of purified pp60$^{c\text{-}src}$. Scheme shows phosphopeptides of pp60$^{c\text{-}src}$ identified by comparison with published fingerprints of pp60$^{c\text{-}src}$.

The role of the PKC-dependent phosphorylation at Ser-12 also remains to be elucidated. Ser-12 is unphosphorylated in resting platelets and may be phosphorylated up to 100% by biologically active agents that activate PKC (9). The fact that a direct activator of PKC like PMA produces a pattern of tyrosine-phosphorylated proteins that is similar to that produced by thrombin and other agonists suggests that the onset of tyrosine phosphorylation occurs downstream from PKC activation and is a late step in platelet activation (10). This coincides with the observed phosphorylation of pp60[c-src] at Ser-12 by PKC and its increased substrate affinity on thrombin stimulation (7). As a consequence of thrombin and PMA stimulation of platelets, pp60[c-src] becomes translocated from the membrane fraction to the cytoskeleton (11). This translocation could bring the kinase closer to putative substrates, allowing the formation of new enzyme-substrate complexes or association with potential modulatory proteins. Our observation that the substrate affinity of pp60[c-src] is increased upon platelet stimulation is consistent with this idea. The mechanism by which pp60[c-src] becomes translocated to the cytoskeleton is unknown. We investigated the effect of the PKC inhibitor, Ro-31-8220, on the ability of pp60[c-src] to translocate to the cytoskeleton.

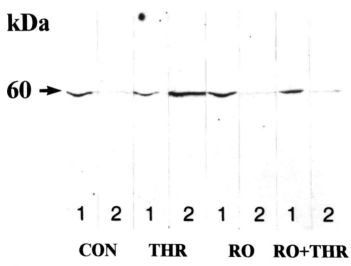

Figure 3. Association of pp60[c-src] with the detergent-soluble and -insoluble fraction of platelets. Fractions were obtained from an equal number of platelets and detergent-soluble (lane 1) and detergent-insoluble, cytoskeletal fractions (lane 2) were prepared as originally described (12). Samples were analysed for pp60[c-src] with a monoclonal antibody, the second antibody was an ^{125}I-labeled anti-mouse-IgG antibody. The resulting autoradiogram is shown. CON, buffer; THR, thrombin (1 U/ml); Ro, PKC inhibitor (Ro-31-8220; 100μM); Ro+THR, preincubation for 20 min with the PKC inhibitor followed by stimulation with thrombin for 3 min.

As can be seen in Fig. 3 the PKC-inhibitor blocked thrombin-induced translocation of pp60^{c-src} to the cytoskeleton. The inhibitory effect of Ro-31-8220 suggests that modulation of pp60^{c-src} is consecutive to PKC activation. The association of pp60^{c-src} with cellular membranes is mediated by myristoylation of Gly-2; both this residue and Ser-12 are located within the fourteen amino acid membrane-binding domain (9). The surrounding basic amino acids, which are thought to be important for the substrate specificity of PKC, and the spatial relationship of Ser-12 to the plasma membrane are similar to the PKC-phosphorylation site of the epidermal growth factor receptor (13). Phosphorylation of the epidermal growth factor receptor at this site promotes receptor internalization. A mechanism for PKC-dependent phosphorylation and translocation of the MARCKS-protein, a myristoylated membrane-bound protein, has also been reported (14). We propose that the effect of a negatively charged phosphoserine displaces pp60^{c-src} from the plasma membrane, which might be important for either translocation to the cytoskeleton or for physical interactions between pp60^{c-src} and other proteins. These protein-protein interactions, mediated by the src-homology domains SH2 and SH3 that were originally found as a common motif in the members of the src-family, may also exist in platelets (15). It has been reported that upon platelet stimulation pp60^{c-src} and p59^{fyn} associate with the phosphatidylinostiol 3-kinase (16). Recently it was shown that phosphatidylinositol 3-kinase redistributes to the cytoskeleton upon thrombin stimulation (17). Our preliminary studies have shown that p59^{fyn} and p62^{c-yes} also translocate to the cytoskeleton in a stimulation-dependent manner. Additionally, an interaction with the cytoskeleton has been demonstrated for pp60^{c-src}, phospholipase C, inositol lipid kinase and diacylglycerol kinase (18). An association of p59^{fyn}, p62^{c-yes} and p54/58^{lyn} with the p21^{ras} GTPase-activating protein has also been reported (19). An additional role for the protein tyrosine kinases of the src-family in platelet signal transduction was suggested by the finding that p59^{fyn}, p62^{c-yes}, and p54/58^{lyn} associate with the thrombospondin receptor (CD36) (20). In other hematopoietic cells the src-protein tyrosine kinases have been shown to be involved in the signalling pathways of different surface receptors, suggesting that the non receptor protein tyrosine kinases of the src-family are involved in surface receptor-induced signal transduction (for review see ref. 2). Integrin receptors associate with proteins of the cytoskeleton, and these complexes might serve to anchor the src-family protein tyrosine kinases and other signalling molecules. It would therefore be of great interest to identify the cytoskeletal targets of pp60^{c-src} and the other members of the src-family.

Acknowledgments:
We are grateful to Roche Products Ltd. for their kind gift of the PKC-inhibitor and to Prof. F. Hofmann, München, for providing purified, A-kinase from bovine heart.

References:

1. Halbrügge, M. and Walter, U. (1993) in Protein Kinases and Blood Cell Function. (Huang, C. K. and Shàafi, R. I., eds.), pp. 245-298, CRC Press, Boca Raton, FL
2. Shattil, S. J. and Brugge, J. S. (1991) Curr. Opin. Cell Biol. 3, 869-879
3. Lerea, K. M., Tonks, N. K., Krebs, E. G., Fischer, E. H. and Glomset, J. A. (1989) Biochemistry 28, 9286-9292
4. Smilovitz, H. M., Aramli, L., Xu, D. and Epstein P. M. (1991) Life Sci. 49, 29-37
5. Lipfert, L., Haimovich, B., Schaller, M. D., Cobb, B. S., Parsons, J. T and Brugge, J. S. (1992) J. Cell Biol. 119, 905-912
6. Clark, E. A. and Brugge, J. S. (1993) Mol. Cell. Biol. 13, 1863-1871
7. Liebenhoff, U., Brockmeier, D. and Presek, P. (1993) Biochem. J., in press
8. Reuter, C., Findik, D. and Presek, P. (1990) Eur. J. Biochem. 190, 343-350
9. Cooper, J. A. (1990) in Peptides and Protein Phosphorylation. (Kemp, B. E., ed.) pp.85-113, CRC Press, Boca Raton, FL
10. Bachelot, C., Cano, E., Grelac, F., Saleun, S., Druker, B. J., Levy-Toledano, S., Fischer, S. and Rendu, F. (1992) Biochem. J. 284, 923-928
11. Horvath, A. R., Muszbek, L. and Kellie, S. (1992) EMBO J. 11, 855-861
12. Fox, J. E. B., Boyles, J. K., Berndt, M. C., Steffen, P. K. and Anderson, L. K (1988) J. Cell Biol. 106, 1525-1538
13. Hunter, T., Ling, N. and Cooper, J. A (1984) Nature 311, 480-483
14. Thelen, M., Rosen, A., Nairn, A. C. and Aderem, A. (1991) Nature 35, 320-322
15. Pawson, T. and Schlessinger, J. (1993) Curr. Biol. 3, 434-442
16. Gutkind, J. S., Lacal, P. M. and Robbins, K. C. (1990) Mol. Cell. Biol. 10, 3806-3809
17. Zhang, J., Fry, M. J., Waterfield, M. D., Kaken, S., Liao, L., Fox, J. E. B. and Rittenhouse, S. E. (1992) J. Biol. Chem. 267, 4686-4692
18. Grondin, P., Plantavid, M., Sultan, C., Breton, M., Mauco, G. and Chap, H. (1991) J. Biol. Chem. 266, 15705-15709
19. Cichowski, K., McCormick, F. and Brugge, J. S. (1992) J. Biol. Chem. 267, 5025-5028
20. Huang, M.-M., Bolen, J. B., Barnwell, J. W., Shattil, S. J. and Brugge, J. S. (1991) Proc. Natl. Acad. Sci. USA 88, 7844-7848

Biology of Vitronectins and their Receptors
K.T. Preissner, S. Rosenblatt, C. Kost, J. Wegerhoff and D.F. Mosher, editors

The Role of αv Integrins in Adenovirus Infection

G.R. Nemerow, T.J. Wickham and D.A. Cheresh

Department of Immunology, The Scripps Research Institute, 10666 North Torrey Pines Road, La Jolla, California 92037, USA

INTRODUCTION

Cell receptors involved in the initial attachment of both enveloped and non-enveloped viruses to host cells have been identified (1). Although these receptors play an important role in virus tropism, it is clear that additional host cell factors are required for subsequent steps in virus infection following attachment.

Adenovirus (Ad), a major cause of respiratory and gastrointestinal disease of man (2,3), has provided a useful model for analyzing the mechanisms of virus internalization (4,5). Adenovirus binds to specific receptors which are widely distributed on various types of host cells (6). The identity of the initial attachment receptor has not yet been established. Following attachment to cells, adenovirus undergoes receptor-mediated internalization into clathrin-coated endocytic vesicles (4). Virions subsequently cause disruption of the cell endosome by a pH-dependent mechanism (5) which is still poorly understood. Virion particles enter the cytoplasm following endosome rupture and then are transported to the nuclear pore complex (4) where the viral genome is uncoated and enters the nucleus thus initiating infection.

Adenovirus is a nonenveloped virus which contains a DNA genome of 35 kb (2). The protein coat or capsid of the virion is composed of 252 subunits (capsomeres), of which 240 are hexons and 12 are pentons (7). Each penton contains a 400 kd penton base (polypeptide III) which is noncovalently linked to a 186 kd fiber protein projecting from the base. The fiber protein mediates high-affinity binding to a cell receptor (8), while the penton base has been suggested to have an important role in virus penetration (9). Each of the five polypeptide III subunits of the penton base contains an Arg-Gly-Asp (RGD) peptide sequence (10) that is also found in a number of cell adhesion molecules including vitronectin and fibronectin. These matrix molecules mediate binding to cell surface integrins (11). The presence of an RGD sequence in Ad penton base as well previous studies indicating that soluble penton base detaches cells in culture (12,13) suggested a role for this protein in interaction with cell surface integrins. In the studies described here, we demonstrate that Ad attachment and uptake into cells are separate but cooperative events that result from the interaction of distinct viral coat proteins with a receptor for attachment and αv integrins for internalization.

RESULTS

Expression of recombinant Ad2 penton base in insect cells

In order to examine its interaction with host cells as well as its role in Ad infection, recombinant Ad2 penton base was overexpressed in Sf9 insect cells using baculovirus. Immunoprecipitation analysis of the insect cell-produced penton base protein indicated that the recombinant protein possessed the same mobility on SDS-gels (85 kd) as the native protein produced by Ad2 infection of mammalian cells (Figure 1). High-resolution transmission electron microscopic studies and ultracentrifugation analysis also indicated that the recombinant protein is pentameric.

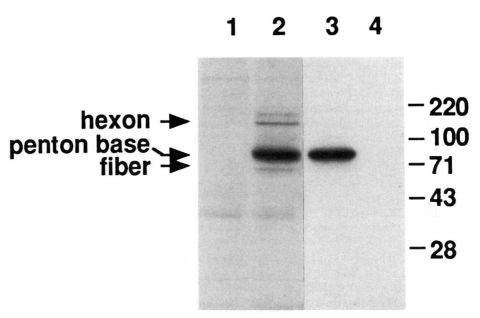

Figure 1. Immunoprecipitation of baculovirus-produced Ad2 penton base. ^{35}S-labeled cell lysates from uninfected (lane 1) or Ad2-infected HeLa cells (lane 2) or from Sf9 insect cells infected with recombinant (lane 3) or wild-type baculovirus (lane 4) were immunoprecipitated with polyvalent anti-penton base antiserum followed by analysis on 7% SDS-gels and autoradiography.

Penton base binds to cell surface integrins and is involved in Ad2 infection

We next examined the possibility that recombinant penton base binds to cell surface integrins on human epithelial cells using two types of cell adhesion assays. In the first assay, we examined whether soluble penton base could compete for matrix proteins in tissue culture media and thereby detach cell monolayers. As shown in Figure 2, addition of soluble penton base to confluent monolayers of HeLa epithelial cells caused the cells to round up and eventually detach from the plastic tissue culture surface.

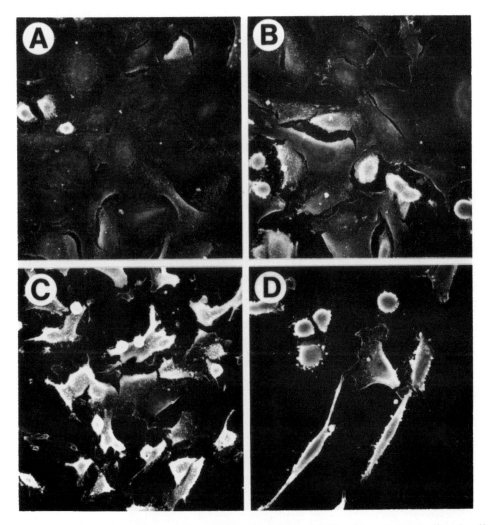

Figure 2. Scanning electron microscopic detection of Ad2 penton base-mediated cell detachment. A549 human epithelial cells were incubated with a control protein (A) or with recombinant penton base for 5 min (B), 30 min (C), 60 min (D).

This result suggested that penton base could compete for cell adhesion molecules by binding to cell integrins. To further address this possibility, we asked whether cells would attach and/or spread on tissue culture wells precoated with penton base. As shown in Figure 3, human epithelial cells attached to wells coated with penton base or with the cell matrix proteins, vitronectin, fibronectin, laminin and collagen but not to a control protein BSA. Cell adhesion to penton base was blocked by soluble RGD peptides and by chelation of divalent metal cations (EDTA) (14). These results strongly indicated that penton base binds to cell surface integrins.

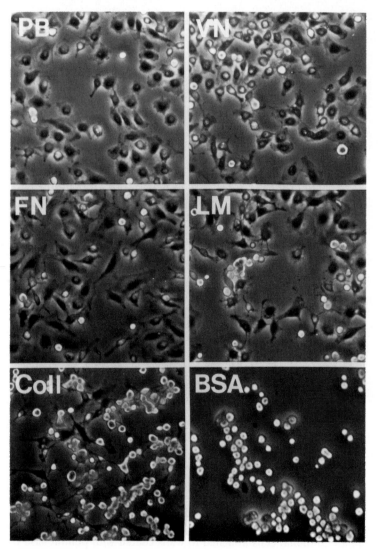

Figure 3. Cell adhesion to recombinant penton base (PB), vitronectin (VN), fibronectin (FN), laminin (LM), collagen (coll) or an irrelevant protein (BSA).

Further studies were performed to determine whether penton base binding to cell integrins was important for adenovirus infection. These studies showed that cells that adhered to vitronectin but not to collagen were less susceptible to Ad2 infection, suggesting that occupancy of vitronectin-binding integrins by cell matrix proteins blocked virus utilization of these cell surface receptors. This hypothesis was further supported by studies showing that soluble RGD but not RGE peptides blocked penton base binding to cells (Figure 4). Soluble penton base or RGD peptides were also capable of causing dose-dependent inhibition of adenovirus infection (14).

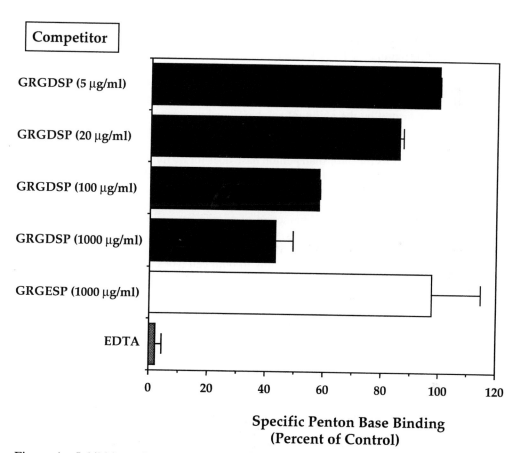

Figure 4. Inhibition of recombinant penton base to HeLa cells by synthetic RGD peptides and EDTA. HeLa cells were incubated with peptides or EDTA prior to the addition of ^{35}S-labeled penton base.

Ad2 penton base binding to αv integrins is required for efficient virus infection

Studies were next performed to identify the cell integrins to which penton base binds. Since cells adhered to vitronectin and/or fibronectin were less susceptible to Ad infection, this suggested that αv integrins such as αvß3 and αvß5 were involved in penton base binding. To address this possibility we examined adenovirus interaction with a defined pair of human melanoma cell lines (M21) which either lack αv integrins (M21-L12) or express αvß3 and αvß5 following transfection with αv cDNA (M21-L4) (15). M21-L4 but not M21-L12 cells adhered to immobilized penton base (14). M21-L4 cell binding to penton base was blocked by RGD peptides and also by a combination of function blocking monoclonal antibodies to αvß3 and αvß5. These studies indicate that αv integrins represent the primary integrin recognized by adenovirus penton base.

Further studies were performed to investigate the role of αv integrins in adenovirus infection. M21-L4 cells were 5-10 fold more susceptible to adenovirus infection than M21-L12 cells indicating that expression of αv integrins is required for efficient adenovirus infection.

αv integrins mediate adenovirus internalization but not attachment

High-affinity virus binding to cell receptors is mediated by the adenovirus fiber coat protein as indicated by the ability of soluble fiber but not penton base to block virion absorption to cells (14). Moreover, Scatchard analyses of the saturation binding data performed with purified fiber and penton base indicate that the fiber possesses a 30-fold higher affinity for its receptor than penton base does for αv integrins (14). These studies strongly indicate that the penton base interaction with αv integrins is not required for the initial high-affinity interaction of Ad2 with cells.

To determine whether the interaction of penton base with αv integrins played an important role in subsequent steps in virus entry, we examined internalization of radiolabeled Ad2 in αv-expressing M21-L4 cells or in M21-L12 cells lacking these integrins. Virus particles bound efficiently to both cell types. In contrast, virions were rapidly internalized into M21-L4 cells but not into M21-L12 cells. These studies indicate that penton base interaction with αv integrins is required for virus uptake into cells. Purified labeled fiber, in contrast to penton base, was also not internalized into cells, further indicating that the fiber receptor is unable to mediate virus uptake.

DISCUSSION

A number of cell surface integrins have recently been identified as participating in the entry of microorganisms into host cells. For example, the collagen/laminin-binding integrin, α2ß1 has been identified as a receptor for echovirus 1 attachment (16). Certain bacteria including *Yersinia* (17) and *Bordetella pertussis* (18), have been shown to bind to and internalize into host cells via ß1 and ß2 integrins, respectively. An RGD integrin-binding motif which is present in two picornaviruses, foot-and-mouth disease virus (19) and coxsackievirus A9 (20), has also been implicated in attachment to cell integrins based on the ability of RGD peptides to inhibit infection.

In these and previous studies (14), we have demonstrated that the vitronectin-binding integrins αvß3 and αvß5 play a crucial role in the uptake of adenovirus type 2 into host cells following high-affinity virus binding to the fiber receptor. The identification of αv integrins as receptors for virus uptake represents a unique example of a nonenveloped virus that uses a distinct cell receptor for internalization rather than for attachment. Certain integrins have been previously shown to be rapidly internalized into cells (21). Relatively little is known of the cell trafficking of αv integrins. The ß subunit of αvß3 and αvß5 contains an NPXY sequence known to be involved in the ligand-stimulated endocytic pathway of other receptors, suggesting that this sequence may be involved in αv integrin uptake into cells (22).

Adenovirus uptake into cells is mediated by the pentameric complex of polypeptide III (penton base), each of which contains 5 copies of an RGD sequence. This peptide sequence is responsible for virus attachment to αvß3 and αvß5, since synthetic peptides containing RGD block cell adhesion to the penton base and also inhibit virus uptake and infection. A question arising from these studies is whether any of the other 41 known serotypes of adenovirus also use the penton base RGD sequence to mediate virus uptake. Although the answer to this question is still unknown, at least one other major subgroup of human adenovirus, Ad3, appears to use this motif since we have found that Ad3 infects αv integrin-deficient cells less efficiently than αv-expressing cells. Whether other serotypes of adenovirus also use cell integrins for entry into cells remains to be determined; however, Wadell and Norbby (23) have noted that many adenovirus serotypes possess the ability to detach cells in culture suggesting that the RGD/penton base protein may be highly conserved.

The possibility that other integrins may also participate in Ad2 infection also needs to be considered. In this regard, we have observed a low level of infection of the αv-deficient cell line as well as a low level of adhesion to penton base coated surfaces (14). This low-level infection/adhesion was blocked by a monoclonal antibody to ß1 integrins, suggesting that fibronectin receptors such as α3ß1 or α5ß1 may also mediate virus internalization to a limited extent.

The identification of αv integrins as receptors for adenovirus internalization may represent a potential new target for antiviral therapy against numerous adenovirus serotypes. Cyclic RGD peptides that selectively block αv integrins but not other integrins (24) may represent such compounds. In addition, the elucidation of the precise molecular events involved in virus entry should facilitate the use of adenovirus as a vector for gene therapy (25).

ACKNOWLEDGEMENTS
The authors express their gratitude to Patricia Mathias for excellent technical assistance and to Catalina Hope for preparation of the manuscript. This work was supported in part by a NIH postdoctoral training grant AI07354.

184

REFERENCES

1. Nemerow GR, Cooper NR. Sem Virol 1992; 3: 117-124.
2. Horwitz M (1990) The adenoviridae and their replication. In: Fields B, Knipe D (eds) Virology. Vol.2. Raven Press, New York, pp 1679-1721
3. Brandt CD, Kim HW, Vargosko AJ, Jeffries BC, Arrobio JO, Rindge B, Parrott RH, Chanock RM. Am J Epidemiol 1969; 90: 484-500.
4. Chardonnet Y, Dales S. Virology 1970; 40: 462-477.
5. Fitzgerald DJP, Padmanabhan R, Pastan I, Willingham MC. Cell 1983; 32: 607-617.
6. Hennache B, Boulanger P. Biochem J 1977; 166: 237-247.
7. van Oostrum J, Burnett RM. J Virol 1985; 56: 439-448.
8. Philipson L, Lonberg-Holm K, Pettersson U. J Virol 1968; 2: 1064-1075.
9. Seth P, Fitzgerald D, Ginsberg H, Willingham M, Pastan I. Mol Cell Biol 1984; 4: 1528-1533.
10. Neumann R, Chroboczek J, Jacrot B. Gene 1988; 69: 153-157.
11. Hynes RO. Cell 1992; 69: 11-25.
12. Everett SF, Ginsberg HS. Virology 1958; 6: 770-771.
13. Boudin M-L, Moncany M, D'Halluin J-C, Boulanger PA. Virology 1979; 92: 125-138.
14. Wickham TJ, Mathias P, Cheresh DA, Nemerow GR. Cell 1993; 73: 303-313.
15. Felding-Habermann B, Mueller BM, Romerdahl CA, Cheresh DA. J Clin Invest 1992; 89: 1-5.
16. Bergelson JM, Shepley MP, Chan BMC, Hemler ME, Finberg RW. Science 1992; 255: 1718-1720.
17. Isberg RR. Science 1991; 252: 934-938.
18. Relman D, Tuomanen E, Falkow S, Golenbock DT, Saukkonen K, Wright SD. Cell 1990; 61: 1375-1382.
19. Fox G, Parry NR, Barnett PV, McGinn B, Rowlands DJ, Brown F. J Gen Virol 1989; 70: 625-637.
20. Roivaninen M, Hyypiä T, Piirainen L, Kalkkinen N, Stanway G, Hovi T. J Virol 1991; 65: 4735-4740.
21. Bretscher MS. EMBO J 1992; 11: 405-410.
22. Chen W-J, Goldstein JL, Brown MS. J Biol Chem 1990; 265: 3116-3123.
23. Wadell G, Norrby E. J Virol 1969; 4: 671-680.
24. Aumailley M, Gurrath M, Müller G, Calvete J, Timpl R, Kessler H. FEBS Lett 1991; 291: 50-54.
25. Curiel DT, Agarwal S, Wagner E, Cotten M. Proc Natl Acad Sci USA 1991; 88: 8850-8854.

Extracellular matrix interactions and alterations

Biology of Vitronectins and their Receptors
K.T. Preissner, S. Rosenblatt, C. Kost, J. Wegerhoff and D.F. Mosher, editors   187

Drosophila Development and Cellular Adhesion

J.H. Fessler[a], F.J. Fogerty[a], R.E. Nelson[a], D. Gullberg[b], T. Bunch[c] and L.I. Fessler[a]

[a]Molecular Biology Institute and Biology Department, UCLA, Los Angeles CA 90024-1570, USA

[b]current address: Department of Medical and Physiological Chemistry, BMC Box 575, University of Uppsala, S-751 23 Uppsala, Sweden

[c]Department of Molecular and Cellular Biology, University of Arizona, Tucson, AZ 85721, USA

INTRODUCTION AND OVERVIEW

The proteins of vertebrate extracellular space are partitioned into blood proteins and those of the interstitial space. In contrast, *Drosophila melanogaster*, a genetically most useful animal, has an open circulatory system and a joined space called the hemocoel. This cavity is filled with hemolymph, whose cells and proteins serve dual purposes and are in direct contact with the extracellular matrix components that line this space.

In the classical physiological sense of Claude Bernard, the interstitial fluid provides the *milieu intérieur* of cells, and one might expect that some of its key aspects may be evolutionarily shared between man and flies. There is a shared need to stop fluid loss at the site of accidental lesions and for protection against bacterial or fungal infections. Both organisms require a specialized, permeable form of extracellular matrix, called a basal lamella or basement membrane, to which epithelia and some other cells can attach. Portions of the cytoskeleton of these cells thereby become orientated and anchored to the immediate cellular environment.

Some key proteins of basal lamella have been evolutionarily conserved, in particular basement membrane collagen IV and laminin. The principal cells of hemolymph, hemocytes, are prime producers of these two materials and of several other extracellular matrix (ECM) components [1,2]. While the *Drosophila* equivalents of fibronectin and vitronectin remain elusive, their corresponding receptors, integrins, are clearly established in *Drosophila* at mechanically strategic sites.

Beyond maintaining the *status quo* of the internal environment, lies the interesting challenge of the dynamic role of ECM during the changes of differentiation and development. We have studied the differentiation of primary *Drosophila* embryo cells cultured in defined media on substrates of purified ECM components. The differentiation of *Drosophila* myocytes involves different cytoskeletal intermediate states when the cells are attached to a substrate of *Drosophila* laminin as opposed to one of vertebrate vitronectin. The integrins of some *Drosophila* cells bind vertebrate vitronectin well.

Differentiating hemocytes sequentially make, and deposit, several basement membrane components. In whole mounts of successive stages of embryos special

ECM deposits appear at sites of invagination of cell layers, and hemocytes are among the first identifiable mesodermal cells. Hemocytes also have other functions: they phagocytose specific embryonic cells that have undergone programmed cell death, also called apoptosis. At the same time they synthesize and secrete a novel protein X. Its sequence has homology with vertebrate peroxidases, and contains additional motifs found in ECM and CAM proteins. The isolated protein has peroxidase activity. At later stages hemocytes can also surround, and either engulf or segregate, foreign particulate materials.

In summary, not only amino acid sequence homologies, but also parallels in cellular functions, show that useful comparisons can be made between the superficially quite different *Drosophila* and man, and, hopefully, highlight some basic biological problems.

MUSCLE INSERTIONS

The transmission of contractile force from striated muscle cells to their points of attachment is through specialized ECM structures in both vertebrates and *Drosophila*. The body plan of *Drosophila* is subdivided into a succession of segments from the anterior to the posterior end. Attachment sites for the body wall muscles of the embryo and larva are provided at segmental intervals. Each anchorage site, or apodeme, is formed by a specialized epidermal cell, called a tendon cell, that is connected to the overlying cuticular exoskeleton. Several longitudinal and oblique muscles are anchored to such a tendon cell. The connecting processes, or insertions, transmit the contractile force from the muscle cells to the tendon cell. This requires effective cell adhesion between the component parts. The insertions are specializations of the basement membranes that envelop the myotubes and contain several of the ECM proteins that we isolated from *Drosophila*.

The insertions stain particularly well with antibodies against a new ECM protein and form a pattern of stripes that correspond to the segmental repeats of the apodemes in embryos. We suggest the name Tiggrin for this protein, in view of a striped, tiger-like mythical animal named Tigger that is prone to erratic muscular contractions, readily makes contact with a variety of materials, but has highly selective taste buds that greatly restrict its intimate association with foods and confine its dietary intake to strength-promoting extracts [3].

The apposing muscle and tendon cells at each of the myo-epithelial junctions contain a different *Drosophila* integrin: $\alpha_{PS2}\beta_{PS3}$ at the muscle cell surface and $\alpha_{PS1}\beta_{PS3}$ in the tendon anchor cell [4]. In experiments, that will be described in detail elsewhere, we demonstrated that tiggrin is a ligand for the $\alpha_{PS2}\beta_{PS3}$ integrin.

Among the cells that differentiate in primary cell cultures prepared from *Drosophila* embryos there are clusters of epidermal cells that stain with antibodies against tiggrin and against $\alpha_{PS1}\beta_{PS3}$, but do not contain $\alpha_{PS2}\beta_{PS3}$. This makes it likely that tiggrin also binds to $\alpha_{PS1}\beta_{PS3}$.

We suggest that tiggrin is a key component of muscle insertions and helps to connect muscle cells bearing $\alpha_{PS2}\beta_{PS3}$ integrin with the $\alpha_{PS1}\beta_{PS3}$ receptors of tendon cells. Antibodies to tiggrin also stain the basement membranes and insertions of larval skeletal muscles. In sections of pupal muscles tiggrin and $\alpha_{PS2}\beta_{PS3}$ integrin locate immunologically to the muscle Z-bands. Thus tiggrin participates in several aspects of *Drosophila* muscle structure. However, tiggrin is not confined to muscles, e.g. it also occurs in the basement membrane under the epidermal cells which form the insect's cuticle.

The sequence of tiggrin's nearly 2200 amino acids contains a single RGD sequence near its COOH end, and this functions in tiggrin's binding to integrin. There is only one Cys residue, near the NH_2 end of tiggrin. The central portion of tiggrin, some 1000 amino acid residues, is a contiguous set of 14 copies of a novel motif. The 72-75 amino acids sequence of this motif indicates a marked likelihood for α-helical folding and coiled-coil formation of amphipathic α-helices. Preliminary biophysical calculations are consistent with some form of elongated, thread-like molecule. A distinctly asymmetric molecule was indicated by the frictional ratio of tiggrin, which was calculated from its sedimentation coefficient and its amino acid sequence molecular weight. It will be interesting to learn more about the supramolecular associations of tiggrin with itself and with neighboring molecules of the *Drosophila* extracellular matrix.

CELL CULTURE AND CELL CULTURE SUBSTRATES

Embryonic *Drosophila* cells, obtained by dissociation of late gastrulae, are readily cultured on a substrate of *Drosophila* laminin [5]. The cells spread and differentiate on this substrate, and begin to secrete ECM components [6]. If the concentration of these newly made ECM components is kept low by changing the culture medium, then the differentiation of myocytes proceeds as far as fused myotubes. This process proceeds equally well with cells that genetically lack $\alpha_{PS2}\beta_{PS3}$ integrin, and were obtained from *lethal (1) myospheroid* mutants that can not synthesize the β_{PS3} chain. Thus adhesion of *Drosophila* cells to laminin does not require integrins. However, the further differentiation of myotubes to a striated, sarcomeric form does require integrin, and fails to proceed in *myospheroid* cell cultures or embryos.

Vertebrate vitronectin also acts as an excellent substrate for the culture and differentiation of *Drosophila* myocytes, and for some other cell types, though the spectrum of cells is narrower than for *Drosophila* laminin [6]. In contrast to culture on laminin, cell spreading and differentiation on vertebrate vitronectin requires the presence of $\alpha_{PS2}\beta_{PS3}$ integrin, and does not occur with the mutant, *myospheroid* cells. Furthermore, Bunch and Brower [7] showed that *Drosophila* S2 cells, which normally express very little $\alpha_{PS2}\beta_{PS3}$ integrin and fail to spread on vitronectin, do spread on this substrate after transfection with DNA coding for this pair of integrin chains and expression of the integrin.

We now find that tiggrin is also an excellent substrate for *Drosophila* myocyte differentiation, consistent with its ability to bind to $\alpha_{PS2}\beta_{PS3}$ integrin. Spreading of *Drosophila* embryo cells on either a vitronectin or a tiggrin substrate is accompanied by the formation of focal adhesions rich in $\alpha_{PS2}\beta_{PS3}$ integrin. In contrast, these focal adhesions and their accompanying cytoskeletal stress fibers do not appear when *Drosophila* cells spread on laminin. The above S2 cells that have been transfected with DNA constructs to express $\alpha_{PS2}\beta_{PS3}$ integrin also spread on a substrate of tiggrin. This spreading is inhibited by addition of an RGD peptide.

Prominent among the cells that differentiate in primary cultures from *Drosophila* embryos is a type of large cell which corresponds to the embryonic hemocytes. These cells manufacture and secrete, in culture and in the embryo, tiggrin, collagen IV, laminin, the proteoglycan papilin and another new protein provisionally named protein X.

PROTEIN X AND HEMOCYTES

As will be described in detail elsewhere, protein X occurs in the conditioned cell culture media of some *Drosophila* cell lines, from which it was purified. Antibodies made against this material stained hemocytes, fat body cells, and portions of basement membrane. The protein is a disulfide-linked homotrimer of a 170 kd polypeptide. The cDNA sequence that codes for it was determined. It suggests an interesting hybrid molecule.

The central portion has strong homology with the family of vertebrate peroxidases [8]. To the NH_2-side of this portion is a region that contains four motifs of the super- immunoglobulin family, with some similarity to IgA. Closer to the NH_2-end is a region containing six leucine-rich proteins. Near each end of the translated amino acid sequence are a number of cysteine residues.

Micro-amino acid sequencing of portions of the isolated protein proved correspondence of amino acid and cDNA sequences. The purified, isolated protein X has H_2O_2-dependent peroxidase activity and it oxidizes standard substrates for peroxidases. Its absorption spectrum is consistent with it containing a heme group. When supplied with H_2O_2 and $Na^{125}I$ it readily radio-iodinates added proteins.

A biological question is the function of this peroxidase. Protein X is the only ECM component that we can identify by immunostaining of embryos that is not extractable with detergents after reduction. Biosynthetic studies with cultured cells indicate that it is not simply a cytoplasmic enzyme that appears in the culture medium due to cell lysis. Its ability to oxidize tyrosine to dityrosine suggests that protein X may stabilize ECM by forming tyrosine-tyrosine cross-bridges.

Partial hydrolysis of embryos leads to release of protein fragments which react with antibodies to protein X. This suggests that protein X may cross-link itself into the adjacent extracellular matrix. The leucine-rich and Ig-like regions of protein X might direct its interaction with a variety of ECM components, and these interactions might be subsequently stabilized by oxidatively formed cross-bridges.

Thus protein X seems to be a matrix component that brings with itself the enzymatic ability for cross-linking the matrix into which it incorporates. Taking this speculation a step further, protein X might also be an important component produced by hemocytes to wall off foreign objects.

The myeloperoxidases of vertebrate white cells are associated with the killing and removal of pathogens and damaged cells. Protein X is also likely to participate in such functions. Programmed cell death, apoptosis, occurs early in embryogenesis and is important in removing supernumerary neurons and other cells. In the head region of *Drosophila* embryos hemocytes that immunostain for protein X arise early during mesoderm formation and later spread to the remainder of the embryo. Staining with acridine orange, or toluidine blue, shows that many of these hemocytes are associated with dead cells which they engulf [9]. Thus protein X may serve several functions. The hemocytes that manufacture protein X are also versatile, as double immunostaining shows that the same cells also elaborate other ECM components at different times.

Controlled destruction, as well as synthesis, is important during embryogenesis. Not only cells, but also previously used ECM needs to be removed or modified. Vertebrate collagenases were originally sought for such an action. Recently it was found that the metamorphosis of *Drosophila* from larva through pupa to adult, is associated with cleavage of basement membranes at specific, strategically located sites [10]. We regard the extracellular matrix of *Drosophila* as a dynamic rather than a static structure, exhibiting problems of synthesis, modification such as cross-linking, and controlled degradation.

In this brief selection we have not mentioned accompanying studies at the mRNA and gene levels, with ongoing work on modifying ECM production and properties with the help of mutants and recombinant DNA methodologies. The persistent, and sometimes unexpected, homologies of *Drosophila* and vertebrate macromolecules remind one that this interplay of cells with their immediate ECM environment seems to be a fundamental aspect of metazoan existence.

ACKNOWLEDGMENTS

We thank our colleagues for valuable discussion and advice, and gratefully acknowledge support from the Muscular Dystrophy Association and the US Public Health Service, grant AG-02128. D.G. held a postdoctoral fellowship of the Swedish Natural Science Research Council and F.J.F. was supported by the National Research Service Awards 5-T-32-CA09056-16 and 5-F32-HD07476-02.

REFERENCES

1 Fessler JH, Fessler LI. Ann. Rev. Cell Biol. 1989: 5: 309-339.
2 Hortsch M, Goodman CS. Ann. Rev. Cell Biol. 1991: 7: 505-557.
3 Milne AA, House at Pooh Corner, London, Methuen, 1928; 19-35.
4 Leptin M, Bogaert T, Lehman R, Wilcox M. Cell 1989: 56: 401-408.
5 Volk T, Fessler LI, Fessler JH. Cell 1990: 63: 525-36.
6 Gullberg D, Fessler LI, Fessler JH submitted for publication
7 Bunch T, Brower D. Development 1992: 116: 239-247.
8 Kimura S, Ikeda-Saito M. Proteins 1988: 3: 113-120.
9 Abrams JM, White K, Fessler LI, Steller H. 1993: 117: 29-43.
10 Fessler LI, Condic ML, Nelson RE, Fessler JH, Fristrom JW. Develop. 1993: 117: 1061-1069.

Biology of Vitronectins and their Receptors
K.T. Preissner, S. Rosenblatt, C. Kost, J. Wegerhoff and D.F. Mosher, editors

EPISIALIN MODULATES CELL-CELL AND CELL-MATRIX ADHESION, PROMOTES INVASION IN MATRIGEL AND INHIBITS CYTOLYSIS BY CYTOTOXIC EFFECTOR CELLS.

J. Hilkens[a], J. Wesseling[a], H.L. Vos[a], S.L. Litvinov[a], M. Boer[a], S. van der Valk[a], J. Calafat[a], E. van de Wiel van Kemenade[b] and C. Figdor[b].

[a]Divisions of Tumor Biology and [b]Immunology, The Netherlands Cancer Institute (Antoni van Leeuwenhoekhuis), Plesmanlaan 121, 1066 CX Amsterdam, The Netherlands.

Introduction

Metastasis is a complex multi-step process in which cell surface molecules, in particular cellular adhesion molecules, are involved (1). Adhesion may either decrease or increase during metastasis. An increasing number of molecules mediating cell-cell and cell-extracellular matrix (ECM) adhesion have been identified that might play a role in this process, but to relate these molecules in a simple model to metastasis appears to be very difficult. During certain stages of the metastatic process adhesion should be d iminished (in particular during the initial stages when the cells disseminate from the primary tumor), whereas in other stages firm adhesion may be necessary (to attach to the vascular endothelium). For invading the tissue parenchyma cell motility should be high, which requires low affinity interactions of the adhesion receptors with the ECM and probably specialized linkage of the receptors with the cytoskeleton.

Two types of adhesion can be distinguished: cell-matrix and cell-cell adhesion. Integrins are the most important adhesion receptors that mediate cell-matrix interactions. Changes in expression level, affinity and type of integrin accompany metastasis. Cadherins are the main intercellular adhesion molecules and play an important role in tissue morphology and cellular sorting out. E-cadherin is mainly present in epithelial tissues, and has been extensively reviewed (2). Other cadherins have been identified in, for instance, liver and neural tissues. E-cadherin is considered as a metastasis suppressor molecule since lowering the effective amount of this molecule is one of the steps that can increase invasion (3, 4, 5).

Here, we show that episialin, a cell membrane mucin-like glycoprotein (also designated EMA, PEM, CA 15-3, etc.), has anti-adhesive properties and strongly decreases cell-cell and cell-matrix interactions. Overexpression of this molecule can have a similar effect as down regulation of adhesion receptors. Consequently, the molecule has a strong effect on adhesion mediated cellular properties such as invasion in gels of ECM components and may be involved in tumor progression.

The structure of episialin

Episialin is a heavily glycosylated molecule with an apparent molecular mass of over 400 kDa encoded by the MUC1 gene. The molecule has been classified as mucin-like, since more than half of the molecule consists of O-linked glycans. It is a transmembrane molecule with a relatively large extracellular domain and a cytoplasmic domain of 69 amino acids. The extracellular domain consists mainly of a region of nearly identical repeats of 20 amino acids. As a result of genetic polymorphism the number of repeats, and thus the size of the molecule, is highly variable in the human population. The repeats, together with adjacent degenerated repeats, contain many potential attachment sites for O-linked glycans and constitute the mucin-like domain which comprises more than half of the protein backbone.

The mucin domain of episialin contains many prolines and other helix-breaking amino acids resulting in molecules with an extended structure and many β-turns (6). The extended structure becomes very rigid by the addition of numerous sialylated O-linked glycans to the molecule. As a result, the mucin domain of episialin reaches an extreme length. According to Jentoft (6), an extensively O-linked glycosylated polypeptide of 28 amino acids is approximately 7 nm long. This means that the mucin-like domain of episialin extends at least 200 to 500 nm above the cell membrane. A model of the molecule is shown in Fig. 1. Electron microscopical studies on purified episialin revealed that it appears indeed as a long thread-like structure as predicted. Episialin also shows this structure at the plasma membrane of mammary tumor cells, in particular the microvilli are covered by elongated structures similar to the ones present in the purified episialin preparations.

Localization of episialin in normal tissues

Episialin is mainly localized at the apical side of glandular epithelial cells as determined by immunohistology. We also observed a strong expression of episialin in the canaliculi of for instance the salivary gland and the parietal cells of the stomach; canaliculi are minute secretory channels which can extend deep into the cytoplasm of the cell and are connected with the extracellular environment. These channels usually contain numerous short microvilli. Immunohistochemical studies using various monoclonal antibodies (mAbs) revealed that episialin is also present at the luminal surfaces of certain cell layers lining other body cavities such as the mesothelium. Moreover, certain hematopoietic cells show episialin expression: a small percentage of the plasma cells as detected by mAbs, and eosinophilic granulocytes as detected by *in situ* hybridization.

The expression level of episialin is enhanced in cancer cells

The expression level of episialin in breast and colon carcinomas is often higher than in the normal epithelia. We have shown that at least 3 different epitopes of episialin are not or only at a very low level present in normal colonic epithelium and low dysplastic adenomas. However, in 61% of the severely dysplastic adenomas and cells with premalignant cellular changes, and in almost all carcinomas of the

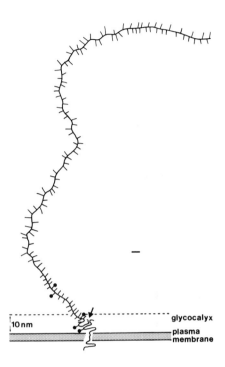

Figure 1. Proposed structure of a mature episialin molecule containing about 40 repeats. The vertical bar indicates the approximate length of one glycosylated repeat (5 nm, 20 amino acids). The symbols ⟍ and | indicate N- and O-linked carbohydrates, respectively. The precise number of O-linked glycans is not known; the depicted number of O-linked glycans is only a very rough estimate. The arrow indicates the site at which the precursor is cleaved in the endoplasmic reticulum. However, both fragments remain non-covalently associated (7). For comparison, the thickness of the glycocalyx is indicated (10 nm). Reproduced from (8).

colon more of these epitopes were present, suggesting that the expression of episialin in these tissues is increased (9). Episialin is also expressed in a number of non-epithelial cancers such as 30% of the sarcomas and brain tumors, and 10% of the lymphomas, whereas none of the tissues of origin of these cancers showed any episialin expression by immunohistochemical means. Apparently, episialin expression is stimulated in these cancers during tumorigenesis. However, the enzymatic detection methods commonly used in immunohistology, do not allow to quantitatify the expression level. Moreover, due to differential glycosylation of episialin in normal and tumor tissues mAb staining does not necessarily represent the true expression level. Therefore, we have quantified the apparent overexpression of episialin in carcinomas at the mRNA level by *in situ* hybridization using an antisense RNA probe representing a non-repetitive region of the episialin cDNA. Episialin mRNA expression in almost all breast carcinomas is at least ten times higher than in adjacent normal glandular epithelium.

It should be stressed that episialin is mainly present at the apical surface of normal glandular epithelial cells, whereas in carcinoma cells episialin is present at the entire cell surface including the plasma membrane facing the stroma or adjacent cells as a result of the absence of a normal polarized architecture.

Episialin expression and cellular adhesion

To investigate whether overexpression of this very long and rigid negatively charged molecule affects cellular adhesion we have transfected episialin cDNA under the control of the CMV promotor into A375 melanoma cells and HBL-100 cells (SV-40 immortalized normal breast cells having a very low endogeneous episialin expression). In addition, we generated episialin-negative revertant cells of these transfected clones by bulk selection using the cell sorter. Bulk selection was chosen over cloning to avoid clonal selection of cell types that accidentally have altered adhesion properties.

Surprisingly, a proportion of the transfected cells of several clones that express episialin to levels also found on carcinoma cell lines grow in suspension. To investigate this, we have tested the capacity of several transfectants and revertants of the same clones to bind to ECM components (fibronectin, laminin, collagen I and collagen IV). Fig. 2 shows the adhesion to these molecules of ACA 19$^+$ (one of the A375 transfectants) and ACA 19$^-$ (the episialin negative revertant) as an example. The adhesion capacity of these and other transfectants is much lower than that of the revertants, indicating that episialin expression can reduce cell-matrix interactions. Since the adhesion of the revertants as well as the residual adhesion of the transfectants to these ECM components can be blocked by mAbs directed against the β1 and β3 chain of the integrins, we conclude that episialin directly interferes with the binding of these integrins to the ECM. A similar but less pronounced effect was also observed with HBL-100 transfectants.

We have also compared capacity of the transfectants to aggregate homotypically with that of their parental cells or revertants that have lost episialin expression and found that cell surface episialin inhibits cellular aggregation (10). The presence of episialin on only one of a pair of aggregating cells is already sufficient to interfere with this process. E-cadherin is the most important molecule on glandular epithelia that mediates homotypic adhesion. Therefore, we have investigated whether episialin could abrogate E-cadherin mediated adhesion. Mouse L cell double transfectants expressing both E-cadherin and episialin showed that episialin could indeed prevent intercellular adhesion mediated by E-cadherin.

The anti-adhesion properties of episialin are probably the result of the unique extended and rigid structure of the molecule. As we have discussed above, episialin is towering 200-500 nm above the plasma membrane whereas most proteins at the cell surface remain inside the boundaries of the glycocalyx which is approximately 10 nm thick. Therefore, we postulate that episialin present at high

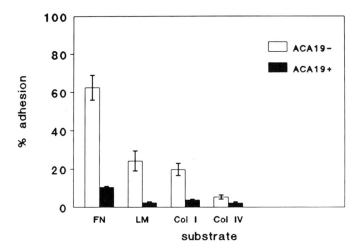

Figure 2. Adhesion of ACA19[+] transfectant and its revertant (ACA19[−]) on ECM components. Microtiter plates were coated with varies ECM components at concentrations that provided optimal binding of the cells. [51]Cr labeled cells were allowed to adhere to the wells for 30 min. Non-adherend cells were removed and the radioactivity in each well, representing the adherend cells, was counted. The percentage of the total amount of the added cells is indicated. FN, fibronectin; LM, Laminin; Col, II collagen II; Col IV, collagen IV.

density, as we observed in many carcinomas, prevents cellular adhesion by masking adhesion molecules, and thus can severely disturb the interaction of cell surface proteins with macromolecules on adjacent cell membranes. In addition to its size, the abundance of sialic acid residues contributes to some extent to the anti-adhesion properties of episialin (10). Since sialic acid residues result in a strongly negative charge of the glycoprotein, the anti-adhesion effect could, at least in part, also be caused by charge repulsion. Alternatively, the sialic acids could merely contribute to the rigidity of episialin. Anti-adhesion effects, induced by charge repulsion, have been postulated before. For instance, the polysialylated embryonic form of N-CAM modulates adhesion of certain malignant cells and of neural cells during embryogenesis by charge repulsion (11).

The adhesion studies described above were carried out with tissue culture cell lines that do not grow in a polarized fashion. On normal differentiated glandular epithelial cells, episialin is only expressed at the apical side. Therefore, episialin can normally not interfere with adhesion molecules, which are only present at the basolateral side of the cell. At the apical side however, episialin will prevent interactions between glycoproteins on opposite membranes, which could cause adhesion of the apical membranes. In this way, episialin might facilitate the formation and maintenance of the lumen present in glandular structures and other tissues.

Episialin expression on target cells affects susceptibility to allogeneic stimulated T cell and LAK cell mediated cytolysis.

Because episialin prevents cellular aggregation, we assumed that the molecule may also prevent conjugate formation between cytotoxic effector cells and the target cell. We first investigated the capacity of episialin transfected A375 melanoma cells and revertants to form conjugates with allogeneic T lymphocytes, stimulated with A375 cells (CTL) and rIL-2 activated large lymphocytes (LAK cells). A high percentage of conjugates (60%) was formed between the allogeneically stimulated lymphocytes and parental A375 cells, whereas the percentage conjugates with episialin positive cells was significantly diminished. The episialin transfected melanoma cells formed hardly any conjugates with the LAK cells, whereas 20% of the reverted cells formed conjugates with these large lymphocytes.

A375 cells can be lysed by allogeneically stimulated CTL and by LAK cells. We have measured the lysis of episialin transfected A375 melanoma cells and episialin-negative revertants by the CTL and LAK cells with time in a ^{51}Cr release assay. The CTLs lysed episialin positive A375 transfected cells only at a very slow rate, whereas the revertants were efficiently killed. The kinetics of lysis of episialin-negative cells by the LAK cells was comparable to that of a standard target of LAK cells, whereas lysis of the episialin-positive A375 transfectants was much slower. Maximal lysis of episialin-negative target cells was reached after 120 min incubation, whereas the maximal lysis of the episialin-positive transfectants was reached only after 180 min. The differences between the rate of lysis of episialin-positive transfectants and episialin-negative control cells became even larger when LFA-1/ICAM1 adhesion pathway was blocked (12).

From these results we can conclude that episialin transfected cells are less susceptible to lysis. The effect of episialin in immune reactions proves that episialin expression has important implications for the cell that go beyond just inhibition of intercellular adhesion.

Episialin and tumor progression

The episialin levels found in transfectants that were not aggregating were similar to those in many carcinoma cells. Therefore, the anti-adhesion effect of episialin as observed with the transfectants is likely to occur also *in vivo*. As long as the carcinoma cells remain polarized, episialin would probably not have a major effect on the biological properties of the cell. However, in many adenocarcinomas the polarization is lost and the molecule is also found at those parts of the plasma membrane that are facing the stroma and adjacent cells where it can interfere with adhesion. Moreover, in at least some types of carcinomas the expression of episialin is at least tenfold higher than in normal cells. Electron microscopy studies of thin sections of breast carcinomas showed that in the regions of the plasma membrane where the molecule is abundantly present, the adjacent membranes make no direct contacts (13, Calafat, Jansen and Hilkens, manuscript in preparation). This implies that the presence of episialin destabilizes cell-cell interactions also *in vivo* and may be an important cofactor in metastasis.

Indeed, an increasing amount of evidence indicates that lowering the effective amount of the homotypic adhesion molecule E-cadherin is one of the steps that can increase invasion (3, 4, 5). To investigate whether episialin indeed promotes invasion we have seeded aggregates of episialin transfected HBL-100 cells in Matrigel (a gel consisting of the major ECM components). Transfected cells were able to invade the Matrigel while the episialin revertants and the parental cells could not invade the gel, suggesting that episialin indeed promotes invasion, possibly by lowering the E-cadherin mediated cell-cell adhesion.

Hence episialin overexpression has a similar effect on invasion *in vitro* as lowering the E-cadherin expression. As shown above, episialin also deminishes cellular adhesion to ECM components, which is known to be another important event during metastasis. We conclude that the balance between various adhesion processes can be disturbed when cellular polarization is lost and episialin is (over)expressed at the entire cell surface, and thus this mucin-like glycoprotein may enhance invasiveness and increase the efficiency of the metastatic process. In addition, the less efficient killing of episialin expressing cells might be crucial to the survival of metastasizing cells in the circulation.

Other membrane associated mucin-like molecules involved in tumor progression

Two other membrane associated mucin-like molecules, epiglycanin and ASGP-1, have been shown to affect cellular functions or to be involved in adhesion and metastasis. Epiglycanin is produced by TA3 Ha mouse mammary carcinoma cells as one of the major components of their cellular surface. This long rod-like molecule has been implicated in masking the histocompatibility antigens and in rendering the cells allotransplantable (14). The amino acid composition of epiglycanin and murine episialin suggests that both glycoproteins are not related. Moreover, we could not demonstrate episialin mRNA in cells that express high levels of epiglycanin.

ASGP-1,2 is a membrane associated mucin present on rat mammary carcinoma cell lines. The molecule shares many properties with episialin with respect to its structure, biosynthesis and processing (15). However, ASGP-1,2 shows no sequence homology with episialin (16). ASGP-1,2 is abundantly present on the surface of the rat mammary carcinoma cell line 13762. Resistance of these cells to lysis by spleen lymphocytes and NK cells was shown to correlate with the expression level of this sialomucin (17, 18). Moreover, studies on different metastatic clones of this rat mammary tumor cell line have established a positive correlation between their metastatic potential and the expression level of ASGP-1 (19).

Recently, it has been shown that a third molecule with a large mucin-like domain, CD43 or leukosialin (specifically expressed on T cells), affects cellular adhesion. Targetted disruption of the CD43 gene enhanced T Lymphocyte adhesion, whereas cells transfected with CD43 cDNA show a reduced cell-cell adhesion (20, 21). Thus, in addition to episialin there are at least three other, genetically unrelated molecules, that share structural properties which can lead to masking of cell surface molecules. It is very likely that cancer cells can employ these molecules to invade, metastasize and escape immune destruction.

References

1. Albelda SM, Lab. Invest. 1993; 68: 1-17.
2. Takeichi M, Science 1991; 251: 1451-1455.
3. Behrens J, Mareel MM, VanRoy FM, and Birchmeier WJ, Cell Biol. 1989; 108: 2435-2447.
4. Vleminckx K, Vakaet L, Mareel M, Fiers W. and Van Roy F. Cell 1991; 66, 107-119.
5. Frixen UH, Behrens J, Sachs M, Eberle G, Voss B, Warda A, Lochner D, and Birchmeier W, J. Cell Biol. 1991; 113: 173-185.
6. Jentoft N, TIBS 1990; 15: 291-294.
7. Ligtenberg MJL, Kruijshaar L, Buijs F, Van Meijer M, Litvinov SV, and Hilkens J, J. Biol. Chem. 1992; 267: 6171-6177
8. Hilkens J, Ligtenberg MJL, Vos HL, and Litvinov S, TIBS 17; 1993: 359-363
9. Zotter S, Lossnitzer A, Hageman PC, Delemarre JF, Hikens J, and Hilgers J. 1987; 57: 193-199.
10. Ligtenberg MJL, Buijs F, Vos HL, and Hilkens J, Cancer Res. 1992; 52: 2318-2324.
11. Yang P, Rutishauser U, J. Cell Biol. 1988; 116: 1487-1496.
12. Van de Wiel van Kemenade E, Ligtenberg MJL, de Boer AJ, Buijs F, Vos HL, Melief CJM, Hilkens J and Figdor CG, J. Immunol. 1993; 151: 767-776.
13. Hilkens J, Buys F, Hilgers J, Hageman Ph, Calafat J, Sonnenberg A, and van der Valk M, 1984; 34: 197-206.
14. Codington J.F. and Haavik S, Glycobiology 1993; 2: 173-180.
15. Carraway, KL and Hull SR, Glycobiology 1991; 1: 131-138.
16. Sheng Z, Wu K, Carraway KI and Fregien N, J. Biol. Chem. 1992; 267: 16341-16346.
17. Sherblom AP and Moody CE, Cancer Res. 1986; 46: 4543-4546.
18. Bharathan S, Moriarty J, Moody CE, and Serblom AP, Cancer Res. 1990; 50: 5250-5256.
19. Steck PA and Nicolson GI, Exp. Cell Res. 1983; 147: 255-267
20. Manjunath M, Johnson RS, Staunton DE, Pasqualini R and Ardman BJ, Immunol. 1993; 151: 1528-1534.
21. Ardman B, Sikorski MA, Staunton DE. Proc. Natl. Acad. Sci. USA 1992; 89: 5001-5004.

Biology of Vitronectins and their Receptors
K.T. Preissner, S. Rosenblatt, C. Kost, J. Wegerhoff and D.F. Mosher, editors

201

Molecular organization and antiproliferative activity of arterial tissue heparan sulfate proteoglycans

E. Buddecke[a], A. Schmidt[b], P. Vischer[b], and W. Völker[b]

[a] Institute for Physiological Chemistry and Pathobiochemistry, University of Münster, Waldeyerstr. 15, D-48149 Münster, Germany

[b] Institute for Arteriosclerosis Research, Waldeyerstr. 15, D-48149 Münster, Germany

INTRODUCTION

In the family of heparan sulfate (HS) containing proteoglycans the syndecan types I-IV (1), perlecan (2), glypican (3), and proteoheparan sulfates (HSPG) of arterial wall cells (4,5) and arterial tissue (6) have been characterized with respect to their chemical and macromolecular properties.

Cell membrane integrated HSPGs bind a variety of molecules in the cellular microenvironment via their HS chains and can act as receptors or coreceptors for ligands (for review see 7,8). Examples for macromolecules of the extracellular matrix interacting with cell associated HSPG are fibronectin, vitronectin (9) laminin, and thrombospondin (10). The ectodomain of all cell membrane integrated HSPG can be cleaved at a protease susceptible site adjacent to the plasma membrane and can provide competitive binding sites in the extracellular matrix.

The functional impact of HSPG isolated from arterial tissue or cultured arterial smooth muscle and endothelial cells is deduced from the finding that vascular endothelial cells (11) and cultured arterial smooth muscle cells produce a HS species that causes inhibition of cell proliferation (12,13). These findings suggest that vascular cells may be able to synthesize HS moieties with an unique structure that could regulate biological systems.

BIOSYNTHESIS AND STRUCTURE OF HSPG

Cultured arterial smooth muscle cells synthesize two HSPG species one is associated with the cells whereas the other is excreted into the medium (4,5). A HSPG isolated from arterial tissue has been characterized as a distinct macromolecule by chemical and physicochemical procedures (6). The cell associated HSPG isolated from arterial smooth muscle cells and endothelial cells have been found to be structurally different macromolecules (Tab. 1).

HS is known to exhibit a broad chemical and configurational variability with respect to the ratio of glucuronic acid/iduronic acid, the number and position of O-sulfate ester groups and the ratio of N-sulfate/O-sulfate (14). The biosynthesis of HSPG involves protein core synthesis, polymerization of a polysaccharide backbone consisting of alternating glucuronic acid and N-acetylglucosamine residues and modification of the glycosaminoglycan chain into a partially sulfated polysaccharide with complex structure (for review see 7,15). The first modificati-

202

on reaction (N-deacetylation) converts about 30 % of N-acetylglucosaminyl residues (Glc-NAc) into glucosamine which is in turn N-sulfated. The subsequent modification reactions, C-5-epimerization of GlcA to IduA and O-sulfation in various positions occur in the Golgi complex, mostly in the vicinity of previously incorporated N-sulfated groups. In the case of HS synthesis in arterial tissue the modification reactions are incomplete, leaving 65 % of the potential target sites unmodified (16, 17).

Table 1
Chemical and macromolecular properties of cell associated HSPG isolated from arterial smooth muscle cells and endothelial cells

Parameter	Smooth Muscle Cell	Endothelial Cell
Molecular weight (M_r)	≈ 200	$> 10^6$
M_r of protein core	92	420
M_r of HS chains	30-40	120
Number of HS chains	3-4	6-8

Reprinted from (4,5) and unpublished results

In contrast to other glycosaminoglycans HS has a heterogenous molecular organization where sulfate-rich domains containing mono-, di- or trisulfated disaccharide units are separated by longer sulfate-poor or sulfate-free sequences consisting of GlcAß(1-4)GlcNAcα(1-4) disaccharide units. A similar structural model has been proposed by Turnbull and Gallagher

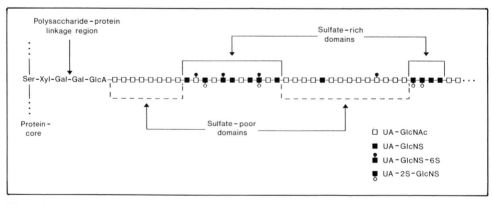

Figure 1
Macromolecular organization of HS (18)

(19) for human skin fibroblast HS. The heterogeneity of the HS molecule is deduced from enzymatic and chemical disintegration experiments by means of heparitinase I and II (18,19), hydrazinolysis and HNO_2 degradation at pH 1.5 and 3.9. From the data obtained a hypothetical model of the macromolecular organization of the HS molecule can be designed (Fig. 1) which consists of 4-5 clusters of highly sulfated hexa- to tetradecasaccharides separated by non- or low sulfated sequences with an average of 10 disaccharide units in length.

ANTIPROLIFERATIVE ACTIVITY: IN VITRO STUDIES

For analyzing the antiproliferative effect of HS and HS oligosaccharides growth arrested cultured arterial smooth muscle cells were plated at a density of 2500 cells/cm² and allowed to attach for 6 h. The plating medium was then removed and an experimental medium (with inhibitor) or control medium (without inhibitor) was added. After 4 days of incubation the cell number was determined from the ratio of growth in the presence of inhibitor/net growth in controls.

Fig. 2 shows that HS is capable of inhibiting the proliferation of arterial smooth muscle cells in a dose-dependent manner when added to cultured cells at a concentration of 5 - 100 μg/ml. Heparin which is not synthesized by arterial smooth muscle cells has a higher antiproliferative effect but other arterial-specific glycosaminoglycans such as chondroitin sulfate, dermatan sulfate, and hyaluronate exhibit no growth inhibitory activity.

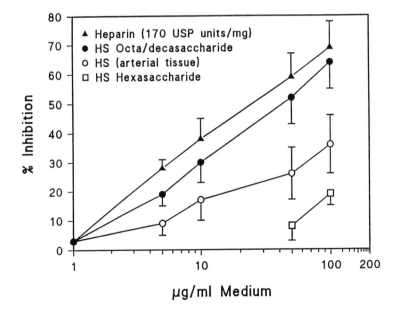

Figure 2
Antiproliferative activity of HS and HS-derived oligosaccharides and heparin (12)

In order to characterize the structural determinants of antiproliferative activity the sulfate-rich domains of HS were isolated after either enzymatic or chemical disintegration. Heparitinase I an microbial endoglucosaminidase, which preferabely splits α-glycosidic linkages of N-acetyl/N-sulfonyl glucosamine if the adjacent uronic acid does not bear a 2-O-sulfate ester group, produces more than 50 % low- or non-sulfated disaccharides in addition to a mixture of oligosaccharides ranging from 6 to 14 monosaccharide residues. The resulting oligosaccharides might be resolved into defined subfractions by gel filtration chromatography on calibrated Biogel P-6 columns. Fig.2 and Tab.2 show that a hexasaccharide

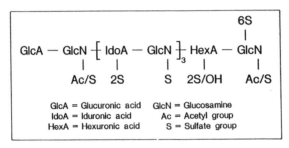

Figure 3
Potential structure of a cell proliferation inhibitory decasaccharide of arterial heparan sulfate

Table 2
Constituent disaccharides of arterial tissue HS and HS derived hexa- and octa/decasaccharide (12)

DISACCHARIDES	ART. HS	HEXA-S	OCTA/DECA-S
△ DiHS-0S	66.4	25.6	18.1
△ DiHS-NS	18.3	24.6	10.2
△ DiHS-6S	3.9	9.1	5.1
△ DiHS-US	0.44	0.75	0.51
△ DiHS-DI(6,N)S	1.3	6.0	6.1
△ DiHS-DI(U,N)S	6.6	13.8	45.5
△ DiHS-TRI(U,6,N)S	2.5	18.4	12.3
TOTAL	99.4	98.3	97.8

and an octa/decasaccharide fraction differ markedly not only in their disaccharide compositi-on (Tab. 2), but also in their ability to suppress arterial smooth muscle cell growth. At a concentration of 50 μg/ml medium the octa/decasaccharide fraction has approximatively about ten times the antiproliferative activity of the hexasaccharide fraction in the in vitro assay system (Tab. 2). The antiproliferative activity of octa/decasaccharide is in the same order as the inhibitory activity of commercial heparin (170 USP units/mg). Fig.3 shows the potential structure of a HS-derived antiproliferatively active decasaccharide. In contrast no antiproliferative activity of the hexasaccharide fraction is detectable at a concentration below 50 μg/ml medium (12).

STRUCTURE-FUNCTION RELATIONSHIP

The data of Tab. 2 reveal that the antiproliferative activity of the HS molecule resides in highly sulfated domains which are enriched in hexuronate-2-sulfate residues and have a minimal length of 4 - 5 disaccharide units. Chemical desulfation abolishs the effect complete-ly. In order to get additional information about the structural determinants of the antiprolife-rative activity the sulfate-enriched domains of HS were isolated after hydrazine/nitrous acid treatment. Hydrazine deacetylates N-acetylglucosamine residues, the following action of nitrous acid at pH 3.9 cleaves the formed α-glucosaminidic bonds converting the deacetylated glucosamine into anhydromannose. This procedure converts 2/3 of HS to disaccharides with the structure uronyl(1-4)anhydromannose. The remaining 1/3 is obtained as oligosaccharide mixture containing N- and O-sulfated glucosamine and 2-O-sulfated uronic acid residues as constituents. A comparative analysis of the oligosaccharides reveal that the antiproliferative activity increases with increasing chain length and is maximal at 7 disaccharides.

The antiproliferative effect of HS oligosaccharides is not considered to be solely the functi-on of their charge density. This is concluded from the fact that the charge of the antiprolifera-tively inactive hexasaccharide is only slightly less than that of the highly active octa/decasac-charide and that this difference does not correspond to the difference in their antiproliferative activity. Moreover, the content of trisulfated disaccharides is higher in the hexa- than in the octa/decasaccharide fraction, and the O-sulfate/N-sulfate ratio of the two oligosaccharide fractions is nearly identical. Charge density, chain length and configuration of anionic groups are the structural determinants of antiproliferative activity of HS oligosaccharides.

The accumulation of 2-O-sulfated uronic acids in the octa/decasaccharide fraction raises the question whether hexuronate-2-sulfate residues are essential for antiproliferative activity. Fedarko and Conrad (20) have demonstrated the presence of HS in the medium, the pericellu-lar matrix, the cytoplasm, and in the nucleus of hepatocytes. From their observation that the nuclear pool of HS contained a high proportion of 2-O-sulfate glucuronate residues, it was suggested that the 2-O-sulfate glucuronic acid-containing sequences of HS are important in growth regulation. In contrast Karnovsky et al. (21) have assayed heparin decasaccharide fractions that lack 2-O-sulfate glucuronic acid and found that they have significant antiprolife-rative activity.

ANTIPROLIFERATIVE ACTIVITY IN VIVO

HS preparations with a sulfate content of 0.5 Mol/Mol disaccharide and a 6 - 8 % content of uronic acid-2-sulfate(1-4)N-sulfoglucosamine disaccharides are able to reduce the formation of neointimal plaques after balloon catheter-induced injury of rat carotids when given intravenously by a miniosmotic pump in a dose of 5 mg/rat/day over a period of 14 days (unpublished results).

In view of the structural similarities of the sulfate-rich domains of HS with heparin and the possibility to obtain a non-anticoagulant heparin by antithrombin III Sepharose affinity chromatography, a low affinity heparin (LA-heparin) was tested for its ability to suppress arterial smooth muscle cell proliferation of rat carotid arteries after catheter balloon-induced deendothelialization. In normolipemic rats LA-Heparin reduces the formation of neointimal plaques and stenoses after angioplasty. At a total daily dose of 5 mg/kg body weight given subcutaneously one week before and 2 weeks after ballooning injury the size of the neointimal plaque is reduced to 36 % as compared with a non-treated control group (100 %). This 64 % reduction is statistically highly significant. Since both the LA-heparin and the high affinity fraction of heparin (bound to AT III) were effective in the in vitro assay system for antiproliferation on smooth muscle cells, the antiproliferative activity appears clearly separated from the anticoagulant activity.

MECHANISM OF ACTION

The mechanism by which HS influences cell proliferation is not known. However, the observation that the dose response curve of HS and HS-derived octa/decasaccharide (Tab. 2) tends to saturation (when the x-axis is converted to a linear scale) suggests that a HS oligosaccharide binding protein at the cell surface or a high affinity receptor could be involved in the mechanism of action. Indeed at doses > 100 μg/ml no significant increase of inhibiton was observed. This could suggest that a HS binding protein at the cell surface might be involved in the mechanism of antiproliferative activity. This assumption is supported by the finding that immunoprecipitation of smooth muscle cell lysates with monospecific antibodies against a heparin/heparan sulfate binding glycoprotein (M_r 78 000) from bovine uteri (22) shows that arterial smooth muscle cells also express the HS binding glycoprotein 78 000 (unpublished results).

Studies on the inhibitory effect of heparin on 3T3 fibroblasts and calf arterial smooth muscle cells suggested that heparin inhibits a protein kinase C-dependent pathway for cell proliferation and suppresses the induction of c-fos and c-myc specific mRNA at a site distal to activation of the protein kinase C (23). The protooncogenes c-fos and c-myc appeared to be essential for proliferation. Furthermore the significance of c-myb for cell proliferation was demonstrated by the observation that an antisense oligonucleotide of c-myb mRNA was able to inhibit smooth muscle cell proliferation in vitro and in vivo (24). Detailed cell cycle analysis indicated that heparin must be present during the last 4 h before S phase for an antiproliferative effect, suggesting a mid-to-late G_1 heparin block. In the continued presence of heparin

G_1 blocked cells slowly moved back into a quiescent growth state (25). A further potential mechanism of action is related to the negative charge which enables HS to bind many substances including basic fibroblast factor (bFGF) and other growth factors (7). A heparin-derived hexasaccharide having high affinity to bFGF effectively inhibits the binding of bFGF to cell membrane integrated syndecan and can act as antagonist of bFGF-mediated cell mitogenesis. On the other hand the hexasaccharide can restore the bFGF-dependent proliferative response to cells grown in the presence of sodium chlorate (26) which suppresses the sulfation of the endogeneous HS precursor polysaccharide.

CONCLUSIONS AND PERSPECTIVES

In addition to the biological interest of characterizing the antiproliferative features of HS and HS oligosaccharides in structural-functional terms there is considerable interest to elucidate the mechanism of inhibitory activity of HS and heparin and the role of arterial tissue HS as endogenous inhibitor of vascular smooth muscle cell proliferation, an early event in the pathogenesis of arteriosclerosis. Moreover the physiological relevance of the inhibitory activity of sulfate enriched domains of HS lies in the fact that HS inhibits cell growth at a concentration that is of the same order as its physiological concentration in native arterial tissue and that in human arterial tissue the HS concentration decreases with increasing degree of arteriosclerois (27). Moreover, the impact of HS or non-anticoagulant heparin and their antiproliferative domains lies in their potential clinical use following percutaneous transluminal coronary angioplasty, vascular surgery or diagnostic angioscopy. Further applications are the accelerated arteriosclerosis characterized by massive vascular smooth muscle cell proliferation as seen in heart transplants and in pulmonary hypertension where cell proliferation is the dominant feature of the lesion.

REFERENCES

1   Bernfield M, Kokenyesi, R, Kato M, Hinkes MT, Spring J, Gallo RL, Lose EJ.
     Annu Rev Cell Biol 1992;8:365-393.
2   Pansson M, Yurchenco GC, Ruben R, Engel I, Timpel R. J Mol Biol
     1987;197:297-305
3   David G, Lovies, V, Decock B, Marynen P, Cassiman JJ, Van den Berghe H.
     J Cell Biol 1990; 111:3165-3172
4   Schmidt A, Buddecke E. Exp Cell Res 1988;178:242-253
5   Schmidt A, Buddecke E. Exp Cell Res 1990;189:269-275
6   Schmidt A, Schäfer E, Buddecke E. Eur J Biochem 1988;173:661-666
7   Kjellén L, Lindahl V. Annu Rev Biochem 1991;60:443-475
8   Zhon F, Höök T, Thompson JA, Höök M. Heparin protein interactions
     in Lane DA, Björk I, Lindahl V (eds) Heparin and related polysaccharides
     1992;Plenum Press New York and London 141-153

9   Preissner KD, Müller-Berghaus G. J Biol Chem 1987;262:12247-12252
10   Schön P, Vischer P, Völker W, Schmidt A, Faber V. Eur J Cell
     Biol 1992; 59:329-339
11   Benitz WE, Kelly RT, Anderson CM, Lovant DE, Bernfield M. Am J Respir Cell
     Mol Biol 1990;2:13-24
12   Schmidt A, Yoshida K, Buddecke E. J Biol Chem 1992;287:19442-19447
13   Fritze LMS, Reilly CF, Rosenberg RD. J Cell Biol 1985;100:1041-1049
14   Gallagher JT, Lyon M, Steward WP. Biochem J 1986;236:313-325
15   Lindahl V, Kjellén L. Biosynthesis of heparin and heparan sulfate
     in Wight TN, Mecham RP (eds) Biology of proteoglycans. Academic Press
     New York 1987;59-104
16   Göhler P, Niemann R, Buddecke E. Eur J Biochem 1984;138:301-308
17   Schmidt A, Buddecke E. Eur J Cell Biol 1990;52:229-235
18   Schmidt A, Lemming G, Yoshida K, Buddecke E. Eur J Cell Biol
     1992;59:322-328
19   Turnbull JE, Gallagher JT. Bioch J 1990;265:715-724
20   Fedarko NS, Conrad HE. J Cell Biol 1986;102:587-599
21   Karnovsky MJ, Wright TC, Castellot JJjr, Choay J, Lovmeau JC, Petitou M.
     Ann.N.Y Acad Sci 1989;556:268-281
22   Lankes W, Griesmacher A, Grünwald J, Schwartz-Albiez R, Keller R.
     Bioch J 1988;251:831-842
23   Wright TCjr, Pukac LA, Castellot JJjr, Karnovsky MJ, Levine RA,Kim-Park HY
     Hy,Campisi J. Proc Natl Acad Sci 1989;86:3199-32039
24   Simons M, Edelman ER,Dekeyser JL, Langer R, Rosenberg RD. Nature
     1992;359:67-70
25   Castellot JJjr, Pukac LA, Caleb BL, Wright TCjr, Karnovsky MJ. J Cell Biol
     1989;109:3147-3155
26   Tyrell DJ, Ishihara M, Rao N, Horne A, Kiefer MC, Stauber GB, Lam LH,
     Stack RJ. J Biol Chem 1993;268:4684-4689
27   Hollmann J, Schmidt A, v. Bassewitz D-B, Buddecke E. Arteriosclerosis 1989;
     9:154-158

Biology of Vitronectins and their Receptors
K.T. Preissner, S. Rosenblatt, C. Kost, J. Wegerhoff and D.F. Mosher, editors

THROMBIN INTERACTION WITH THE VASCULAR SYSTEM

R. Bar-Shavit , Y. Eskohjido, M. Benezra and I. Vlodavsky

Department of Oncology, Hadassah-University Hospital, Jerusalem 91120, Israel

INTRODUCTION

Cell interactions with the extracellular matrix (ECM) play important roles in determining cellular morphology, growth and differentiation (1,2). As cells adhere to a substratum, contacts are established mainly via members of the integrin superfamily (3,4). During this process, integrins become localized on the ventral surface of the cells in focal contacts, providing a transmembrane link between the ECM and the cytoskeletal cell machinery (5,6). The vascular endothelium has emerged as a highly dynamic environment maintaining its integrity by factors that mediate the attachment and growth of endothelial cells (ECs). The integrity of the vascular endothelium is an essential requirement governing the vascular tone and permeability, as well as preventing the vessel wall from platelet deposition and thrombus formation (7,8). Most EC receptors for ECM proteins belong to the integrin superfamily recognizing an Arg-Gly-Asp (RGD) motif found in many ECM proteins and the Leu-Asp-Val (LDV) sequence in fibronectin (1,2). Adhesive interactions play a leading role in the progression of vascular thrombosis and subsequent wound healing. Following vascular injury, the fibrin dependent aggregation of platelets adherent to the exposed subendothelium (8,9) contributes to the formation of thrombus that initially seals the vessel to prevent excessive blood loss. Subsequently, a local repair mechanism is initiated, involving EC attachment, migration and proliferation to renew the damaged vessel. Therefore, to understand the molecular events involved in thrombus formation and wound healing, it is necessary to delineate structural interactions between cells and molecules participating in this process. The serine protease thrombin is unique among the enzymes participating in the clotting cascade, by virtue of its cell activation effects induced via the enzymatic pocket and or via functional domains located throughout the molecule. In this overeview, we elaborate on thrombin interactions with the vessel wall and its induction of EC adhesion and smooth muscle cell proliferation.

The role of thrombin, which circulates mainly in the form of prothrombin, and its regulation of the vascular system depends upon its distribution between the extra- and intravascular space. We have demonstrated that thrombin may be present within the vascular system, immobilized to the subendothelial ECM through a short anchorage binding site, leaving the majority of the molecule functional and available for cellular interactions (10). ECM-immobilized thrombin was found to be protected from inactivation by the circulating inhibitor antithrombin III (ATIII). Thus, thrombin when sequestered by the ECM may exhibit a long-acting and localized stimulation of surrounding tissues (11). Thrombin can exert these effects when either present in a fluid phase or immobilized to ECM. An internal source for thrombin in the vascular system may be provided by the thrombus, from which it can be released intact and active during fibrinolysis, or by transendothelial passage through gaps formed between adjacent EC (12-14).

THROMBIN INDUCED CELL ADHESION

Amino acid analysis of thrombin B-chain reveals the presence of an Arg-Gly-Asp-Ala sequence at residues 187-190 (15). Because thrombin is ubiquitous to injury sites, we wondered whether it can participate in EC adhesion and thus contributes to repair mechanisms of vascular lesions. A variety of thrombin species that were chemically modified to alter thrombin procoagulant or catalytic functions were tested for their ability to promote bovine and human endothelial cell adhesion. NO_2-α-thrombin was found to be the most potent adhesive thrombin analogue, inducing attachment of ~80% of the cells within 2 h. Modified thrombin preparations inactivated by the fibrinopeptide exosite affinity label (Exo-α-thrombin) and the mesyl-conjugated active site serine (MeSO$_2$-α-thrombin) were highly active in promoting EC attachment (29). γ-Thrombin, lacking the procoagulant site of thrombin, was also active. In contrast, the native enzyme (α-thrombin) and the esterolytically inactive forms, N-α-tosyl-L-lysylchoromethyl-ketone (TLCK)- α-thrombin and diisopropylfluorophosphate(DIP) -α-thrombin, were poorly active and the level of cell attachment was only slightly above that observed on noncoated surfaces (29).

The fact that native α-thrombin does not exhibit adhesive properties, is consistent with crystal structure analysis of thrombin. Stereo views of α-thrombin displayed with the "connolly dot surface" method (16) show that of the RGD motif only Arg 187 is surface exposed while Gly 188 and Asp 189 are completely buried. Therefore, exposure of the functionally active, cryptic RGD domain in thrombin is necessary in order to convert the protein to an adhesive molecule. NO_2-α-thrombin obtained by nitration of the tyrosine residue adjacent to the RGD sequence, induced spreading of EC in a manner comparable to the subendothelial ECM. This tyrosine (Tyr-185) apparently plays a significant role in altering the conformation of α-thrombin, resulting in exposure of the RGD sequence at an exosite domain. The degree of nitration (eight or three nitrotyrosines per mole) did not affect significantly the attachment-promoting activity of the modified thrombin. EC attachment to NO_2-α-thrombin exhibited a typical distribution of the microfilamentous cytoskeleton with formation of stress fibers concomitant with vinculin streaks at their endings, corresponding to focal contacts (29). Qualitatively, this type of cellular response was comparable to those obtained after seeding the cells on characteristic matrix proteins such as vitronectin, fibronectin or von Willebrand factor. Immunofluorescence staining using anti-α_V and anti-β_3 antibodies gave a peculiar pattern of oval and arrowhead-shaped spots usually located at stress fiber endings. The amount of β_3 was reduced in comparison to EC adhering to vitronectin.

EC attachment was completely abolished in the presence of ATIII, forming a complex with NO_2-α-thrombin, suggesting that ATIII masks the RGD region in thrombin. In contrast, hirudin, a leech-derived protein (M_r ~7,000) with high affinity to thrombin, did not inhibit its attachment promoting activity. Hirudin has a negatively charged tail (17,18), thought to bind to a region neighboring the catalytic site located up-stream to the RGD sequence at the frontal side of thrombin. The RGD domain is therefore an unlikely candidate for interaction with hirudin (19).

PHYSIOLOGICAL CONVERSION OF THROMBIN TO AN ADHESIVE PROTEIN

The question arises as to the possible physiological significance of the cryptic RGD motif in thrombin. During tissue repair and wound healing the hemostatic plug is dissolved by active fibrinolysis, where plasmin is generated to degrade the fibrin mesh, releasing functionally intact and active thrombin (20). The plasminogen activation system has been implicated in a variety of physiological and pathophysiological events, such as cell migration, fibrinolysis, tissue remodeling, inflammation and tumor metastasis (21). The presence of plasminogen in appreciable quantities in extracellular fluids and extravascular compartments is a prerequisite for the concerted action of plasminogen activators and their inhibitors, resulting in generation of plasmin (22). Moreover, cell surface receptors for these activators and plasminogen (13,24), as well as the fibrin clot, may serve to localize plasmin formation during pericellular proteolysis and fibrinolysis.

We investigated whether activation of the fibrinolytic system may act on the procoagulant protein thrombin in a manner that exposes its RGD domain. Plasmin was found to convert thrombin to an adhesive protein for EC in a time and dose dependent manner. To better characterize this conversion, thrombin was preincubated with plasmin and cleavage fragments were subjected to gel filtration analysis on FPLC Superdex 75 column (Fig. 1a). Four distinct cleavage products were obtained and the pooled fractions of each peak were used to coat the surfaces of 4 well plates. EC adhesion was induced by a low Mr (~3,000 daltons) cleavage product (peak IV) and was completely inhibited in the presence of the synthetic peptide GRGDSP (Fig. 1a). In an attempt to identify and sequence the RGD containing fragment, the active peak (#IV, Fig. 1b) was subjected to reverse phase HPLC (C_{18} column). From these results we concluded that exposure of thrombin generated a 31 amino acid peptide, containing the RGD sequence (Arg_{187}- Arg_{221}, Fig. 1c). This is due to formation of a disulfide bond (Cys_{191} & Cys_{219}) between two fragments, taking into account that a six amino acid peptide (Lys_{202} - Arg_{206}) is cleaved away (Fig. 1c).

EFFECT OF SOLUBLE AND CELL SURFACE HEPARAN SULFATE ON PLASMIN STIMULATED THROMBIN INDUCED EC ATTACHMENT

Conversion of thrombin to an adhesive molecule was significantly enhanced in the presence of heparin or heparan sulfate, while other glycosaminoglycans (GAGs) (e.g. dermatan sulfate, keratan sulfate, chondroitin sulfate) had no effect. We investigated the potential involvement of cell membrane associated proteoglycans in plasmin cleavage of thrombin and its subsequent ability to promote EC attachment. For this purpose, we have used CHO cell mutants (25-28), expressing specific defects in various aspects of GAG biosynthesis. We compared the abilities of CHO wild type and mutant cells to promote plasmin mediated conversion of thrombin to an adhesive molecule, as monitored by the stimulation of EC attachment. Mutant 745 CHO cells, fail to produce all types of proteoglycans (25,26). The CHO mutant 677, synthesizes all GAGs except heparan sulfate (HS) due to a defect in the enzymatic activity required to form the repeating disaccharide unit characteristic of HS (27). Exposure of thrombin to plasmin was carried out by incubation on the surface of wild type and mutant CHO cells. In order to prevent secretion of adhesive proteins by the CHO cells, monolayers were pretreated with monensin (29), or were fixed with 3% formaldehyde. Under these conditions no secretion of cellular adhesive proteins was observed, as indicated by the lack of EC adhesion when supernatants of the CHO cells were used to coat plastic surfaces. We co-incubated thrombin and plasmin on the surface of fixed wild-type and mutant CHO cells and used the supernatants to coat dishes prior to monitoring the levels of EC attachment. A marked enhancement of EC attachment was observed following incubation of thrombin and a suboptimal concentration of plasmin on top of wild type CHO cells, as compared to minimal levels of cell attachment obtained following a similar incubation with mutant cells lacking all GAGs or only heparan sulfate. Mutant 677 cells exhibiting a 3 fold higher level of chondroitin sulfate than wild type cells and mutant 803 cells deficient in HS failed to promote EC attachment to thrombin. The residual extent of cell adhesion and spreading observed following incubation of plasmin and thrombin on top of HS deficient CHO mutants may be due to very limited exposure of the RGD sequence in the presence of a suboptimal amount of plasmin (0.2 μg/ml) alone and residual HS (<5%) expressed by the mutant cells. These results indicate that cell surface associated HS is involved in plasmin mediated conversion of thrombin to an adhesive molecule. Conversion of thrombin to an adhesive molecule was also accelerated upon incubation of thrombin and plasmin on top of a fixed subendothelial ECM. Similar to the results observed with the CHO cells, exposure of thrombin to a low concentration plasmin (0.2 μg/ml) resulted in its conversion to an adhesive molecule only in the presence of intact ECM, or when thrombin was first incubated with ECM. This accelerated conversion was mediated by the ECM-HS since it was markedly inhibited when the ECM-HS was first extensively degraded by bacterial

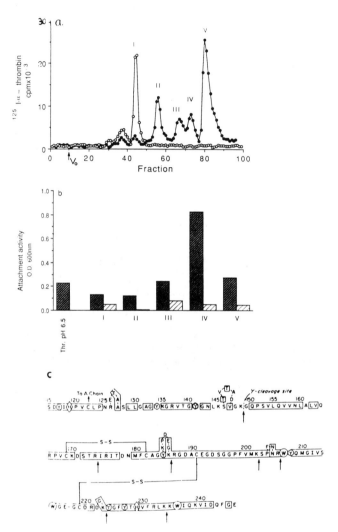

Figure 1: Gel filtration analysis of thrombin derived fragments cleaved by plasmin.
a. ^{125}I-α–thrombin (50μg, 5x10[5] cpm) was incubated (4h, 10mM Tris-HCl pH 6.5) in the absence (-o-) or presence (-●-) of 10 μg/ml plasmin. The reaction mixture was then subjected to gel filtration analysis on FPLC Superdex 75 column (Pharmacia, Uppsala). Fractions (0.4 ml) were collected (1min/ml) and counted in a γ-counter. **b. Attachment activity of thrombin fragments generated by plasmin and separated by gel filtration.** Peak fractions were pooled and each peak (I,II,III,IV, V) was used to coat 4-well plates for measurments of EC attachment in the absence (▨) and presence (▨) of GRGDSP peptide. **c. Amino acid sequence of the RGD containing region of human thrombin B chain.** The residues are aligned with residus 115 through 245 of bovine chymotrypsin. Individual residues are identified by the single letter code. Plasmin cleavage sites are indicated by arrows.

heparitinase (0.5 U/ml, 2 h, 37°C). Use of metabolically sulfate ($Na_2{}^{35}SO_4$) labeled ECM revealed that about 90% of the total incorporated radioactivity was released by treatment with the heparitinase enzyme. EC attachment to thrombin following exposure to plasmin on fixed ECM was inhibited in the presence of GRGDSP peptide or anti thrombin/prothrombin antibodies.

HS proteoglycans (HSPG) are a diverse group of macromolecules containing at least one covalently bound HS chain, with N-linked and O-linked sulfate groups. They are widely distributed throughout animal tissues associated with cell surfaces and basement membranes and are becoming increasingly recognized as mediators of the binding and function of heparin-binding growth factors, adhesive proteins, plasma proteins, enzymes, enzyme inhibitors and viruses (30-32). The structural diversity of HS is particularly suited for generating specific domain structures that can be utilized for biological recognition of a specific protein. Thrombin mediated acceleration of cell adhesion may function in maintaining the integrity of the vessel wall. Experiments utilizing antibodies directed against the RGD epitope of thrombin will hopefully clarify whether such a role is indeed fulfilled by thrombin. Our results indicate that under certain conditions thrombin may acquire adhesive properties and hence be actively engaged in supporting EC adhesion. Adoption of this unique function of thrombin is accelerated by heparin and HS, when present on cell surfaces and in ECM. Chemical modifications of thrombin revealed that Lys_{240} and Lys_{169}, positioned near Arg_{187}, render thrombin inaccessible to heparin (16), suggesting that these positively charged groups are crucial for the conversion of thrombin to an adhesive protein. Altogether these results ascribe a new physiological significance to the heparin binding property of thrombin.

THROMBIN INDUCED CELL PROLIFERATION WITHIN THE VASCULAR SYSTEM

I. THROMBIN IMMOBILIZED TO ECM IS A GROWTH FACTOR FOR SMOOTH MUSCLE CELLS (SMC)

Vascular medial SMC are normally screened from blood-borne components by subendothelial ECM and the intact barrier of EC. Inappropriate migration and proliferation of SMC in the intima is associated with blood vessel wall pathology, mainly atherogenesis. It is becoming evident that atherosclerotic plaque formation *in vivo* may occur within the intact vessel wall through intrinsic mediators, independent of the delivery of exogenous growth factors from the blood (33,34). ECM-immobilized thrombin retains its functional properties (e.g. induction of platelet activation and SMC proliferation) and is protected from inhibition, making the molecule a potential regulator of events occurring in the vessel wall (11,35). In view of the possibility that active thrombin is firmly bound to the subendothelial ECM, we investigated the ability of thrombin immobilized to ECM to promote vascular SMC proliferation. When SMC were seeded on ECM that was first exposed to thrombin a 3-4 fold stimulation of proliferation was obtained as compared to cells maintained on ECM alone. A 5-6 fold stimulation of DNA synthesis was obtained when soluble native α-thrombin was added to quiescent SMC. Thrombin induced SMC proliferation was effectively inhibited in the presence of the leech derived thrombin inhibitor hirudin.

Binding of $[^{125}I]$α-thrombin to SMC monolayers revealed the existence of specific cell surface receptors with an apparent K_d of 6 nM and an estimated 5.4×10^5 binding sites per cell. Vu and co-workers simultaneously with Rassmusen and co-workers adopted an expression cloning approach to isolate a functional human thrombin receptor DNA (36,37). The deduced amino acid sequence of the thrombin receptor revealed that it is a member of the seven transmembrane domain receptor family. It appears that thrombin cleaves its receptor next to Arg_{41} at the amino terminal extension, exposing a new amino terminus that acts as a ligand and activates the receptor. In fact, by using a synthetic peptide that mimicks this region, one can bypass the requirement for proteolysis, and activate the receptor directly, although the signal transduction pathway has yet to be elucidated (38). Cross-linking analysis using an iodinated 14 amino acid synthetic peptide that mimics the internal ligand within the cloned thrombin receptor, revealed the presence of an ~80 kDa receptor on monolayers of SMC.   . The biosynthesis of this receptor and its mode of activation is currently being studied.

II. THROMBIN INDUCED RELEASE OF MITOGENIC ACTIVITY FROM ECM

Identification of ECM-bound growth factors, enzymes and plasma proteins (39,40), indicate that the ECM provides a storage depote for active molecules and that some of the effects of ECM can be attributed to the combined action of structural components and of ECM-immobilized molecules that are thereby protected and stabilized (41). This may allow a more localized, regulated and persistent mode of action, as compared to the same molecules in a fluid phase. Among these growth factors is bFGF which was identified as a complex with HSPG in the subendothelial ECM produced *in vitro* (39) and in basement membranes of diverse tissues and blood vessels (40,42,43). Members of the FGF family exhibit a high affinity for heparin and are among the most potent inducers of neovascularization and mesenchyme formation (44, 45). ECM-bound bFGF, unlike soluble recombinant bFGF, is stable to heat and acid inactivation and to proteolytic degradation (46, 47). Release of bFGF from its storage in ECM by heparin-like moleculs, HS-degrading enzymes (i.e. heparanase) (48), or proteases (i.e. plasmin)(49), was suggested to elicit a localized neovascularization in processes such as wound healing, inflammation and tumor development.

Exposure of ECM to thrombin resulted in release of a high Mr HSPG-bFGF complex, as indicated by its immunoprecipitation with anti bFGF antibodies, susceptibility to degradation by bacterial heparitinase and inhibition of its mitogenic activity in the presence of neutralizing anti-bFGFantibodies (50). The ECM-resident bFGF-HSPG complex was not released by catalytically blocked thrombin preparations, or by thrombin in the presence of hirudin or anti-thrombin III (50). Our results indicate that the large reservoir of proteolytic activity present in plasma in the form of prothrombin may participate in release of biologically active bFGF-HSPG complexes from the subendothelial ECM, depending on the accessibility of thrombin.

FUTURE PERSPECIVES:

Our finding that thrombin binds to the subendothelial ECM retaining its functional properties and exhibiting long-acting and localized functions suggests that thrombin may play a significant role in vessel wall physiology and pathology. The fact that advanced atherosclerotic plaques show evidence of repeated thrombotic/coagulative events led to the suggestion that platelet derived factors play key roles in the development of atherosclerotic plaques (51). Thrombin may be a better candidate, especially in relation to restenosis after angioplasty. It was shown that thrombin activity remains elevated in balloon-injured vessels in animals for weeks after injury (52). In contrast, the platelet response is transient and disappears after one or two days (53). In fact, thrombin may regulate indirectly the expression of PDGF-A and the synthesis of PDGF-B by ECs and macrophages within the atherosclerotic plaques (53). Specific inhibitors of thrombin activities are being developed. Of particular interest is hirudin, shown to inhibit intimal thickening following balloon injury (53).The recently cloned thrombin receptor (36-38) revealed a new member of the seven transmembrane domain receptor family with a large amino terminal extracellular extension site. Activation of thrombin receptor requires the proteolytic unmaslking of an internal ligand. This observation led to the development of a novel agonist peptide that activates thrombin receptor regardless of the presence of thrombin. It is possible now to develop thrombin receptor agonists that may provide the basis for a new class of antiproliferative pharmaceuticals. In a search for potent anti-thrombotic drugs, the design of receptor inhibitors alone may not be sufficient. For example, release of matrix embedded growth factors (e.g. bFGF-HSPG) (50) by ECM-immobilized thrombin will still take place, resulting in stimulation of vascular SMC proliferation. The fact that thrombin molecule possesses a functional RGD site suggests that it may be involved in wound healing and repair processes in the vascular system. Our studies indicated that conversion of thrombin to an adhesive protein is mediated by plasmin and accelerated by the cell surface HSPG. Moreover, it appears that this cell sueface associated activity is primarily attributed to glypican since it can be released by treatment of cell monolayers with glycosyl-phosphatidylinositol- specific phospholipase C (PI-PLC). This enzyme releases a unique cell surface associated HSPG anchored via a covalently linked-glycosyl-phophatidylinositol residue that has its fatty acyl chains buried in the lipid bilayer. Thus, it is conceivable that RGD containing fragments of thrombin may be present in the circulation and play a role in processes (e.g. thrombosis, metastasis) involving cell-cell and cell-substrate interactions mediated by the RGD motif. With the aid of antibodies that recognize specifically thrombin-RGD, but fail to interact with other RGD containig adhesive proteins (e.g. fibronectin, vitronectin, fibrinogen, von-Willebrand factor) it may be feasible to illustrate a role for the RGD sequence in thrombin. Thus, thrombin may activate cells through a new binding site, recognizing the RGD.

Acknowledgment

This work was supported by grants from the GSF (Forschungszentrum fuer umwelt und gesundheit) and the Israeli Ministry of Health awarded to R.B.

References

1. Albelda SM, and Buck CA. FASEB J 1990; 4: 2868-2880.
2. Hynes RO. Cell 1992; 69: 11-25.
3. McClay DR, and Ettensohn CA. Ann Rev Cell Biol 1987; 3: 5357-5363.
4. Streuli CH, and Bissell MJ. J Cell Biol 1990;110: 1405-1415.
5. Burridge K, Faith K, Kelly T, Nuckolls G, and Turner C. Ann Rev Cell Biol 1988; 4: 487-525.
6. Otey CA, Pavalko FM, and Burridge K. J Cell Biol 1990; 111: 721-729.
7. Fuster V, Badimon L, Badimon JJ, and Chesebro JH. N Engl J Med 1992; 326: 424-251.
8. Marguerie GA, Plow EF, and Edington TS. J Biol Chem 1979;254: 5357-5363.
9. Bennet JS, and Vilaire G. J Clin Invest 1979; 64: 1393-1401.
10. Bar-Shavit R, Eldor A, and Vlodavsky I. J Clin Invest 1989;84: 1096-1104.
11. Bar-Shavit R, Benezra M, Eldor A, Hy-Am E, Fenton II JW, Wilner GD, and Vlodavsky I. Cell Reg 1990;1: 453-463.
12. Garcia JGN, Siglinger-Birnboim A, Bizios R, Del Vecchio PJ, Fenton II JW, and Malik AB. J Cell Physiol 1986; 128: 96-104.
13. Laposada M, Dovuarsky DK, and Solkin HS. Blood 1983; 62: 549-556.
14. Wilner GD, Danitz MP, Mudd MS, Hsieh KH, and Fenton II JW. J Lab Clin Med. 1981; 97: 403-411.
15. Furie B, Bing DH, Feldmann RJ, Robinson DJ, Burnier JP, and Furie BC. J Biol Chem 1982; 257: 3875-3882.
16. Bode W, Turk D, and Karshikov A. In: Protein Science 1992; 1: 426-471.
17. Markwardt F. Methods Enzymol 1970; 19: 924-932.
18. Dodt J, Muller HP, Seemuller V, and Chang J-Y. 1 FEBS (Fed. Eur. Biochem. Soc.) Lett 1984; 165: 180-183.
19. Fenton II JW, and Bing DH. Sem Thromb Haemostasis 1986; 12: 200-208.
20. Blasi F, Stopelli MP, and Cubellis MV. J Cell Biochem 1986;32: 179-186.
21. Dano K, Andearsen PA, Grodahl-Hansen J, Kristensen P, Nielsen LS, and Skriver L.. Adv Cancer Res 1985; 44: 139-266.
22. Plow EF, Freaney DE, Plesica J, and Miles LA. J Cell Biol 1986; 103: 2411-2420.
23. Saksela O, and Rifkin DB. Annu Rev Cell Biol 1988; 4: 93-126.
24. Blasi F. Fibrinolysis 1988; 2: 73-84.

216

25. Esko JD, Stewart TE, and Taylor WH. Proc Natl Acad Sci USA 1985; 82: 3197-3201.
26. Esko JD, Rostand KS, and Weinke JL.Science (Wash. DC) 1988; 241:1092-1096.
27. Esko JD. Curr Opin Cell Biol. 1991; 3: 805-816.
28. Lidholt K, Weinke JL, Kiser CS, Lugemwa FN, Bame KJ, Cheifetz S, Massague J, Lindahl U, and Esko JD. Proc Natl Acad Sci USA 1992; 89: 2267-2271.
29. Bar-Shavit R, Sabbah V, Lampugnani MG, Marchisio PC, Fenton II JW, Vlodavsky I, and Dejana E. J Cell Biol 1991; 112: 335-344.
30. Ruoslahti E, and Yamaguchi Y. Cell 1991; 64: 867-869.
31. Jackson RL, Busch SJ, Cardin AD. Physiol Rev 1991; 71: 481-539.
32. Kjellen L, Lindahl U. Biochemistry 199; 60: 443-675.
33. Ross R. N Engl J Med 1986; 314: 488-499.
34. Barrett TB, and Benditt EB. Proc Natl Acad Sci USA 1988; 85: 2810-2814.
35. Bar-Shavit R, Benezra M, Sabbah V, Bode W, and Vlodavsky I. Am J Respir Cell Mol Biol 1992; 6: 123-130.
36. Vu TH, Hung DT, Wheaton VI, and Coughlin SR. Cell 1991; 64: 1057-1068.
37. Rasmussen UB, Vouret-Craviari V, Jallat S, Schlesinger YM, Pages G, Pavirni A, Lecocq JP, Pouyssegur J, and Van-Obberghen-Schilling E. FEBS 1992; 288: 123-128.
38. Coughlin SR, Vu TH, Hung DT, and Wheaton VL. J Clin Invest 1992; 89: 351-355.
39. Vlodavsky I, Folkman J, Sullivan R, Fridman R, Ishai-Michaeli R, Sasse J, and Klagsbrun M. Proc Natl Acad Sci USA 1987; 84: 2292-2296.
40. Folkman J, Klagsbrun M, Sasse J, Wadzinski M, Ingber D, and Vlodavsky I. Am J Pathol 1988; 130: 393-400.
41 Vlodavsky I, Bar-Shavit R, Ishai-Michaeli R, Bashkin P, and Fuks Z. TIBS 1991; 16: 268-271.
42. Cardon-Cardo C, Vlodavsky I, Haimovitz-Friedman A, Hicklin D, and Fuks Z. Lab Invest 1990; 63: 832-840.
43. Gonzalez AM, Buscaglia M, Ong M, and Baird A.. J Cell Biol 1990; 110: 753 -765.
44. Burgess WH., and Maciag T. Ann Rev Bichem 1989; 58: 575-606.
45. Folkman J, and Klagsbrun M. Science 1987; 235: 442- 447.
46. Gospodarowicz D, and Cheng J. J Cell Physiol 1986; 128: 475-484.
47. Saksela O, Moscatelli D, Sommer A, and Rifkin DB. J Cell Biol 1988; 107: 743-751.
48. Bashkin P, Klagsburn M, Doctrow S, Shvan C-M, Folkman J, and Vlodavsky I. Bichemistry 1989; 28: 1737-1743.
49. Saksela O, and Rifkin DB. J Cell Biol 1990; 110: 767-775.
50. Benezra M, Vlodavsky I, Neufeld G, and Bar-Shavit R. Blood 1993; 81: 3324-3332.
51. Ross R, Masuda J, and Raines E. Ann. NY Acad Sci 1990; 598: 102-112.
52. Hatton MW, Moar SL, and Richardson M. Am J Pathol 1989; 135: 499-508.
53. Schwartz MS. J Clin Invest. 1993; 91: 4.

Biology of Vitronectins and their Receptors
K.T. Preissner, S. Rosenblatt, C. Kost, J. Wegerhoff and D.F. Mosher, editors

217

Basic fibroblast growth factor expression in human bone marrow cells and phospholipase C release of biologically active growth factor-heparan sulfate proteoglycan complexes

G. Brunner[a] and H. Nguyen[b].

[a]Department of Cell Biology and [b]Department of Surgery, New York University Medical Center, 550 First Avenue, New York, NY 10016, USA

INTRODUCTION

Hematopoiesis in human bone marrow is dependent on the continuous supply of growth factors such as the colony-stimulating factors (CSFs) [1] as well as on cellular interactions of hematopoietic progenitor cells with the bone marrow stroma [2]. It is thought that stromal cells express adhesion molecules ("anchor factors") as well as growth factors essential for adhesion, proliferation and differentiation of the progenitor cells [3, 4]. The hematopoietic growth factors appear to be localized in the stromal cell microenvironment and presented to the progenitor cells in a biologically active form [3, 4]. This hypothesis is supported by the fact that granulocyte-macrophage CSF and interleukin-3 have been shown to bind to heparan sulfate in bone marrow stroma in a biologically active form [5, 6].

Basic fibroblast growth factor (bFGF), a growth factor known for its role in angiogenesis, wound healing, and mesoderm induction [7], has also been found to be a potent hematopoietic growth factor. bFGF stimulates stem cell [8] and stromal cell growth [9], as well as myelopoiesis [10] and megakaryocytopoiesis [11]. bFGF is deposited into the extracellular matrix of cells [12] and matrix-bound bFGF provides a reservoir of biologically active growth factor [13]. bFGF is released from these sites as a biologically active growth factor-heparan sulfate proteoglycan (HSPG) complex by the serine proteinase plasmin [14] or by heparanases [15]. In a complex with the HSPG, bFGF is protected from proteolytic inactivation [16] and its diffusion within a cell monolayer is enhanced [17]. Therefore, the bFGF-HSPG complex can be considered the biologically active form of this growth factor.

In order to further evaluate the significance of bFGF as a hematopoietic growth factor, we studied bFGF expression in human bone marrow cells *in vitro* and *in vivo* as well as the interactions of this cytokine with the bone marrow stromal cell matrix.

bFGF EXPRESSION IN HUMAN HEMATOPOIETIC CELLS

Human bone marrow cells as well as peripheral blood cells were examined for the presence of bFGF by immunofluorescence. In bone marrow smears megakaryocytes and platelets stained strongly for bFGF, whereas weaker staining was observed in immature and mature cells of the granulocyte series [18]. Similar results were obtained with peripheral blood smears. The amount of bFGF in platelets was quantified by ELISA and platelet-associated bFGF in the circulation was estimated to be 1.6 ng/ml of blood. These results suggest that platelets are a potentially relevant source of bFGF in the circulation. Examination of primary human long-term bone marrow cultures by immunofluorescence combined with metabolic labeling of proteins followed by immunoprecipitation revealed bFGF synthesis by bone marrow stromal cells *in vitro* [18]. Because bFGF is a potent mitogen for human bone marrow stromal cells [9], it may act as an autocrine growth factor for these cells as it has been reported for endothelial cells [19]. Alternatively, bFGF might be produced by the bone marrow stromal cells and act on hematopoietic progenitor cells in a paracrine fashion.

bFGF BINDING TO STROMAL HEPARAN SULFATE PROTEOGLYCANS

Because bFGF is known to bind to HSPGs [7], we hypothesized that proteoglycans in the bone marrow stromal cell microenvironment might serve as a reservoir for biologically active bFGF. We therefore examined primary human bone marrow cultures for heparin-like low-affinity binding sites for bFGF expressed on the cell surface or in the extracellular matrix. bFGF binding sites were identified by incubation of the cells with [^{125}I]bFGF followed by quantification of bound bFGF in cell and matrix extracts (Figure 1). The results demonstrate the presence of heparin-like binding sites in the cultures, which are not saturable up to a bFGF concentration of 1 μg/ml. Approximately 13% of bFGF was recovered in the cell extract and 27% in the matrix extract resulting in a total recovery of 40%. The bFGF binding capacity of the bone marrow cultures exceeded 70 ng/10^5 cells. The majority of the bFGF binding sites were heparin-like in nature, because heparin competed for more than 95% of the cellular and 98% of the matrix binding sites.

Thus, primary human bone marrow cultures express a large number of heparin-like low-affinity binding sites for bFGF suggesting that these bFGF binding sites provide a reservoir of this growth factor in the stromal microenvironment.

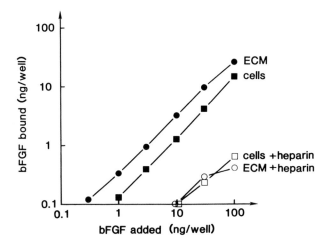

Figure 1. bFGF binding to heparin-like low-affinity sites expressed in primary human bone marrow cultures. Cells were incubated with 0-1,000 ng/ml of bFGF containing 7.9-79 nCi/ml [125I]bFGF (7.9 mCi/mg) in the absence or presence of 100 µg/ml of heparin in serum-free medium supplemented with 0.15% gelatin and 25 mM Hepes buffer, pH 7.2. After incubation on a shaker for 2 h at 4°C, excess bFGF was removed by extensive washing of the cell layer with PBS. Cells were lysed with 0.5% Triton X-100 in PBS, and the extracellular matrix was extracted with SDS-PAGE sample buffer. The amount of radioactivity in the cell and matrix extracts was determined using a gamma counter.

ENZYMATIC RELEASE OF bFGF-HEPARAN SULFATE PROTEOGLYCAN COMPLEXES

Using bFGF ligand blotting and ligand precipitation techniques, the major bFGF binding site on the stromal cell surface was identified as a HSPG of an apparent molecular weight in SDS-PAGE of approximately 200 kDa [20]. Because the soluble bFGF-HSPG complex appears to represent the biologically active form of this growth factor [16, 17], bFGF is most likely released from the stromal reservoir as a growth factor-HSPG complex which exerts its biological effects. We therefore investigated whether treatment with various enzymes releases bFGF-HSPG complexes from the bone marrow stromal cell surface.

Primary human bone marrow cultures were incubated with bFGF followed by treatment with heparinase, plasmin, phosphatidylinositol-specific phospholipase C (PI-PLC), or phosphatidylcholine-specific phospholipase C (PC-PLC). The enzymatically released bFGF-binding HSPG was detected by a bFGF

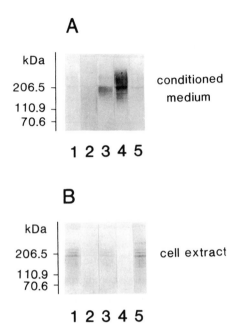

A

kDa

206.5 –

110.9 –
70.6 –

conditioned medium

1 2 3 4 5

B

kDa

206.5 –

110.9 –
70.6 –

cell extract

1 2 3 4 5

Figure 2. Enzymatic release of the bFGF-binding HSPG from bone marrow cultures. Cells were incubated for 2 h at 37°C with serum-free medium containing no enzyme (lane 1), 0.5 µg/ml heparinase (lane 2), 1 µg/ml plasmin (lane 3), 0.5 µg/ml PI-PLC (lane 4), or 0.5 µg/ml PC-PLC (lane 5). Cells were lysed with 0.5% Triton X-100 in PBS and aliquots of the cell extracts (B) and of the conditioned media (A) were analyzed by SDS-PAGE in 3-16% gradients under reducing conditions. The bFGF-binding HSPG was detected by ligand blotting using bFGF (1 µg/ml), affinity-purified anti-bFGF IgG (5 µg/ml), and anti-rabbit IgG coupled to alkaline phosphatase (1:10,000).

ligand blotting technique (Figure 2). The HSPG was most efficiently released by incubation with PI-PLC (lane 4), whereas PC-PLC did not release detectable amounts of the HSPG (lane 5) compared to control (lane 1). Incubation of the cells with plasmin also resulted in release of the HSPG, although, on a molar basis, PI-PLC appeared to be more efficient. Enzymatic release of the HSPG into the medium was accompanied by a decrease in the amount of the HSPG in the cell extract (panel B compared to panel A), and PI-PLC treatment of the cells almost completely transferred the HSPG from the cell extract into the medium (lane 4). Heparinase digested the HSPG in the medium as well as in the cell extract abolishing its bFGF binding capacity (lane 2). Consequently, when primary bone marrow cultures were incubated with bFGF, cell-associated bFGF was most efficiently released by incubation with PI-PLC [20]. bFGF was released as a soluble growth factor-HSPG complex and retained its biological activity.

The highly efficient release of the bFGF-binding HSPG by PI-PLC suggested that the proteoglycan is linked to the stromal cell surface via a glycosylphosphatidylinositol (GPI) anchor. This was confirmed by metabolic labeling of the HSPG with the GPI anchor constituents, ethanolamine and palmitic acid, indicating the presence of a GPI anchor attached to the HSPG [20].

CONCLUDING REMARKS

The hematopoietic growth factor, bFGF, is produced by primary human bone marrow cells *in vitro*. bFGF is also expressed in bone marrow *in vivo* and was found in cells of two hematopoietic lineages, megakaryocytes/platelets and granulocytes. bFGF binds to a GPI-anchored HSPG on the stromal cell surface providing a reservoir of this growth factor in the stromal cell microenvironment. Biologically active bFGF-HSPG complexes can be released from this reservoir by the action of PI-PLC and, to a lesser extent, by the serine proteinase plasmin. The soluble bFGF-HSPG complex might act in an autocrine and/or paracrine fashion to support stromal and hematopoietic progenitor cell growth (Figure 3). Therefore, enzymes releasing biologically active bFGF-HSPG complexes such as GPI-specific phospholipases or the proteolytic cascade of plasminogen activation might play a crucial role in regulating bFGF-supported hematopoiesis.

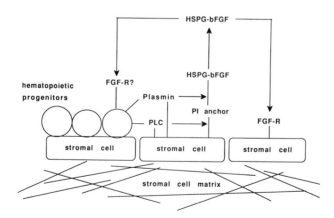

Figure 3. Biological action of bFGF in the bone marrow stromal microenvironment.

ACKNOWLEDGEMENTS

This work was supported by grants from the Deutsche Forschungsgemeinschaft (DFG; Br 1125/1-1), from the National Institute of Health (CA 49419, CA 34282, and CA 20194), and from the American Cancer Society (DHP-41).

REFERENCES

1   Clark SC, Kamen R. Science 1987; 236:1229-1237.
2   Dexter TM. J Cell Physiol Suppl 1982; 1:87-94.
3   Quesenberry P, Temeles D, McGrath H, Lowry P, Meyer D, Kittler E, Deacon D, Kister K, Crittenden R, Srikumar K. J Cell Biochem 1991; 45:273-278.
4   Lowry PA, Deacon D, Whitefield P, McGrath HE, Quesenberry PJ. Blood 1992; 80:663-669.
5   Gordon MY, Riley GP, Watt MS, Greaves MF. Nature 1987; 326:403-405.h
6   Roberts R, Gallagher J, Spooncer E, Allen TD, Bloomfield F, Dexter TM. Nature 1988; 332:376-378.
7   Rifkin DB, Moscatelli D. J Cell Biol 1989; 109:1-6.
8   Huang S, Terstappen LWMM. Nature 1992; 360:745-749.
9   Oliver LJ, Rifkin DB, Gabrilove J, Hannocks M-J, Wilson EL. Growth Factors 1990; 3:231-236.
10   Wilson EL, Rifkin DB, Kelly F, Hannocks M-J, Gabrilove JL. Blood 1991; 77:954-960.
11   Bruno E, Cooper RJ, Wilson EL, Gabrilove JL, Hoffman R. Blood 1993; 82:430-435.
12   Vlodavsky I, Folkman J, Sullivan R, Fridman R, Ishai-Michaeli R, Sasse J, Klagsbrun M. Proc Natl Acad Sci USA 1987; 84:2292-2296.
13   Flaumenhaft R, Moscatelli D, Saksela O, Rifkin DB. J Cell Physiol 1989; 140:75-81.
14   Saksela O, Rifkin DB. J Cell Biol 1990; 110:767-775.
15   Ishai-Michaeli R, Eldor A, Vlodavsky I. Cell Regulation 1990; 1:833-842.
16   Saksela O, Moscatelli D, Sommer A, Rifkin DB. J Cell Biol 1988; 107:743-751.
17   Flaumenhaft R, Moscatelli D, Rifkin DB. J Cell Biol 1990; 111:1651-1659.
18   Brunner G, Nguyen H, Gabrilove J, Rifkin DB, Wilson EL. Blood 1993; 81:631-638.
19   Sato Y, Rifkin DB. J Cell Biol 1988; 107:1199-1205.
20   Brunner G, Gabrilove J, Rifkin DB, Wilson EL. J Cell Biol 1991; 114:1275-1283.

Biology of Vitronectins and their Receptors
K.T. Preissner, S. Rosenblatt, C. Kost, J. Wegerhoff and D.F. Mosher, editors

VITRONECTIN IN INFLAMMATORY CONDITIONS: LOCALIZATION IN RHEUMATOID ARTHRITIC SYNOVIA

Bianca R. Tomasini Johansson
Dept. of Medical and Physiological Chemistry, University of Uppsala, BMC Box 575, S-75123, Uppsala, Sweden.

Rheumatoid Arthritis and Vitronectin

Rheumatoid arthritis is a common chronic inflammatory disease characterized by damage to peripheral joints. No clear mechanism has been elucidated explaining its development or progression (1). It is characterized by an inflammation of the synovial membrane surrounding the joint, increased numbers of mononuclear cells, synovial fibrosis, blood vessel proliferation, and destruction of the underlying tissue (1). It is known that an up-regulation of cytokines and proteases contributes to the progression of damage to the affected tissue (2). In addition, coagulation abnormalities have also been observed and it has been postulated that the deposition of fibrinogen/fibrin contributes to the pathogenesis of the disease (3).

Vitronectin has been characterized as possessing several activities, *in vitro*, which, *in vivo*, could be associated with inflammatory conditions. In addition to its adhesive properties (4), vitronectin has been shown to act as a scavenger of excess membrane attack complexes of complement (C5b-9), thereby protecting innocent bystander cells from lysis (5). Vitronectin has also been shown to be a scavenger for heparin in the heparin-dependent inactivation of thrombin and factor Xa by antithrombin III (6,7) and to act as a stabilizer molecule for the activity of plasminogen activator inhibitor-1 (PAI-1). The protection of thrombin and the inhibition of plasminogen activators (with the consequent inhibition of plasmin generation) could be postulated to result in the preservation of fibrin deposition.

Vitronectin circulates in plasma at a relatively high concentration (200-400 µg/ml), its site of synthesis being the liver (8,9). However, vitronectin has been localized extravascularly in association with various tissues (10). Thus, vitronectin could affect fibrin formation not only in blood but also at sites of extravascular deposition. In reference to the localization of vitronectin in tissues, it appears that except for adult dermal elastic fibers in the skin

(11) and blood vessels in the kidney (12), vitronectin is not found in normal tissues. Mostly, it has been found in amyloid deposits, sclerotic and fibrotic lesions associated with diseased skin, kidney, and vessel walls (12-15). Localization studies that have utilized the monoclonal antibody 8E6 (16) can be misleading, since it is now known that this monoclonal antibody cross-reacts with a matrix protein unrelated to vitronectin (17). It would be of interest to re-evaluate localization data obtained with this monoclonal antibody in order to ascertain that the molecule being recognized is vitronectin.

In order to investigate a potential role for vitronectin in inflammation, rheumatoid arthritic synovia (RAS) was analyzed by immunohistochemistry for the presence of vitronectin and haemostatic components with which it is known to interact, particularly in relation to fibrin deposition.

Deposition of vitronectin in RAS

Frozen sections from rheumatic synovia removed from patients in conjunction with synovectomy were processed for immunohistochemistry using described double-immunofluorescence techniques(17). Staining of vitronectin was carried out using affinity purified rabbit polyclonal antibodies and verified with a previously characterized monoclonal antibody (17). Vitronectin was found in all RAS studied mostly in association with "pannus" or granulation tissue areas (18). The amount of vitronectin present could be correlated to the development of granulation tissue. In some cases, where obvious granulation tissue was nil, vitronectin was present in small amounts in loose connective tissue and was absent in areas that appeared undamaged, as ascertained by examination of sequential sections after haematoxylin-eosin staining. This is in disagreement with a study reporting the presence of vitronectin in normal synovia (19). A reason for this discrepancy might be the use of the monoclonal antibody mentioned above that recognizes a matrix molecule different from vitronectin (17,19).

Mechanism of vitronectin deposition

Vitronectin has been shown to undergo a conformational change upon binding of thrombin-serpin complexes (20) or C5b-9 complement complex (21) that exposes its heparin-binding region. The exposure of this region might result in vitronectin binding to heparan sulfate proteoglycans in the stroma or on cell surfaces. These interactions could be involved but do not completely explain the mechanism of deposition of vitronectin onto granulation tissue areas, as vitronectin did not always co-localize with complement (stained with a monoclonal antibody to C5b, CBL192, Cymbus Bioscience) or thrombin (stained using a rabbit polyclonal antiserum to prothrombin, Dakopatts). Vitronectin has been shown to bind intracellular proteins in human and mouse systems (22-24). It is possible that extravasated vitronectin binds to intracellular components left in granulation tissue areas after cell death. In this respect, it has recently been shown that the $\alpha_v\beta_5$ integrin (a vitronectin receptor) mediates the uptake and degradation of conformationally changed vitronectin by fibroblasts (25). It had also been reported that another vitronectin receptor, the $\alpha_v\beta_3$ integrin on macrophages, mediates the uptake of apoptotic cells (26). It could be postulated that binding to cytoskeletal components or adsorption onto granulation tissue might change the conformation of vitronectin so that it and its scavenged products are taken up and degraded by phagocytes. Thus, vitronectin might also have a role in promoting the phagocytosis of cell debris in granulation tissue areas.

Vitronectin and fibrin deposition

The deposition of fibrin in RAS has been previously demonstrated (3), although the mechanisms involved in its deposition and turnover are unclear as yet. Staining of fibrin was performed with monoclonal antibody (E8, Immunotech) that recognizes fibrin and not fibrinogen. Of the RAS examined in this study, only a minority showed extensive fibrin deposition. In these cases, there was a strict co-localization of vitronectin and fibrin deposition (18) present in granulation tissue areas. In other RAS there was little fibrin deposition when vitronectin was found in loose connective tissue or small areas of granulation tissue. Thus, in the observed cases, a clear deposition of fibrin coincided with vitronectin in extensive granulation tissue areas.

Does vitronectin affect fibrin deposition?

In order to determine whether the observed fibrin deposition could be promoted by vitronectin via either of the pathways mentioned above (protection of thrombin or binding of PAI-1 by vitronectin) staining for these two components was carried out. Staining with a rabbit polyclonal antibody to prothrombin indicated the presence of cell-associated thrombin/prothrombin but striking co-localization with fibrin-rich areas was not detected. Staining for PAI-1 (monoclonal antibody MAI-11, Biopool) was mostly cell associated and rather extensive throughout all the RAS tissues examined. PAI-1 was clearly present in the endothelial lining of blood vessels whereas vitronectin was absent in these areas. No deposits of PAI-1 either in stroma or granulation tissue were observed. However, PAI-1-positive cells were found in granulation tissue areas that were also positive for vitronectin and fibrin (18). Vitronectin has been shown to be the binding protein for PAI-1 in matrices of cultured cells (27) and blood (28). Binding of PAI-1 by vitronectin has been shown to cause a decrease in cell-associated plasmin activity (29). It is possible that vitronectin stabilizes the cell-asociated PAI-1 activity in granulation tissue areas, thus inhibiting fibrinolysis at these sites.

Cell-associated vitronectin in RAS

Interestingly, in most of the RAS studied, small groups of cells with associated vitronectin were found (18). Some of these cells were also stained with a macrophage and leukocyte markers (monoclonal antibody anti-CD68, Dakopatts and monoclonal antibody against the β2 chain of leukocyte integrins). However, most macrophages and other leukocytes were negative for vitronectin. It is interesting to note that, in contrast to other adhesion-promoting molecules, such as fibronectin, laminin, or collagen, vitronectin is not synthesized by connective tissue cells, in culture (30) or *in situ* (31). Only hepatocytes or hepatoma cells have been reported to synthesize vitronectin in culture, as shown by metabolic labeling and immunoprecipitation (30,32). Peritoneal macrophages and blood monocytes were reported to synthesize vitronectin in culture as detected by ELISA and immunoblotting of culture media (33) and glioblastoma cells were reported to synthesize vitronectin *in situ*, but not in culture (34). Isolation of adherent cells from RAS followed by metabolic labeling and immunoprecipitation with polyclonal and monoclonal antibodies to vitronectin demonstrated the synthesis of vitronectin by these cells (18). Isolated cells producing vitronectin were mesenchymally-derived as they were positive for vimentin, some cells being recognized by leukocyte markers as well. Demonstration of *in situ* synthesis awaits further investigation. However, it is clear that cell populations present in RAS are capable of producing vitronectin in culture. This finding could enable further analysis of the regulation of vitronectin synthesis by inflammatory or "activated" cells and its potential role(s) in inflammation.

Summary

The presence of vitronectin in RAS might promote the preservation of fibrin at sites such as granulation tissue by binding to cell-associated PAI-1 present in these areas. PAI-1 present in other areas devoid of fibrin, such as blood vessels, did not apparently co-localize with vitronectin. The observed association of vitronectin with cell populations in RAS is probably due to vitronectin synthesis by these cells, and not only binding to the cell surface by extravasated vitronectin.

References

1. Krane, S.M. and Simon, L.S. Med. Clin. North Am. 1986 70:263-284.
2. Hamilton, J. A., Campbell, I.K., Wojta, J., and Cheung, D. Ann. N. Y. Acad. Sci. 1992 667:87-100.
3. Zacharski, L.R., Brown, F.R., Memoli, V.A., Kisiel, W., Kudryk, B.J., Rousseau, S.M., Hunt, J.A., Dunwiddie, C., and Nutt, E.M. Clin. Immunol. Immunopathol. 1992 63:155-162.
4. Hayman, E. G., Engvall, E., A'Hearn, E., Barnes, D., Pierschbacher, M. and Ruoslahti, E. J. Cell Biol. 1982 95:20-23.
5. Podack E.R., Kolb. W.P., Müller-Eberhard, H.J. J. Immunol. 1978 120:1841-1848.
6. Podack, E.R., Dahlbäck, B., and Griffin, J.H. J. Biol. Chem. 1986 261:7387-7392.
7. Preissner, K.T., Müller-Berghaus, G. J.Biol. Chem. 1987 262:12247-12253.
8. Kemkes-Matthes, B., Preissner, K.T., Langenscheidt, F., Matthes K.J. and Müller-Berghaus, G. Eur. J. Haematol. 1987 39:161-165.
9. Conlan, M.G., Tomasini, B.R., Schultz, R.L. and Mosher, D.F. Blood 1988 72:185-190.
10. Tomasini, B.R. and Mosher, D.F. Prog. Hemost. Thromb., Coller, B., editor, 1991 10:269-305.
11. Dahlbäck, K. Löfberg, H., Alumets, J. and Dahlbäck, B. J. Invest. Dermat. 1989 92:727-733.
12. Dahlbäck, K., Löfberg, H., and Dahlbäck, B. Histochem. 1987 87:511-515.

228

13. Falk, R.J., Podack, E., Dalmasso, A., Jenette, J.C. Am. J. Pathol. 1987 127:182-190.
14. Guettier, C. Hinglais, N. Bruneval, P., Kazatchkine, M., Bariety, J., Camilleri, J-P. Virchows Archiv. A. Pathol. Anat. 1989 414:309-313.
15. Niculescu, F., Rus, H.G., Porutiu, D., Ghiurca, V., and Vlaicu, R. 1989 78:197-203.
16. Hayman, E.G., Pierschbacher, M.D., Ohgren, Y., and Ruoslahti, E. Proc. Natl. Acad. Sci. 1983 80:4003-4007.
17. Tomasini-Johansson, B.R., Ruoslahti, E., and Pierschbacher, M.D. Matrix 1993 13:203-214.
18. Tomasini-Johansson, B.R. and Rubin, K., manuscript in preparation.
19. Demaziere, A. and Athanasou, N.A. J. Pathol. 1992 168:209-215.
20. Tomasini B.R. and Mosher, D.F. Blood 1988 72:903-912.
21. Hogåsen, K., Mollnes, R.E., and Harboe, M. J. Biol. Chem. 267:23076-23082.
22. Hintner, H., Stanzl, U., Dahlbäck K., Dahlbäck B., and Breathnach, SM. J. Invest. Dermatol. 1989 93:656-661.
23. Podor, T.J. Joshua, P., Butcher, M., Seiffert, D., Loskutoff, D., and Gauldie, J. Ann. N.Y. Acad. Sci. 1992 667:173-177.
24. Dahlbäck, K., Wulf, H-C., and Dalbäck, B. J. Invest. Dermatol. 1993 100:166-170.
25. Scalise Panetti, T. and McKeown-Longo P.J. J. Biol. Chem. 1993 268:11492-11495.
26. Fadok, V.A., Savill, J.S., Haslett, C., Bratton, D.L., Doherty, D.E., Campbell,P.A. and Henson, P.M. J. Immunol. 1992 149:4029-4035.
27. Mimuro, J., and Loskutoff, D.J. J. Biol. Chem. 1989 264:936-939.
28. Wiman, B., Almqvist, A., Siguardardottir, O., and Lindahl, T. FEBS Lett 1988 242:125-128.
29. Ciambrone, G.J., and McKeown-Longo, P.J. J. Cell Biol. 1990 111:2183-2195.
30. Barnes D.W., and Reing. J. J. Cell Physiol. 1985 125:207-214.
31. Seiffert, D., Keeton, M., Eguchi, Y., Sawdey, M., and Loskutoff, D.J. Proc. Natl. Acad. Sci. 1991 88:9202-9206.
32. Jenne, D., Hille, A., Stanley, K.K., and Huttner, W.B. Eur. J. Biochem. 1989 185:391-395.
33. Hetland, G., Pettersen, H.B., Mollnes, T.E. and Johnson, E. Scand. J. Immunol. 1989 29:15-21.
34. Gladson, C.L. and Cheresh, D.A. J. Clin. Invest. 1991 88:1924-1932.

Biology of Vitronectins and their Receptors
K.T. Preissner, S. Rosenblatt, C. Kost, J. Wegerhoff and D.F. Mosher, editors

Tenascin and DSD-1-proteoglycan - extracellular matrix components involved in neural pattern formation and remodeling

Andreas Faissner, Albrecht Clement, Bernhard Götz, Angret Joester, Claudia Mandl, Christiane Niederländer, Oliver Schnädelbach and Angela Scholze

Department of Neurobiology
University of Heidelberg
INF 364
69120 Heidelberg
Germany (FRG)

Introduction

The development of the central nervous system of vertebrates unfolds as a finely tuned sequence of individual steps. Several key events have been distinguished, such as proliferation of epithelial stem cells, migration of neuronal precursors from the ventricular zone to target areas, formation of neuronal nuclei, innervation of distal territories, neuronal cell death and, finally, pruning and rearrangement of connections. These developmental stages are tightly controlled by several regulatory mechanisms including specific intercellular, membrane-mediated interactions, the presence of limiting amounts of growth factors and electrical activity [1]. In this context, the interplay between neurones and astrocytes might play a pivotal role, for example for the guidance of migrating neuronal precursors or of advancing growth cones and for the segregation of forming neuronal aggregates and nuclei [2, 3]. Several adhesion and extracellular matrix (ECM) molecules have been identified which play a role in neural cell interactions. Thus, members of the Ig-superfamily underlie Ca^{++}-independent and cadherins Ca^{++}-dependent adhesion processes [4, 5] and integrins mediate the interactions with ECM glycoproteins, e.g. laminin and fibronectin [6]. Also, an increasing number of proteoglycans has been detected, but the functions of this class of ECM components in neural tissues are poorly characterized [7]. Recently, tenascin glycoproteins have attracted increasing attention because they are transiently expressed by astrocytes during neural development and exhibit definable effects on neural cell behaviour [8, 9].

Distribution and structure of tenascin glycoproteins

Tenascin glycoproteins have been independently discovered by several workers in various developing organs [10]. In most of the cases, tenascin expression is confined to circumscribed areas in a "site-restricted" manner, which has been proposed as one prominent feature of the glycoprotein [11]. In particular, tenascin glycoproteins show a striking transient expression pattern in several defined areas of

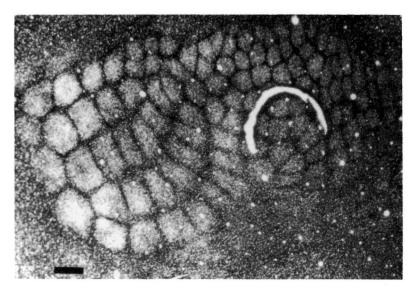

Figure 1. Tenascin boundaries in the cerebral cortex. A postnatal day 6 (P6) flattened tangential section of the mouse somatosensory cortex stained with tenascin antibodies is shown. Note that transient tenascin immunoreactivity delineates the prospective barrels, processing units for sensory afferent information. Bar: 200 µm, picture courtesy by Dr. Steindler, see text and [12] for details.

the nervous system, for example in the barrel field of the somatosensory cortex where they delineate the borders of prospective barrels, processing units of sensory afferent information originating in the periphery (Fig. 1, and [3, 12, 13]). The emergence of tenascin in discrete glial boundaries rises the possibility that tenascin contributes to the patterning of neural structures into functional units [3, 9]. In adult neural and non-neural tissues tenascin is not detectable or present at low levels, but displays enhanced expression after infliction of leasion and in tumor tissues [14-16]. Derived primary amino acid sequences have been reported for several species, including chicken, mouse and human tenascin. In the mouse, the glycoprotein consists of a cysteine-rich segment at the aminoterminus, subsequent 141/2 EGF-type repeats followed by 8 fibronectin type III (FNIII) repeats and homologies to fibrinogen β and γ at the carboxyterminus. Isoforms of tenascin are generated by insertion of additional FNIII repeats between the 5th and the 6th FNIII repeat of the basic structure (Fig. 2, [17]). Mouse tenascin glycoproteins display molecular weights ranging from 190 kD to 260 kD apparent Mr and are linked to multimers through disulfide-bridges at their amino terminal ends. Tenascin hexamers appear as so called hexabrachia by electron microscopy of rotary shadowed preparations [10, 18, 19].

Functional properties of tenascin in neural tissue
In short term cell-binding assays polyclonal antibodies to tenascin reduce the attachment of neurones to astrocyte surfaces but do not affect neurone-neurone or

astrocyte-astrocyte adhesion, suggesting that tenascin glycoproteins specifically contribute to the molecular basis of neurone-glia interactions [8, 20]. In view of the asserted role of glia in guiding neuronal precursor cells [2], the influence of tenascin on neuronal migration has been examined in more detail. Both poly- and monoclonal antibodies to tenascin retard granule cell migration from the external to the internal granule cell layer in explant suspension cultures of the developing mouse cerebellum, where tenascin is clearly expressed by Bergmann glia. The use of a panel of monoclonal antibodies with known binding sites on mouse tenascin allowed to circumscribe FNIII repeats 3-5 as critically important in regulating the migration process (Fig. 2, [19, 21, 22]). This region contains an RGD-motive in chicken and human tenascin and endothelial cells attach to it by an integrin-dependent mechanism [23]. A second cell binding site has been attributed to the carboxyterminal end which contains a heparin binding site and proteoglycan(s) may be required for cell-tenascin interactions with this domain [24-27]. Yet, although tenascin contains at least two cell binding sites the purified glycoprotein forms a poor substrate for cell culture in most cases examined so far [9, 28]. This is particularly true for neurones, where tenascin exerts pronounced anti-adhesive effects. In fact, both cell bodies and neurites are specifically repulsed from tenascin-containing territories of patterned tenascin/poly-ornithine substrates [29]. Studies with bacterially-derived fusion proteins suggest that the anti-adhesive properties of tenascin are encoded by domains separate from those involved in cell binding and migration [25, 26]. Hence, it is conceivable that cell-tenascin interactions evolve as a temporal sequence of singular events, where cell binding would precede recognition of the anti-adhesive site(s) of the molecule. The latter may cause anti-spreading by launching secondary events which might eventually impinge onto cytoskeletal organisation. The signal transduction pathways implied by this model are presently unknown [30, 31], for detailed discussion see [9, 28]. Interestingly, even though tenascin is anti-adhesive for neurons and their processes, the molecule promotes neurite outgrowth from several types of PNS and CNS neurons cultivated on homogeneous tenascin-containing substrates [18, 32]. In mouse, the FNIII domains D or 6 are implicated in this phenomenon, as demonstrated with the function-blocking antibody J1/tn2 [18]. It is noteworthy in this context that fiber outgrowth from retinal explants is prevented by tenascin under comparable conditions, indicating that enhanced neurite growth in response to tenascin depends on the neuronal lineage [33, 34]. Increased neurite lengths are presumably caused by stimulation of growth cone velocity, which rises by 50% on homogeneous tenascin-containing substrates [34]. Nevertheless growth cones, comparable in this regard to neurites and neuronal cell bodies, are deflected from tenascin-containing territories in choice situations on patterned substrates where tenascin has been combined with conducive components like polycations, laminin or fibronectin [9, 34]. In conclusion, tenascin glycoproteins display both inhibitory, repulsive or conducive, growth promoting properties for neurones, depending on the mode of presentation. This is consistent with a role of tenascin in mediating both supporting and segregating glial properties in differentiating neural tissues [3, 9]. In view of its influences on neuronal cell behaviour, tenascin could also play a role in tissue remodeling after injury. Indeed, tenascin is up-regulated both in the PNS and in the CNS after lesioning [14, 16, 35]. For example, it has been proposed that tenascin prevents regeneration of sensory axons into the spinal cord after damage to the dorsal root [35]. However, it has to be kept in mind that axons cruise freely through

232

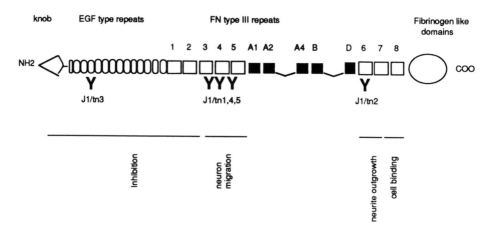

Figure 2. Functional domains of tenascin. The drawing shows the structural elements of the largest known isoform of mouse tenascin. Monoclonal antibodies J1/tn1-5 have been used to map distinct tenascin functions to separate domains. The figure condenses data from [17-19, 25]. For detailed discussion see text and [9, 28].

tenascin-containing territories in the lesioned peripheral nerve [14]. The functional role(s) of tenascin in the context of pathology have not yet been clarified. Several distinct receptors probably respond to the different functional domains of tenascin glycoproteins, but information relating to neural tissues is scarce. Recently, a *neuronal* chondroitin sulfate proteoglycan receptor called CTBP has been documented [36, 37], reviewed in [9]. For these reasons, the chondroitin sulfate proteoglycan DSD-1-PG, which co-localizes with tenascin in several areas of the developing nervous system, e.g. the boundaries of the somatosensory barrel field [13], has been investigated in more detail.

Distribution, structure and function of DSD-1-PG
DSD-1-PG has been identified with the help of the monoclonal antibody DSD-1 (for dermatan-sulfate-dependent Nr.1, formerly 473, [38]) generated against L2/HNK-1-positive glycoprotein fractions purified from mouse brain (rest-L2, [8, 38]). In immunofluorescence double labeling experiments performed on mouse cerebellar cultures of various ages, the antibody stains the surface of subsets of glial cells. Use of glial markers such as vimentin and glial fibrillary acidic protein (GFAP) for immature and mature astrocytes and O4 and O1 for immature and mature oligodendrocytes, respectively, in comparison with DSD-1 has clearly assigned the DSD-1-epitope to the surfaces of early differentiation stages of these glial lineages [38]. This conclusion is corroborated by immunohistological localisation studies which document expression of DSD-1 in immature glia, in particular of glial boundaries,

where DSD-1 co-localizes with tenascin in several cases [13]. The antigen recognized by DSD-1-PG in neural tissues migrates as polydisperse smear of 1000 kD Mr according to Western blot experiments [38]. Material of similar size can be immunoprecipitated from detergent-extracts and supernatants of glial cultures biosynthetically labeled with ^{35}S-SO$_4$ and ^{3}H-fucose. The immunoprecipitates are degradable by chondroitinases ACII and ABC, but resistant to heparinase, heparitinase and to keratanase, consistent with the notion that DSD-1-PG is a chondroitin sulfate proteoglycan [38]. In order to deepen the understanding of DSD-1-PG, the molecule has been purified from postnatal mouse brains using a combination of immunoaffinity and ion-exchange chromatography. The purified DSD-1-PG possesses an Mr of 1000 kD and a major core glycoprotein of 350-400 kD which appears after digestion with chondroitinases ACII and ABC, confirming the results obtained in biosynthetic labeling studies. Interestingly, the purified DSD-1-PG promotes neurite outgrowth from rat embryonic day 18 hippocampal neurons in a concentration-dependent manner [39, and Faissner et al., submitted]. Thus, DSD-1-PG could be one of the constituents of immature glial surfaces which are supportive of neuronal process extension and constitutes a potentially important ECM component of developing neural tissues in its own right. It might be one of the proteoglycans described in bulk preparations from rat brains [7]. The question whether the molecule interacts with tenascin is presently under scrutiny.

Acknowledgements

We thank the Deutsche Forschungsgemeinschaft for grant support (SFB 317/A2 and Fa 159/5-1 to A. Faissner), the Graduiertenkolleg Zelluläre und Molekulare Neurobiologie and the Studienstiftung des Deutschen Volkes for graduate training grants to A. Joester, B. Götz, and A. Scholze, Prof. Dr. W. Huttner for encouragement and support and Dr. J. Trotter for stimulating, ongoing discussion. A. Faissner is tenant of an endowed professorship of the H.-L.-Schilling-Stiftung of the Stifterverband für die Deutsche Wissenschaft.

REFERENCES

1   Jacobson M. Developmental Neurobiology. 3rd Edition, New York, London: Plenum Press, 1991; 1-776.
2   Rakic P. Science 1988; 241: 170-176.
3   Steindler DA. Annu Rev Neurosci 1993; 16: 445-470.
4   Rathjen FG, Jessel TM. Semin Neurosci 1991; 3: 297-307.
5   Takeichi M. Science 1991; 251: 1451-1455.
6   Reichardt LF, Tomaselli FK. Annu Rev Neurosci 1991; 14: 531-570.
7   Herndon ME, Lander AD. Neuron 1990; 4: 949-961.
8   Kruse J, Keilhauer G, Faissner A, Timpl R, Schachner M. Nature 1985; 316: 146-148.
9   Faissner A. Perspectives Dev Neurobiol 1993; 1: 155-164.
10 Erickson HP, Bourdon MA. Annu Rev Cell Biol 1989; 5: 71-92.
11 Crossin KL, Hoffman S, Grumet M, Thiery J-P, Edelman GM. J Cell Biol 1986; 102: 1917-1930.
12 Steindler DA, Cooper NGF, Faissner A, Schachner M. Dev Biol 1989; 131: 243-260.

234

13   Steindler DA, O`Brien TF, Laywell E, Harrington K, Faissner A, Schachner M. J Exp Neurol 1990; 109: 35-56.
14   Martini R, Schachner M, Faissner A. J Neurocytol 1990; 19: 601-616.
15   Koukoulis GK, Gould VE, Bhattacharyya A, Gould JE, Howeedy AA, Virtanen I. Hum Pathol 1991; 22: 636-643.
16   Laywell E, Dörries U, Bartsch U, Faissner A, Schachner M. Proc Natl Acad Sci USA 1992; 89: 2634-2638.
17   Weller A, Beck S, Ekblom P. J Cell Biol 1991; 112: 355-362.
18   Lochter A, Vaughan L, Kaplony A, Prochiantz A, Schachner M, Faissner A. J Cell Biol 1991; 113: 1159-1171.
19   Husmann K, Faissner A, Schachner M. J Cell Biol 1992; 116: 1475-1486.
20   Grumet M, Hoffman S, Crossin KL, Edelman GM. Proc Natl Acad Sci USA 1985; 82: 8075-8079.
21   Chuong C-M, Crossin KL, Edelman GM. J Cell Biol 1987; 104: 331-342.
22   Bartsch S, Bartsch U, Dörries U, Faissner A, Weller P, Ekblom P, Schachner M. J Neurosci 1992; 12: 736-749.
23   Sriramarao P, Mendler M, Bourdon M. J Cell Sci (in press).
24   Faissner A, Kruse J, Kühn K, Schachner M. J Neurochem 1990; 54: 1004-1015.
25   Spring J, Beck K, Chiquet-Ehrismann R. Cell 1989; 59: 325-334.
26   Prieto AL, Andersson-Fisone C, Crossin KL. J Cell Biol 1992; 119: 663-678.
27   Aukhil I, Joshi P, Yan Y, Erickson HP. J Biol Chem 1993; 268: 2542-2553.
28   Chiquet-Ehrismann R. Curr Opin Cell Biol 1991; 3: 800-804.
29   Faissner A, Kruse J. Neuron 1990; 5: 627-637.
30   Lotz MM, Burdsal CA, Erickson HP, McClay DR. J Cell Biol 1989; 109: 1795-1805.
31   Murphy-Ullrich JE, Lightner VA, Aukhil I, Yan YZ, Erickson HP, Höök M. J Cell Biol 1991; 115: 1127-1136.
32   Wehrle B, Chiquet, M. Development 1990; 110: 401-415.
33   Perez, RG, Halfter W. Dev Biol 1993; 156: 278-292.
34   Taylor J, Pesheva P, Schachner, M. J Neurosci Res 1993; 35: 347-362.
35   Pindzola RR, Doller C, Silver J. Dev Biol 1993; 156: 34-48.
36   Hoffman S, Edelman GM. Proc Natl Acad Sci USA 1987; 84: 2523-2527.
37   Hoffman S, Crossin KL, Edelman GM. J Cell Biol 1988; 106: 519-532.
38   Faissner A. Soc Neurosci Abstr 1988; 14: 920.
39   Faissner A, Lochter A, Streit A, Schachner M. Eur J Neurosci 1992; Suppl 5: 85.

Immune defense and micro-organisms

Biology of Vitronectins and their Receptors
K.T. Preissner, S. Rosenblatt, C. Kost, J. Wegerhoff and D.F. Mosher, editors

Lymphocyte mediated cytotoxicity

Eckhard R. Podack, M.D.[1]

University of Miami School of Medicine, Department of Microbiology and Immunology, 1600 NW 10th Avenue (R-138), Miami, FL 33136

INTRODUCTION

Cytotoxic lymphocytes may be divided into two groups. First, cells with cytolytic activity that do not require prior stimulation, e.g. natural killer (NK) cells; and, second, cells that become cytotoxic only after an inductive event, usually presentation of antigen. In this latter group are found the cytotoxic T-lymphocytes (CTL), belonging to the system of adaptive immunity. Both, natural cytotoxicity and adaptive cytotoxicity are important for removal of virally infected or otherwise parasitized cells. In addition it is postulated that cytotoxic lymphocytes are important in immune surveillance against tumors. Of pathophysiological significance is the role of cytotoxic lymphocytes in transplant rejection and in autoimmune diseases.

The cytotoxic mechanisms used by naturally or adaptively cytotoxic cells are largely identical (For review see Annual Rev. Immunol 1, Ann. Rev. Cell Biol. 2). Differences however appear to exist between subpopulations of T-cells regarding their killing mechanism. For instance a subset of CD4+ T-cells does not express perforin yet is endowed with cytolytic activity.

CYTOTOXIC MECHANISMS

Cytolytic lymphocytes kill target cells by membrane damage, i.e. pore formation (3,4), and by mediating nuclear disintegration of the target cell (5). Lysis is mediated by secretory and by non secretory mechanisms. The latter pathway signals target cell death via membrane protein interaction involving the Fas antigen and other receptors (7). It is evident from this brief summary that multiple mechanisms of target cell lysis exist and are deployed by CTL and NK cells simultaneously. Fig. 1 shows in schematic form the different pathways of CTL mediated cytolysis. It is likely that normally more than one pathway is activated and that target cell lysis results from multiple, partly independent insults by the killer cell. For instance, it is probable that the secretion of granules containing perforin and granzymes may be triggered concomitantly with the secretion of TNF and lymphotoxin and with the engagement of Fas, mediating target suicide. On the other hand the presence of multiple pathways may be critical in situations where the target cells are resistant to certain killing pathways. Absence of TNF receptors or of Fas on the target, for instance, renders the lytic mechanisms relying on these receptor interactions ineffective and other effector mechanisms must be used.

238

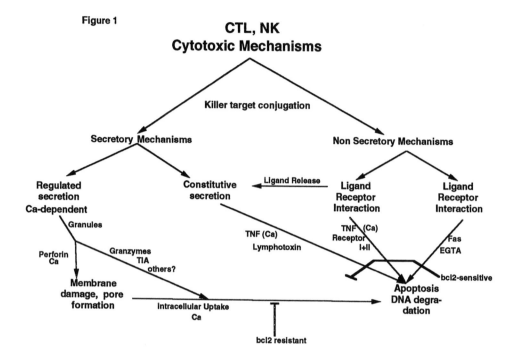

Figure 1

CTL, NK Cytotoxic Mechanisms

Killer target conjugation

Secretory Mechanisms

Non Secretory Mechanisms

Regulated secretion Ca-dependent

Constitutive secretion

Ligand Release

Ligand Receptor Interaction

Ligand Receptor Interaction

Granules

Perforin Ca

Granzymes TIA others?

TNF (Ca) Lymphotoxin

TNF (Ca) Receptor I+II

Fas EGTA

bcl2-sensitive

Membrane damage, pore formation

Intracellular Uptake Ca

Apoptosis DNA degra- dation

bcl2 resistant

NON SECRETORY KILLING MECHANISMS. FAS AND MEMBRANE TNF MEDIATED LYSIS

Engagement by the CTL of Fas on the target cell causes target cell lysis by DNA degradation. Lysis is rapid and proceeds in the presence of EGTA. Cells from LPR mice which are deficient in Fas do (8) not exhibit CTL lysis in the presence of EGTA (6), yet show normal CTL killing in the presence of Ca. This indicated that Fas, although contributing to the cytolytic activity of CTL, is not essential for cytotoxicity. Quantitative measurements indicate that Fas in the presence of EGTA can produce maximally 30% of the lysis seen normally in Ca (9) within 4h.

How Fas is triggered by the CTL is not known. It is possible that CTL express the Fas ligand on their surface or that the Fas ligand is assembled upon killer target conjugate formation. It cannot be ruled out currently that Fas is activated by mechanisms that do not rely on receptor - ligand interaction. However, a genetic defect in gld mice causing the identical lymphoproliferative syndrome as Fas deficiency, has generally been interpreted to indicate a deficiency of the Fas ligand.

Fas mediated cell death has been suggested also for the elimination of T-cells during thymic selection; the deficiency of Fas in lpr mice is accompanied by an autoimmune proliferative disorder as a result of the failure to eliminate autoreactive cells. Since Fas mediated T-cell death seems to be counteracted by the expression of bcl2, it appears likely that susceptibility to Fas killing is not only controlled by the availability of Fas and the putative Fas ligand, but also by the level of expression of bcl2 in the target cell (10).

TNFα and LT (lymphotoxin)αß are type two membrane proteins that occur also in soluble form (11, 12). The membrane associated form of these molecules can mediate cytotoxicity of sensitive cells form in a relatively slow (12h to 24h) Ca dependent reaction.

SECRETORY KILLING MECHANISMS. CONSTITUTIVE SECRETION

The prime examples of toxic molecules constitutively secreted by CTL and NK cells are TNFα and ß (13). These two cytokines in addition occur in membrane associated form (see above). Membrane associated TNFα is upregulated upon CTL activation by CD3 and is responsible for the slow (i.e. 18h) lysis of TNF sensitive target cells. TNF does not account for or even significantly contribute to the acute lysis (i.e. 4h) of target cells (7). Although TNF secretion is blocked by protein synthesis inhibitors and cyclosporin A, lysis by membrane associated TNF is not affected by theses agents. Moreover slow TNF mediated lysis is not TCR dependent and requires Ca. TNF mediated cytotoxic mechanisms therefore are restricted to TNF sensitive cells lysed in a TCR independent fashion.

REGULATED SECRETION

CTL and NK cells contain microscopically detectable storage granules. These organelles are especially abundant in cultured cells following repeated antigen stimulation and generous availability of Il2. Interaction with the target cell induces the secretion of these granules in the direction of the killer-target cell contact area in a Ca dependent reaction. The vectorial secretion results in the exocytosis of granules into the interstitial space between killer and target.

MEMBRANE DAMAGE

Granules contain perforin, a Ca dependent pore forming protein, whose receptor is the phospholipid headgroup of bilayer membranes (14). Phospholipid binding and polymerization, accompanied by hydrophobic membrane insertion, of perforin are Ca dependent. Poly-perforin pores vary in size from 1 to 16nm depending on the multiplicity of perforin protomers per complex (15). Since perforin is secreted in granule size packets containing many perforin molecules, perforin pores will tend to occur in clusters on the target membrane. Perforin pores delivered in sufficient numbers are of sufficient size to destroy a cell simply by irreparable damage of the membrane, thus disrupting the osmotic barrier and allowing equilibration of the intra- and extracellular milieu. At sufficiently low dose however perforin pores may be repaired by the target cell through repair endocytosis. The perforin pore in the target cell membrane allows influx of Ca, which in turn represents the signal to endocytose the perforated membrane. In this reaction the membrane is resealed and the pore repaired. However, in the same

Table 1: Effector Molecules of Cytotoxic Lymphocytes

Molecule (Effector)	Localization	Target/ Receptor	Cofactor	Ca-Need	Target Response	Note
Perforin	Granule	Phospholipid	None	Yes	Membrane Pores	Ca-influx
Granzymes	Granule	Substrate unknown	Perforin	Yes	Apoptosis	bcl2-insensitive
TIA	Granule, Cytoplasm	Polyadenylated RNA	Perforin	ND	Apoptosis	RNA binding
Membrane TNF, TNFα	Membrane and secreted	TNF receptor	None	Yes	Apoptosis	slow kinetics
Membrane LT, LT (TNFβ)	Membrane and secreted	TNF receptor	None	ND	Apoptosis	slow kinetics
Fas ligand or activator	ND	Fas	ND	No	Apoptosis	bcl2 sensitive

ND: Not determined

process the target cell is forced to take up extracellular liquid enclosed by the endocytosed membrane vesicle (pinocytosis). If the extracellular liquid contains soluble TNF or granzymes, for instance, rapid DNA degradation will take place. Similarly, other molecules may gain entry into the target cell and cause nuclear break down (16).

APOPTOSIS

In the cytotoxic pathway mediated by regulated secretion of granules (17, 18) the formation of membrane pores by perforin is the prerequisite for apoptosis of the target cell. The main protein component of granules is a group of serine proteases known as granzymes. In the murine system the granzymes have been designated by letters (A-F) (19) and in the human system by numbers (1-3) (20). These granule proteases are highly homologous to each other. Their main function is to mediate target DNA degradation upon gaining entry into the target cell through the presence of perforin pores. The diversity of granzymes has been postulated to be necessary for the induction of apoptosis in diverse cells, endowed with different mechanisms of DNA degradation. DNA degradation is also induced by TIA (21), a protein homologous to a group of RNA binding proteins, and possibly by Ca influx alone in sensitive cells.

The mechanism by which DNA degradation is induced is not known. However in contrast to many forms of developmental apoptosis, granule mediated DNA degradation is not dependent on macromolecular synthesis. It is however dependent on the presence of intracellular Ca in the target cells and is not inhibited by bcl2 (16).

SUMMARY

CTL and NK mechanisms of mediating target cell lysis use an arsenal of cytotoxic molecules, that kill target cells through multiple pathways (Table 1). The use membrane associated proteins on the target cell to signal target suicide obviously requires a compliant target that is not only expressing the respective receptors and but also responds in the appropriate way. Quite frequently such compliance cannot be taken for granted. This is true in particular when cells become virus infected or are oncogenically transformed. Both, viral infection and tumor transformation, cause the down regulation or abrogation of normal growth control pathways including resistance to normal suicide mechanisms. Overexpression of bcl2, or virus encoded homologous proteins, for instance, confer resistance to several forms of apoptosis. It is clear from these examples that CTL and NK mechanisms relying solely on target cell signalling could be easily doomed to failure.

The granule exocytosis mechanism of target cell lysis does not rely on such compliance of the target cell. In this pathway of cytolysis no specific receptor or target cell response is required due to the ability of perforin to bind to phospholipid and polymerize in the membrane in the presence of Ca. This event in itself can destroy the target viability. Survival of the target requires membrane repair, which by necessity and involuntarily leads to the uptake of other toxic granule proteins into the target, setting off the DNA degradation pathway. Significantly, this pathway of DNA degradation is not protected by bcl2.

It would seem that the granule exocytosis pathway of cytotoxicity has evolved precisely to dispose off renegade cells which resist normal growth control pathways due to viral or oncogenic transformation. Since the prime component of this cytolytic pathway is perforin, perforin deficiency is predicted to be particularly serious in virus infection and tumor transformation.

REFERENCES

1. Podack, E.R., Hengartner, H., Lichtenehld, M.G. Ann. Rev. Immunol. 1991; 9:129-157.
2. Podack, E.R. and Kupfer, A. Ann. Rev. Cell. Biol. 1991; 7:479-504.
3. Henkart, P.A. Ann. Rev. Immunol. 1985; 3:31-58.
4. Podack, E.R. Immunology Today. 1985; 6:12.
5. Russell, J.H. Immunol. Rev. 1983; 72:97-117.
6. Rouvier, E., Luciani, M.F., and Golstein, P. J. Exp. Med. 1993; 177:195.
7. Ratner, A., and Clark, W.R. J. Immunol. 1993; 150:4303-4314.
8. Watanabe-Fukunaga, R., Brannan, C.I., Copeland, N.G., et al. Nature 1992; 356:314.
9. Ostergaard, H., Clark, W.R. J. Immunol. 1989; 143:2120-2126.
10. Itoh, N., Tsujimoto, Y., and Nagata, S. J. Immunol. 1993; 151:621-627.
11. Kriegler, M., Perez, C., Defay, K., et al. Cell 1988; 53:45-53.
12. Ware, C.F., Crowe, P.D., Grayson, M.H., et al. J. Immunol. 1992; 149:3881-3888.
13. Ruddle, N.H., and Homer, R. Prog. Allergy. 1988; 40:162.
14. Tschopp, J., Schafter, S., Masson, D. et al. 1989; 337:272-274.
15. Young, J-D.E., Cohn, Z.A., Podack, E.R., et al. Science. 1986; 233:184-190.
16. Hameed, A., Olsen, K.J., Lee, M.-K., et al. J. Exp. Med. 1989; 169:765-777.
17. Shi, L., Kam, C-M., Powers, J.C., et. al., J. Exp. Med. 1992; 176:1521-1529.
18. Shiver, J.W., Su., L., Henkart, P.A. Cell. 1992:71:315-322.
19. Masson, D., Tschopp, J. Cell. 1987; 49:679-685.
20 Hameed, A., Lowrey, D.M., Lichtenheld, M., et al. J. Immunol. 1988; 141:3142-3147.
21. Tian, Q., Streuli, M., Saito, H., et al. Cell. 1991; 67:629-639.

Biology of Vitronectin and their Receptors
K.T. Preissner, S. Rosenblatt, C. Kost, J. Wegerhoff and D.F. Mosher, editors

STRUCTURE & FUNCTION OF CD59

Peter J. Sims[a]

[a]Blood Research Institute of The Blood Center of Southeastern Wisconsin, 1701 W. Wisconsin Ave., Milwaukee, WI 53233, USA

INTRODUCTION

It has been known for many years that human erythrocytes are remarkably resistant to lysis by human serum, whereas these cells can be readily lysed when the sera of rabbit, guinea pig, or other species serves as the complement source. This capacity of human cells to resist lysis by homologous complement is now recognized to be conferred by a plasma membrane protein that expresses the leukocyte CD59 antigen. In this article, I will briefly summarize the known functional and structural properties of this complement regulatory protein, and discuss the potential importance of this component to understanding the biology of complement system in man.

HOMOLOGOUS RESTRICTION FACTOR

The existence of a membrane factor that specifically serves to interrupt assembly of the C5b-9 pore was suggested by evidence that the resistance of human erythrocytes to lysis by complement is related to the species of origin of C8 and C9[1-4]. The identity of this factor was first reported in 1984 as a 60-65 kDa component of the erythrocyte membrane, designated "homologous restriction factor" or "C8-binding protein"[5,6]. This protein was reported to exhibit affinity for C8 or C9, and to confer resistance to human C5b-9 when incorporated into lipid vesicles or the membranes of xenotypic erythrocytes. Furthermore, antibodies directed against this protein were shown to increase the sensitivity of human red cells to lysis by the human C5b-9 proteins. This antigen was shown to be attached to the plasma membrane through a glycosylphosphatidylinsitol (GPI) anchor, and to be missing from the hemolytically-sensitive erythrocytes of patients with the disorder, Paroxysmal Nocturnal Hemoglobinuria, a condition traced to abnormal GPI-anchoring due to defective biosynthesis of the glycosyl moiety of the anchor[7,8]. Although the reported activities of this protein appear to account for the known species-restricted activity of the C5b-9 complex towards human blood cells, peptide or DNA sequence for this protein has not yet been reported, and it remains uncertain how this factor might be related CD59, an 18-21 kDa protein for which peptide, gene structure, and complement regulatory function is now established.

CD59

The identity of another membrane protein that restricted lysis by human C5b-9 was first reported by Sugita et al [9] in 1988, and subsequently confirmed by several other laboratories the following year [10-14]. This protein, which exhibits an electrophoretic mobility of 18-21 kDa in SDS, was shown to be widely distributed on human blood cells, endothelium, and several other cells, and to exhibit many of the properties that had been reported for the 64 kDa "homologous restriction factor"[15-23]. In particular, when purified, this protein can re-incorporate into cell membranes and confer resistance to lysis by the human C5b-9 complex. This inhibitory function appears to be directly related to CD59's capacity to restrict incorporation of polymeric C9 into the target membrane[16,24]. CD59 is cleaved from the surface of many cells with bacterial PI-PLC, and is deficient in the hemolytically-sensitive erythrocytes that arise in PNH[13,14,25]. Amino acid sequence derived by cDNA cloning established the identity of this protein as the antigen associated with leukocyte CD59[10,11,26-28]. Proof that the complement regulatory activity attributed to CD59 actually resided in this protein was provided by the demonstration that transfection with full-length cDNA coding for the CD59 sequence conferred all of the functional properties attributable to the human erythrocyte "homologous complement restriction factor"[29]. A single individual with an inherited CD59 gene defect has been identified in Japan, and this patient exhibited the clinical manifestations associated with PNH (discussed further below)[30,31].

CD59 Protein Structure

CD59 is synthesized as a 128 amino acid propeptide consisting of a 25 amino acid leader sequence, and a 26 amino acid C-terminal membrane spanning domain that conforms to a motif for GPI-anchoring[10,11,26-28]. The residual 77 amino acid polypeptide that remains after transamination of Asn77 to its GPI-anchor contains 10 cysteine residues. The amino acid sequence exhibits 24-30% homology with the members of the murine Ly-6 superfamily, with a high degree of conservation of the number and position of the 10 cysteine residues[11]. Sequence homology with squid glycoprotein 2 and human urokinase plasminogen activator receptor has also been noted[32,33]. The reported disulfide structure of CD59 is also similar to that of the erabutoxin family of snake venoms, for which disulfide assignments, NMR solution and X-ray crystal structures are available[34,35]. X-ray crystal structure for CD59 has not been reported.

In addition to its GPI anchor, CD59 contains two potential glycosylation sites (at Asn8 and Asn18)[11]. Tryptic mapping has established that only Asn18 is glycosylated[36]. There is no detectable O-linked carbohydrate[36]. Removal of the GPI-anchor from the purified protein abrogates its capacity to incorporate into membranes and causes marked reduction in its capacity to protect membranes from C5b-9 damage[11,16]. A reduction in the complement-inhibitory activity of membrane CD59 is also observed after enzymatic deglycosylation at Asn18, or, upon biosynthetic modification of branch-chain structure of the N-linked sugars[36]. The species-selective affinity of CD59 for human C8 and C9 is nevertheless retained after removal of either the GPI anchor or upon deglycosylation, suggesting that functionality of the protein is conferred through the disulfide-bonded peptide structure, independent of subsequent post-translational modification[37,38]. Consistent with this interpretation, disulfide reduction abrogates the complement-inhibitory function of CD59 and destroys the immunodominant epitopes.

Site-directed mutagenesis of the cysteine residues in CD59 has revealed that functional protein is expressed after disruption of loop 4 or the internal loop 1. No protein is detected if the disulfide bonds of either loop 1,2,or 3 is disrupted[39].

CD59 Gene Structure

The gene for CD59 has been assigned to the short arm of chromosome 11 (p14-p13 region) and has been sequenced by two laboratories[40-42]. The 128 amino acid coding region is distributed over 27 kb of genomic DNA. Four exons have been identified: The first codes for 45bp of 5'-untranslated region of mRNA; Exon 2 (85 bp) contains additional 18bp of untranslated region plus 22 amino acids of the signal peptide; Exon 3 (102 bp) codes the remaining 3 amino acids of signal peptide plus 31 amino acids of coding sequence; Exon 4 (215 bp) encodes the remaining 72 amino acids of coding sequence. The 5'-flanking region of the gene is G+C rich and does not contain TATA or CAAT boxes. Duplicate sites for transcription factor Sp1 have been identified. The basis for tissue-specific expression remains to be determined. The organizational structure of the CD59 gene is strikingly similar to that of the Ly-6 genes, particularly ly6-C, in both exon structure and in position of intron-exon boundaries. Major differences are found in intron size (27 kb for CD59 versus 4 kb for ly-6C) and in the promotor region: ly-6C is not G+C rich and contains consensus for interferon-inducibility.

Cell and Tissue Distribution

CD59 antigen is widely distributed on all blood cells and many non-hematopoietic tissues, including virtually all vascular and ductal epithelia, as well as fibroblasts, spermatozoa, and myocardium (reviewed in ref.[15]). The antigen has also been detected in human urine, saliva, cerebro-spinal fluid, seminal fluid, amniotic fluid and breast milk, although it remains unclear whether the soluble protein found in these fluids retains its GPI-anchor.

Function

As was noted above, CD59 serves to regulate assembly of the complement pore by restricting the incorporation of C9 into the membrane attack complex. CD59 does not affect the efficiency with which C5b67 binds to the membrane, nor does its block subsequent binding of C8 to membrane C5b67[16,24,43]. The inhibitory function of CD59 is apparently mediated through its affinity for two distinct epitopes exposed in the nascent C5b-9 complex: a region of the C8 α-chain that is exposed when C8 incorporates into the C5b-8 complex, and a region of C9 that is exposed when the first molecule of C9 incorporates into the C5b-9 complex[24,43,44]. This suggests that CD59 may recognize a common structure that is shared by these two homologous polypeptides (C8 α-chain and C9) and which serves to promote the conformational transition of C9 that is required for its polymerization and intercalation into the plasma membrane. The apparent affinity of CD59 for human C8 or C9 conforms to the species-selectivity that is observed for its complement inhibitory function: CD59 shows little affinity for rabbit or guinea pig C8 or C9, and poorly inhibits hemolysis of red cells exposed to human C5b67 plus C8/C9 of rabbit (or, guinea pig) origin[43,44]. The regions of the human C8 α-chain and human C9 that provide the CD59 binding sites remain to be identified. The site has been shown to reside in the C-terminal fragment of C9 produce by α-thrombin cleavage (residues 245-538) and to be destroyed by tryptic cleavage of C9 at Arg391[44]. The specificity of CD59 as an inhibitor of the C5b-8-initiated conversion of C9 into its membrane-inserted state suggests that this protein may provide a useful tool by which to elucidate the conformational transitions that are required to convert C9 from its globular state in plasma into a membrane-embedded homopolymer with cytolytic activity.

In addition to the complement-inhibitory function exhibited by CD59, there is recent evidence that this membrane protein participates in rosette formation between human erythrocytes and T-lymphocytes, and also augments T-cell activation by antigen presenting cells, activities that appear to be mediated through CD59's capacity to bind to CD2 on T-lymphocytes[45-49]. The CD59 binding site in CD2 has recently been localized to a region that partially overlaps the CD58 binding site in the protein[45]. Unresolved is how the affinity of CD59 for CD2 relates to its affinity for the C8 α-chain and for C9. It should also be noted that antibodies against CD59 and other GPI-anchored proteins have been reported to elicit cell-stimulatory responses, which may be related to activation of intracellular tyrosine kinases[10,50,51]. The physiological significance of these responses, as well as the potential role of CD59 as an accessory ligand for CD2, awaits further investigation.

Biological Importance and Application

Considerable insight into the contribution of CD59 to the regulation of the complement system in vivo has come from studies of PNH blood cells and from cells that are genetically deficient in either decay accelerating factor (DAF) or CD59. At the membrane, assembly of the C5b-9 complex is principally regulated by limiting assembly of the C3/C5 convertases, through the action of membrane cofactor protein (MCP) and DAF, and by limiting activation of C9 by membrane C5b-8, the function of CD59. In PNH, both DAF and CD59 can be deleted from the cells surface due to biosynthetic defect in GPI anchoring[12,52-55]. The CD59-deficient cohort of cells are the most lytically sensitive, and are now thought to be responsible for the episodes of intravascular hemolysis observed in this disorder. Consistent with this interpretation, are the clinical manifestations observed in those patients with genetic deficiency in one of these proteins: while a syndrome of intermittent intravascular hemolysis was described in a patient exhibiting a CD59 gene defect but normal levels of DAF, hemolytic anemia is not observed in those individuals who are genetically deficient in DAF (individuals with erythrocytes displaying the Inab antigen of the Cromer blood group)[30,31,55]. CD59 has also been shown to down-regulate the cell-stimulatory and procoagulant responses evoked from platelets, endothelial cells, and other cells exposed to the C5b-9 complex, and CD59-deficient platelets display increased sensitivity to C5b-9 induced exposure of prothrombinase activity[23,56-58]. This suggests that deletion of this complement regulatory protein may underlie the thrombotic episodes observed in those PNH patients who are abnormal for CD59 expression[58].

The capacity of CD59 to confer cellular resistance to human C5b-9 suggests that this protein and its gene may have therapeutic application in protecting cells and tissue from plasma membrane damage mediated directly by the complement system. As was noted above, addition of CD59 to complement-sensitive cells that lack this inhibitor has been shown to confer resistance to C5b-9 mediated lysis[59,60]. Similarly, xenotypic cells that are stably transfected with cDNA coding for CD59 become resistant to lysis by antibody and human serum[29,61,62]. This raises the possibility that similar strategies may provide the means by which to protect CD59-deficient cells, including the vascular endothelium and other cells of discordant xenotransplants, from complement-mediated injury.

ACKNOWLEDGEMENTS

Supported by grant HL36061 from the Heart,Lung Blood Institute, National Institutes of Health; and by the Blood Center Research Fndn. of The Blood Center of Southeastern Wisconsin.

REFERENCES

1. Shin ML, Hansch G, Hu VW, Nicholson-Weller A. J Immunol 1986; 136: 1777.
2. Hansch GM, Hammer CH, Vanguri P, Shin ML. Proc Natl Acad Sci USA 1981; 78: 5118.
3. Hu VW, Shin ML. J Immunol 1984; 133: 2133.
4. Yamamoto K. J Immunol 1977; 119: 1482-.
5. Zalman LS, Wood LW, Muller-Eberhard HJ. Proc Natl Acad Sci USA 1986; 83: 6975.
6. Schonermark S, Rauterberg EW, Shin ML, Loke S, Roelcke D, Hansch GM. J Immunol 1986; 136: 1772.
7. Zalman LS, Wood LW, Frank MM, Muller-Eberhard HJ. J Exp Med. 1987; 165: 572.
8. Hansch GM, Schonermark S, Roelcke D. J Clin Invest 1987; 80: 7.
9. Sugita Y, Nakano Y, Tomita M. J Biochem (Tokyo) 1988; 104: 633.
10. Okada H, Nagami Y, Takahashi K, et al. Biochem Biophys Res Commun 1989; 162: 1553.
11. Davies A, Simmons DL, Hale G, et al. J Exp Med 1989; 170: 637.
12. Holguin MH, Wilcox LA, Bernshaw NJ, Rosse WF, Parker CJ. J Clin Invest 1989; 84: 1387.
13. Okada N, Harada R, Fujita T, Okada H. Int Immunol 1989; 1: 205.
14. Stefanova I, Hilgert I, Kristofova H, Brown R, Low MG, Horejsi V. Mol Immunol 1989; 26: 153.
15. Walsh LA, Tone M, Thiru S, Waldmann H. Tissue Antigens 1992; 40: 213.
16. Rollins SA, Sims PJ. J Immunol 1990; 144: 3478.
17. Dave SJ, Sodetz JM. J Immunol 1990; 144: 3087.
18. Meri S, Morgan BP, Wing M, et al. J Exp Med 1990; 172: 367.
19. Whitlow MB, Iida K, Stefanova I, Bernard A, Nussenzweig V. Cellular Immunology 1990; 126: 176.
20. Harada R, Okada N, Fujita T, Okada H. J Immunol 1990; 144: 1823.
21. Hamilton KK, Ji Z, Rollins S, Stewart BH, Sims PJ. Blood 1990; 76: 2572.
22. Nose M, Katoh M, Okada N, Kyogoku M, Okada H. Immunology 1990; 70: 145.
23. Sims PJ, Rollins SA, Wiedmer T. J Biol Chem 1989; 264: 19228.
24. Meri S, Morgan BP, Davies A, et al. Immunology 1990; 71: 1.
25. Holguin MH, Wilcox LA, Bernshaw NJ, Rosse WF, Parker CJ. Blood 1990; 75: 284.
26. Sugita Y, Tobe T, Oda E, et al. J Biochem [Tokyo] 1989; 106: 555.
27. Philbrick WM, Palfree RG, Maher SE, Bridgett MM, Sirlin S, Bothwell AL. Eur J Immunol 1990; 20: 87.
28. Sawada R, Ohashi K, Anaguchi H, et al. DNA & Cell Biology 1990; 9: 213.
29. Zhao J, Rollins SA, Maher SE, Bothwell AL, Sims PJ. J Biol Chem 1991; 266: 13418.
30. Yamashina M, Ueda E, Kinoshita T, et al. New Eng J Med 1990; 323: 1184.
31. Motoyama N, Okada N, Yamashina M, Okada H. Eur J Immunol 1992; 22: 2669.
32. Roldan AL, Cubellis MV, Masucci MT, et al. Embo Journal 1990; 9: 467.
33. Williams AF, Tse AG, Gagnon J. Immunogenetics 1988; 27: 265.
34. Tomita M, Tobe T, Choi-Mura N, Nakano Y, Kusano M, Oda E. Complement Inflammation, 1991; 8: 233(Abstract)

35. Inagaki F. Cell Structure & Function 1990; 15: 237.
36. Ninomiya H, Stewart BH, Rollins SA, Zhao J, Bothwell AL, Sims PJ. J Biol Chem 1992; 267: 8404.
37. Hahn WC, Burakoff SJ, Bierer BE. J Immunol 1993; 150: 2607.
38. Nilsson B, Hagstrom U, Englund A, Safwenberg J. Vox Sang 1993; 64: 43.
39. Petranka J, Norris J, Hall S, et al. Mol Immunol (suppl) 1993 (in press).
40. Forsberg UH, Bazil V, Stefanova I, Schroder J. Immunogenetics 1989; 30: 188.
41. Tone M, Walsh LA, Waldmann H. J Molec Biol 1992; 227: 971.
42. Petranka JG, Fleenor DE, Sykes K, Kaufman RE, Rosse WF. Proceed Natl Acad Sci USA. 1992; 89: 7876.
43. Rollins SA, Zhao J, Ninomiya H, Sims PJ. J Immunol 1991; 146: 2345.
44. Ninomiya H, Sims PJ. J Biol Chem 1992; 267: 13675.
45. Hahn WC, Menu E, Bothwell AL, Sims PJ, Bierer BE. Science 1992; 256: 1805.
46. Deckert M, Kubar J, Bernard A. J Immunol 1992; 148: 672.
47. Deckert M, Kubar J, Zoccola D, et al. Eur J Immunol 1992; 22: 2943.
48. Venneker GT, Asghar SS. Exp Clin Immunogen 1992; 9: 33.
49. Korty PE, Brando C, Shevach EM. J Immunol 1991; 146: 4092.
50. Stefanova I, Horejsi V, Ansotegui IJ, Knapp W, Stockinger H. Science 1991; 254: 1016.
51. Stefanova I, Horejsi V. J. Immunol. 1991; 147: 1587.
52. Rosse WF, Hoffman S, Campbell M, Borowitz M, Moore JO, Parker CJ. Brit J Haematol 1991; 79: 99.
53. Wilcox LA, Ezzell JL, Bernshaw NJ, Parker CJ. Blood 1991; 78: 820.
54. Mahoney JF, Urakaze M, Hall S, et al. Blood 1992; 79: 1400.
55. Telen MJ, Rosse WF. Baillieres Clinical Haematology 1991; 4: 849.
56. Sims PJ, Wiedmer T. Immunol Today 1991; 12: 338.
57. Hamilton KK, Ji Z, Rollins S, Stewart BH, Sims PJ. Blood 1990; 76: 2572.
58. Wiedmer T, Hall SE, Ortel TL, Kane WH, Rosse WF, Sims PJ. Blood,1993 (in press).
59. Okada N, Harada R, Taguchi R, Okada H. Biochem Biophys Res Commun 1989; 164: 468.
60. Wing MG, Zajicek J, Seilly DJ, Compston DA, Lachmann PJ. Immunology 1992; 76: 140.
61. Walsh LA, Tone M, Waldmann H. Eur J Immunol 1991; 21: 847.
62. Akami T, Sawada R, Minato N, et al. Transplant Proceed 1992; 24: 485.

Biology of Vitronectins and their Receptors
K.T. Preissner, S. Rosenblatt, C. Kost, J. Wegerhoff and D.F. Mosher, editors
249

MOLECULAR MECHANISMS AND THERAPEUTICAL INTERVENTION STRATEGIES OF THE SEPSIS SYNDROME: INDUCTION-PATTERN AND FUNCTION OF LIPOPOLY-SACCHARIDE BINDING PROTEIN

Ralf R. Schumann[a], C. Kirschning[a], N. Lamping[a], H.-P. Knopf[b], H. Aberle[b] and F. Herrmann[a]

[a]Max-Delbrück-Centrum für Molekulare Medizin (MDC), Robert-Rössle-Str. 10, 13122 Berlin, Germany and Free University, Universitätsklinikum Rudolf-Virchow, Abteilung für Onkologie und angewandte Molekularbiologie (department head: Prof. Dr. F. Herrmann), Lindenberger Weg 80,13122 Berlin, Germany

[b]Max-Planck-Institut für Immunbiologie, Stübeweg 51, 79108 Freiburg, Germany

ABSTRACT

Lipopolysaccharide (LPS) or Endotoxin, a part of the outer membrane of gramnegative bacteria, initiates a cascade of events in the host organism when released into the bloodstream. Moderate activation of immune cells by LPS can be beneficial, in an uncontrolled fashion, however, it often leads to severe malfunctions of the organism. Hypotension, fever, multi-organ-failure, disseminated intravascular coagulation and the full gramnegative shock syndrome can be induced by the entry of even small amounts of LPS into the bloodstream. The sepsis syndrome has a high mortality rate and as to now no therapeutical intervention strategy has been established. With the recent discovery of binding proteins and receptors for LPS insight in the Endotoxin recognition and cellactivation processes has been gained over the last years. Here the LPS Binding Protein LBP is discussed, focussing on its synthesis in the liver and the analysis of the promoterregion of the gene. Understanding of the complex mechanism of Endotoxin recognition might ultimately lead to therapeutical approaches to stop the chainreaction initiated by LPS, that leads to the shock syndrome.

INTRODUCTION

Lipopolysaccharide is a structure found in almost all different gramnegative bacteria and constitutes the major threat for the host, once bacteria have invaded. Recently great advances in elucidating the chemical structures of LPS were made and an important subcompound of LPS, namely Lipid A, was found, that carries all the "endotoxic" capacity, as was proven in experimental systems [1]. Meanwhile the complete structure of LPS is revealed and it consists of the polysaccharide part that is made up of the O-specific chain and the two-part core and on the other side of the lipid A component [2]. Since it became clear that Endotoxins are the cause of toxicity of gramnegative bacteria, many groups have investigated the events triggert in the host by the appearance of Endotoxin. It was found that LPS is able to induce a cascade of events in immune cells, mediated by different soluble factors. Proteins and receptor candidates recently were discovered that recognize Endotoxin, bind and possibly transport it, and mediate the signal for activation of the host defense through compartments and cells. LPS that enters the blood stream will be bound unspecifically by serum lipoproteins, mainly of the "high density" class (HDL), which attenuate its effects [3]. Apolipoproteins also have been shown to bind LPS and inhibit its endotoxic potential [4] and LDL lipoproteins can bind Endotoxin and transport it through endothelial cell layers [5]. Specific binding of LPS in serum occurs to a protein, which is called Lipopolysaccharide Binding Protein or LBP [6-10]. Complexed to LBP, even pikograms of Endotoxin induce numerous cellular responses, and it appears as if LBP katalyzes all of the LPS-mediated effects.

THE LIPOPOLYSACCHARIDE BINDING PROTEIN (LBP)

LBP is synthesized in hepatocytes as a glycosylated 58 kD protein and is constitutively secreted into the bloodstream [8,11]. The protein concentration rises in the "acute phase" with a maximum after 24 hours as was shown in rabbit eperiments (Fig. 1). Here "New-Zealand White" rabbits were injected with silvernitrate which is known to induce the acute phase in these animals. LBP, that constitutively is detected in the animals only at 0.5 ug/ml, rises in the serum within the first twelve hours after induction to maximum levels of up to 50 ug/ml LBP after 24 hours. Within the next 5 days serumlevels of LBP drop to almost constitutive levels. The rise of LBP during the acute phase in humans, as newer data confirm (data not shown), appears to be weaker and constitutively synthesized levels are higher in humans as compared to rabbits.

ug/ml LBP

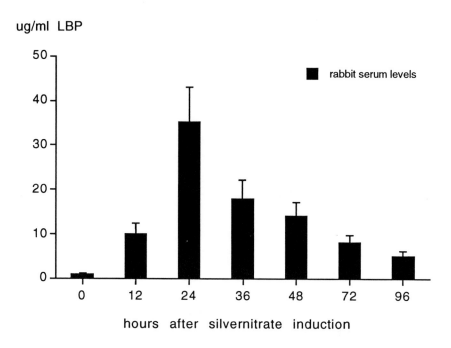

hours after silvernitrate induction

Figure 1. New-Zealand White rabbits were injected with silvernitrate and at the hours indicated, blood was taken and LBP content was measured. Mean value of four experiments.

RNA also was prepared from different tissues of the silvernitrate-induced animals and northern-blot experiments were performed to analyze the amount of LBP transcript induced during the acute phase. Also an in-vitro model, using Hep-G2 cells was set up and induceability of LBP message was investigated. Table 1 shows the induction pattern of LBP mRNA in rabbit liver and in Hep-G2 cells. During the "acute- phase" levels of LBP-RNA rise dramatically in the rabbit liver with a maximum also at 24 hours. A similar induction pattern can be seen, when Hep-G2 cells are stimulated with IL-1b, IL-6 and Dexamethasone. IL-6 alone cannot induce expression of LBP in Hep-G2, but IL-1 is needed, thus LBP appears to be a class I acute phase proteins, according to the definition by Baumann and Gauldie [12].

Table 1
LBP mRNA induction pattern in rabbit liver and Hep-G2 cells.

stimulation (h)	rabbit liver	Hep-G2 IL-1, IL-6	Hep-G2 IL-1, IL-6 + Dex.
0	-	-	-
12	(+)	+	+
24	+++	++	+++
48	+	+	++

Total RNA was prepared from liver or from in-vitro cultures of Hep-G2 cells and northern-blot experiments were performed.

Function of LBP

Upon binding to LPS, LBP does not block, but on the opposite, it enhances the effects of LPS and abrogates cellular responses at subthreshhold LPS levels. LPS-induced TNF-production and TNF-mRNA expresion in rabbit peritoneal macrophages, i.e., is enhanced strongly when LPS is complexed to LBP [13,14]. Rabbit peritoneal macrophages that are rendered unresponsive to LPS stimulation by a process called adaptation can be restored in their ability to produce TNF after LPS stimulation by the addition of LBP [15]. Also Macrophages detect and bind LPS faster and easier when it is complexed with LBP, so it acts as an opsonin for gramnegative bacteria [16]. The LPS-induced response in neutrophilic granulocytes, furthermore, can be enhanced by the addition of LBP to the system [17]. The biological effects of LBP thus can be summarized as complexing to LPS and subsequently enabling the organism to detect small amounts of LPS better and to trigger the defense cascade. Experiments using antibodies to LBP revealed that pretreatment of serum with these antibodies and thus subsequently depleting the serum of LBP, resulted in a much weaker response to LPS-challenge as compared to LBP containing serum [9].

The LBP-promoter

By screening a human genomic library, we were able to isolate, subclone and sequence a clone containing the LBP-promoter. The analysis of 1kB upstream revealed several CAAT-boxes, the IL-6 responsive element H-APF-1 RS, as well as the liver-specific transcription factor HNF-5 and HSTF-hsp 70.5 at -15 to -170. Also the glucocorticoid-responsive elements Oct-2-D and GCRE were found, which correlates with the induction pattern of LBP mRNA in Hep-G2 cells by IL-6, IL-1 and Dexamethasone. Several transcription factors, apparently important in immune regulation also were found, namely Ig-kappa.2, IgNF-A-Igk and several of the IgHC-group. Experiments, elucidating the functional importance of these structures are underway to determine the elements of the promoter that regulate expression of LBP.

THE CELLULAR LPS-RECEPTOR CD 14

The specific cellular responses that, moderately induced, have important defense functions, but when triggerd too strongly are fatal for the host, have been postulated to be mediated on the cellsurface by a specific LPS receptor [18-21]. Several cellular structures have been found that bind LPS but the majority of these compounds like the scavenger receptor or CD11/CD18 most likely are involved in binding LPS for detoxification without being involved in cellactivation. One molecule, that binds LPS-LBP complexes, but also seems to be responsible for cellular activation is the CD14 molecule. It was shown to be able to bind LPS on the cellular surface and mediate effects like TNF production of or opsonic function for macrophages [22]. Soluble CD14 can replace cellular CD14 as was shown recently, when CD14 negative cells could be induced for LPS responsiveness, depending on the presence of soluble CD14 added [23,24]. LPS-induced signaltransduction could not be achieved by CD14 alone yet, so most likely an additional compound is needed to build the LPS receptor, a model that also proved to be right for other cytokine receptors, like the IL-6 / LIF / CNTF / OSM receptor family.

THERAPEUTICAL INTERVENTION STRATEGIES IN GRAM-NEGATIVE SEPSIS

In gramnegative sepsis antibiotics don't work because they usually act too slowly. Also, as we tried to explain in this article, LPS released from killed bacteria still has all the properties to induce the sepsis syndrome. Targets for intervening in septic shock exist at all levels of the development of sepsis. Antibodies to different portions of LPS have been tested and clinical results have been reported with an antiserum directed against the core part of LPS [25]. A problem is the variability in this region of Endotoxin and the lack of cross-reactivity of such antibodies, so that controversies on the usefulness of such antibodies have arisen [26]. Experiments using whole blood were performed, pretreating the blood with anti-LBP antibodies and the TNF secretion in response to LPS could be shifted approx. two logs [9]. Assuming LBP plays a major role in binding LPS in the bloodstream and directing it to the target cell, antibodies or mutated forms of LBP, that still bind LPS, but cannot present it to CD14 or vice versa would be interesting objects to study in prevention of sepsis. The role of soluble CD14 (sCD14) has been controversial, but meanwhile more reports are published, stating that sCD14 augments the effects of LPS in the organism [23,24]. Thus blocking the function of sCD14, i.e. by antibodies could be of clinical value too, and in the experiments leading to the finding that CD14 acts as a LPS receptor, TNF production succesfully was suppressed by anti-CD14 antibodies [22]. Once the monocyte / macrophage is activated, key mediators like IL-1, TNF and IL-6 are targets for therapeutical intervention. Antibodies to the cytokines or blocking of their receptors are strategies in antagonizing their effects. One natural antagonist of the potent proinflammatory cytokine IL-1 has been discovered, that has homology to IL-1 and binds to its receptor, but does not induce cellactivation, this molecule is the IL-1 receptor antagonist (IL-1 RA) [27]. Intervention with this molecule could be of special interest, as a strong synergism between IL-1b and TNF-a was reported, that could be blocked by addition of IL-1 RA [28]. A protein that has 34% homology to LBP and that is found in the granules of neutrophilic granulocytes, termed BPI, binds to LPS also but inhibits its effects. Phase II studies with BPI are underway now to interfer with the development of sepsis in humans and results from animal studies so far are promising.

SUMMARY AND OUTLOOK

Some of the steps required for LPS binding and uptake by the cell that were described here in the future might lead to new concepts of interfering with the development of sepsis. Also studying the induction of LBP during the acute phase response will help understanding the development of the septic shock syndrome. As LPS is the major cause of this syndrome, further insight in every step of the pathogenesis of this phenomenon is needed to be able to understand this complex process and to develop a therapy in the future.

REFERENCES

1. Gmeiner J, Lüderitz O, Westphal O. Eur J Biochem 1969; 7: 370-379.
2. Rietschel ETh, Brade L, Lindner B, Zahringer U. In: Morrison DC, Ryan JL, eds. Bacterial Endotoxic Lipopolysaccharides. CRC Press, London, 1992; 3-40.
3. Ulevitch RJ, Johnston AR, Weinstein DB. J Clin Invest 1979; 64: 1516-1524.
4. Flegel WA, Wolpl A, Männel DN. Infect Immun 1989; 57: 2237-2245.
5. Navab M, Hough GP, van Lenten JA. J Clin Invest 1988; 81: 601-605.
6. Tobias PS, Soldau K, Ulevitch RJ. J Exp Med 1986; 164: 777-793.
7. Tobias PS, Mathison JC, Ulevitch RJ. J Biol Chem 1988; 263: 13479-13781.
8. Tobias PS, Soldau K, Ulevitch R.J. J Biol Chem 1989; 64: 10867-10871.
9. Schumann RR, Leong SR, Flaggs GW, Gray PW, Wright SD, Mathison JC, Tobias PS, Ulevitch RJ. Science 1990; 249: 1429-1431.
10. Schumann RR. Res Immunol 1992; 143: 11-15.
11. Schumann RR, Tobias PS, Mathison JC, Ulevitch RJ. In: Faist E, ed. Host defense dysfunctions in trauma, shock and sepsis, Springer Verlag, Heidelberg, 1993 767-772.
12. Baumann H, Gauldie J. Mol Cell Med 1990; 7: 147-160.
13. Ulevitch RJ, Mathison JC, Schumann RR, Tobias PS. J Trauma 1990; 30: S190-192.
14. Ulevitch RJ, Schumann RR, Mathison JC, Leavesley DI, Martin TR, Soldau K, Kline L, Wolfson E, Tobias PS. In: Baumgartner J-D, Calandra T, Carlet

J, eds. Endotoxin, from pathophysiology to therapeutic approaches, Medecine-Sciences Flammarion, Paris, 1990; 31-41.

15. Mathison JC, Virca GD, Wolfson N, Tobias PS, Glaser K, Ulevitch RJ. J Clin Invest 1990; 85: 1108-1118.

16. Wright SD, Tobias PS, Ulevitch RJ, Ramos RA. J Exp Med 1989; 170: 1231-1241.

17. Vosbeck K, Tobias PS, Müller H, Allen RA, Arfors K-E, Ulevitch RJ, Sklar LA. J Leukocyte Biol 1990; 47: 97-104.

18. Hampton RY, Golenbock DT, Penman M, Krieger M, Raetz CRH. Nature 1990; 352: 342-344.

19. Kirkland TN, Virca GD, Kuus-Reichel T, Multer FK, Kim SY, Ulevitch RJ, Tobias PS. J Biol Chem 1990; 265: 9520-9525.

20. Lei M-G, Stimpson SA, Morrison DC. J Immunol 1991;147: 1925-1932.

21. Couturier C, Haeffner-Cavaillon N, Caroff M, Kazatchkine M. J Immunol 1991; 147: 1899-1904.

22. Wright SD, Ramos RA, Tobias PS, Ulevitch RJ, Mathison JC. Science 1990; 249: 1431-1433.

23. Frey EA, Miller DS, Jahr TG, Sundan A, Bazil V, Espevik T. Finlay BB, Wright SD. J Exp Med 1992; 176: 1665-1671.

24. Pugin J, Schurer-Maly C-C, Leturcq D, Moriarty A, Ulevitch RJ, Tobias PS. Proc Natl Acad Sci USA 1993; 90: 2744-2748.

25. Ziegler EJ, Fisher CJ, Sprung CL and the HA-1A sepsis study group N Engl J Med 1991; 324: 429-436.

26. Baumgartner, JD, Heumann, D, Gerain J, Weinbreck P, Grau GE, Glauser MP. J Exp Med 1990; 171: 889-896.

27. Arend WP, Welgus HG, Thompson RC, Eisenberg SP. J Clin Invest 1990; 85: 1694.

28. Ohlson K, P Bjork, M Bergenfeldt R Hagenau. Nature 1990; 348: 550-552.

Biology of Vitronectins and their Receptors
K.T. Preissner, S. Rosenblatt, C. Kost, J. Wegerhoff and D.F. Mosher, editors

Vitronectin binding surface proteins of Staphylococci and *Helicobacter pylori*

T. Wadström[a], Å. Ljungh[a], J.-I. Flock[b]

[a]Department of Medical Microbiology, University of Lund, Sölvegatan 23,
 S-22362 Lund, Sweden

[b]Center for Biotechnology, S-141 57 Huddinge, Sweden

Introduction

Chhatwal *et al.* [1] and Fuquay *et al.* [2] first described binding of vitronectin (Vn) by groups A, C and G streptococci, *Staphylococcus aureus* and *Escherichia coli*. This has subsequently been confirmed and extended to other microorganisms by others (Table 1). Vn mediated adherence to epithelial cells was proposed as a new mechanism for pathogenic microbes to adhere to and colonize host tissues [3-7]. However, since Vn binds the complement C5-C9 complex and inhibits cell lysis by complement, Vn binding by microorganisms may represent an important protective mechanism from complement-induced cell lysis.

Table 1
Pathogenic microorganisms binding vitronectin

Microorganism	Reference
Staphylococcus aureus	Fuquay *et al.*,1986
Streptococci, *E.coli*, *S.aureus*	Chhatwal *et al.*,1987
S.hæmolyticus	Paulsson and Wadström, 1990
Helicobacter pylori	Ringnér *et al.*,1992
Listeria monocytogenes	Rozalska and Wadström, 1992
Candida albicans	Jakab *et al.*, 1993

Extracellular Matrix Protein binding by *S.aureus* and Coagulase-Negative Staphylococci (CNS).

S.aureus and coagulase-negative staphylococci (CNS) use a number of surface proteins to adhere to cell surfaces and extracellular matrix (ECM) proteins, such as fibronectin (Fn), various collagens, laminin (Ln), thrombospondin and Vn (Table 2; [1,2,8-11]). While binding of all ECM proteins has been detected in CNS strains, immunoglobulin binding and fibrinogen binding seem to be restricted to *S.aureus* [12-14].

258

Table 2
Staphylococci bind the following proteins

Fibronectin (at least two FnBP of *S.aureus*)
Collagens
Vitronectin
Laminin
Elastin
Plasminogen
Thrombospondin
Heparan sulphate
Fibrinogen (only   *S.aureus*)

Recently, we have developed methods to immobilize Vn and other ECM and serum proteins as well as heparin, heparan sulphate and other glycosaminoglycans (GAGs) on various biopolymer surfaces [15-18]. We then found that CNS strains isolated from various surgical and biomaterial-associated infections commonly expressed binding of immobilized proteins (Fig.1), and that CNS bound proteins and GAGs to the same or higher extent than *S.aureus* strains [4,16,19]. Surface immobilized vitronectin as well as heparin and fibrinogen have been proposed to act as acceptor molecules on these surfaces [3,17,18]. Thus, these may be important in a complex interplay *in vivo* when CNS strains colonize on foreign bodies such as wound sutures, vascular grafts, intravascular catheters and orthopædic devices etc. Encouraged by these findings we decided to purify and characterize Vn-binding proteins (VnBPs) of *S.aureus* ISP 546 and compare binding properties of *S.aureus* with that of one CNS species, *S.hæmolyticus*, expressing high binding of Vn [20,21].

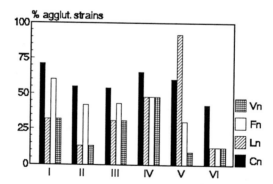

Figure 1. Expression of binding of immobilized collagen type I (Cn), laminin (Ln), fibronectin (Fn) and vitronectin (Vn) by CNS strains isolated from catheter-related septicæmia (I), endocarditis (II), vascular graft infections (III), osteomyelitis (IV), urinary tract infections (V) and normal skin (VI).

Purification of Vitronectin binding proteins (VnBPs).

S.aureus strain ISP 546 was cultured in Todd-Hewitt broth for 24 hrs at 37°C to express optimal Vn binding [20]. After washing, cells were resuspended in 1M LiCl, pH 5.0, for 2 hrs at 45°C to remove cell surface adhered proteins including the VnBPs. LiCl extraction was found to be a mild method which did not cause detectable cell lysis. This treatment removed all cell surface VnBP, and no Vn binding was expressed by the LiCl-extracted cells.

After centrifugation, the supernatant was dialyzed, and used in inhibition studies. This crude extract inhibited Vn binding by *S.aureus* and *S.hæmolyticus* completely [20]. The extract was further purified by affinity chromatogrpahy on heparin Sepharose. SDS-PAGE revealed one major band of M_rs 70 kDa, and two minor bands with M_rs 34 and 36 kDa [20]. Two VnBPs expressing high and moderate affinity binding respectively, were purified from *S.aureus* strain V8 [18]. One 60 kDa protein was identified as a putative high-affinity VnBP with $K_d = 7.4 \times 10^{-10}$ M. The estimated number of binding sites per bacterial cell is 260 [21].

Characterization of vitronectin and heparin binding to staphylococci.

Vn binding by *S.aureus* was highly sensitive to heating and treatment with various proteases [4,20] whereas *S.hæmolyticus* binding of immobilized Vn was heat resistant [14], suggesting that Vn binding of CNS strains is mediated by different surface proteins or other polymers from those mediating binding by *S.aureus* strains. Also fibronectin and immunoglobulin binding to FnBP and protein A on *S.aureus* are mediated by quite heat resistant surface proteins [22]. We earlier showed that Fn binding by the CNS species *S.epidermidis*, *S.hæmolyticus* and *S.lugdunensis* was resistant to heat treatment, and furthermore the gene for FnBP A of *S.aureus* could not be detected in CNS strains [4,16]. In conclusion, *S.aureus* and *S.lugdunensis* bind Vn by similar or identical surface proteins whereas Vn binding as well as Fn binding expressed by *S.hæmolyticus* are mediated by surface molecules which differ from those of *S.aureus* from physico-chemical point of view [4,14].

S.aureus strains expressed optimal Vn binding at pH 6.0-7.2. The binding was significantly inhibited by mannose but not by glucose and galactose. Some characteristics of Vn binding by *S.aureus* are summarized in Table 3.

Table 3
Properties of Vn binding by *S.aureus*

High affinity binding $K_d = 7.4 \times 10^{-10}$ M
Moderate affinity $K_d = 7.4 \times 10^{-8}$ M
About 260 high affinity binding sites per cell
One high affinity binding site binds 6-7 Vn molecules
VnBP of *S.aureus* recognizes the heparin binding domain of Vn
Incubation with heparin oligosaccharides affects binding

Interestingly, Vn binding was inhibited by heparin to the same extent as by Vn. To analyze whether Vn binding and heparin binding was mediated by the same surface molecule we purified heparin and heparan sulphate binding proteins (HeBPs) from *S.aureus* V8 [23]. Two HeBPs with M_rs of 60 and 66 kDa were purified on heparin Sepharose from a LiCl-cell surface extract. The binding was of low affinity ($K_d = 3 \times 10^{-5}$ M) (Fig.2). The complexity of heparin interacting with urea-activated Vn (aVn) and direct heparin binding to staphylococci is illustrated in Figure 2 [21].

260

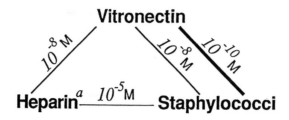

Figure 2. Dissociation constants of intrinsic affinity between human Vn heparin and *S.aureus* cell surface VnBP [21].

The affinity of heparin binding to Vn and the moderate affinity Vn binding by *S.aureus* was the same (Fig.2). Consequently, heparin fragments could only inhibit the moderate affinity Vn binding by *S.aureus* and did not affect the high affinity binding [21]. Further studies revealed that the high-affinity VnBP recognizes the heparin binding domain of Vn. Heparin oligosaccharides of more than 12 monosaccharide units were found to inhibit Vn binding to *S.aureus*, whereas smaller fragments enhanced binding (Figure 3) [21]. Furthermore, 4-12 monosaccharide fragments increased binding of heparin to Vn [21].

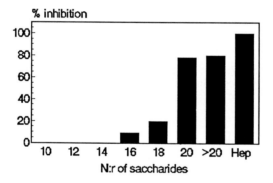

Figure 3. Effect of heparin oligosaccharides and heparin on binding of ^{125}I-Vn and ^{125}I-heparan sulphate to *S.aureus* strains V8 [21].

Studies are in progress to explore if heparin in combination with mannose and other carbohydrates will affect binding to VnBP and HeBP of *S.aureus* and CNS strains.

Studies of Staphylococcal binding to biopolymers.

Vn and Fn are both major serum proteins binding to intravascular catheters and other biopolymer devices. Vn has been shown to adsorb to a higher extent than Fn to glass surfaces, hydrophobic surfaces and to tissue culture polystyrene [24,25]. We recently showed that Vn adsorbs to a high extent to some poly(ether urethanes) and to the same extent as Fn to heparinized polyethylene [17,18,26]. We earlier showed that CNS strains isolated from biomaterial-associated infections commonly express binding of Fn as well as of Vn [12,27]. Binding to serum derived Fn on polymer surfaces has been implicated as a major mechanism in initiating infections caused by CNS as well as *S.aureus* on such devices [28-30]. However, Vn may mediate binding by CNS strains which do not bind Fn as recently shown on heparinized surfaces [18]. These studies emphasize the importance of systematic studies to evaluate the role of Vn binding by *S.hæmolyticus* and other

CNS strains on different kinds of biomaterials [18,24,25].

Vitronectin and laminin binding to *Helicobacter pylori* - lectin-like interactions.

H.pylori was discovered 1983, and has since proven to be one of the most common bacterial pathogens [31]. It's tissue specificity is restricted to the gastric epithelium where it causes the common type of gastritis as well as gastric and duodenal ulcer. We earlier found that *H.pylori* strains expressed binding of collagen type IV but not of Fn [32]. Subsequent studies showed that *H.pylori* strains commonly expressed binding of Vn as well as of Ln [33-35]. Vn binding was shown to be rapid, saturable and reversible [35]. Similar to Vn binding expressed by *S.aureus,* Vn binding by *H.pylori* was sensitive to heat and protease treatment of the bacterial cells [35]. The pH optimum for Vn binding by *H. pylori* is 4.0 whereas that of *S.aureus* cells is 6.0-7.2. Hence, Vn binding may be expressed in the acid environment in the stomach. Vn binding by both species was inhibited by excess of Vn but inhibition experiments with heparin, mannose and sialylated glycoproteins gave differing results (Table 2). Sialic acid specific proteins, but not asialofetuin, inhibited binding expressed by *H.pylori*. This indicates that sialic acid-specific hæmagglutinins may mediate Vn binding [35,36].

Table 4
Vitronectin binding to Staphylococcus aureus ISP 546 and Helicobacter pylori CCUG 17874

Inhibitory substance	Inhibition of Vn binding (%)	
	S.aureus	*H.pylori*
Control (PBS)	0	0
Bovine serum albumin	17	7
Fetuin	10	81
Orosomucoid	4	80
Glucose	6	5
Mannose	46	14
Galactose	4	9
Heparin	88	6
Vitronectin	90	88

Also, laminin binding by *H.pylori* was inhibited by highly glycosylated glycoproteins l like N-acetylneuraminyllactose [33]. Furthermore, neuraminidase treatment of laminin decreased binding by *H.pylori* cells. This suggests that terminal sialic acids on laminin compete for a specific sugar binding protein(s) on *H.pylori* cells. Whether the same structure on the *H pylori* cell surface mediates binding to Ln and Vn has not been elucidated but seems likely.

Interestingly, coccoidal forms of *H.pylori* predominant in gastric tissue of patients of patients with chronic gastritis seem to bind iodine-labelled Vn at similar levels to vegetative spral-shaped forms under our standard *in vitro* test conditions (T Wadström, unpublished).

Future aspects

A number of tissue invasive pathogens seem to interact with various ECM molecules [1,27,34,37]. Vn interacts with complement factors C5b-9 complexes and inhibits cell lysis by complement [38]. By binding to Vn microorganisms may be protected from cell lysis by

262

complement. Binding to different ECM proteins may represent a way of anchoring the microorganism in different tissues. Although the pathogenesis of the different infections reviewed certainly is multifactorial, involving also enzymes and toxins excreted by the microorganism with local effects and which modulate host defense mechanisms [22], it seems that expression of binding of ECM and serum proteins by microorganisms is a major virulence factor in several kinds of infections.

A further role of these binding structures needs to be elucidated in animal experiments with active immunization or administration of specific antibodies to these binding structures [39]. Identification of binding domains on Vn and other ECM proteins may give us possibilities to specifically block binding by microorganisms to Vn.

Acknowledgements. This study was supported by grants from the Swedish Medical Research Council, the Faculty of Medicine of the University of Lund, The Royal Physiographic Society, Lund and Magnus Bergvall Foundation.

References

1　Chhatwal GS, Preissner KT, Müller-Berghaus G, Blobel H. Infect Immun 1987; 55: 1878-1883.
2　Fuquay JI, Loo DT, Barnes DW. Infect Immun 1986; 52: 714-717.
3　Valentin-Weigand P, Grulich-Henn J, Chhatwal GS, Muller-Berghaus G, et al. Infect Immun 1988; 56: 2851-2855.
4　Paulsson M, Wadström T. FEMS Microbiol Immunol 1990; 2: 55-62.
5　Kostrzynska M, Paulsson M, Schmidt KH, Wadström T Microbios 1992; 71: 179-196.
6　Ljungh Å, Emödy L, Steinrück H, Sullivan P, et al. ZBl Bakt 1990; 274, 126-134.
7　Ljungh Å, Emödy L, Aleljung P, Olusanya O, Wadström, T. Curr Microbiol 1991; 22: 97-102.
8　Rydén C, Rubin K, Speziale P, Höök M, et al. J Biol Chem 1983; 258: 3396-3401.
9　Holderbaum D, Hall GS, Erhart LA. Infect Immun 1986; 54: 359-364.
10　Lopes JD, dos Reis M, Brentani RR. Science 1985; 229: 275-277.
11　Herrman M, Suchard SJ, Boxer LA, Waldvogel FA, Lew PD. Infect Immun 1991; 59: 279-288.
12　Wadström T, Paulsson M, Ljungh Å. In: Möllby R, Flock JI, Nord CE, Christensson B, eds. Staphylococcal Infections. Stuttgart: Gustav Fischer Verlag, 1993: (in press).
13　Bodén M, Flock JI. Microb Pathogen 1992;12: 289-298.
14　Switalski LM, Rydén C, Rubin K, Ljungh Å, et al. Infect Immun 1983; 42:628-633.
15　Naidu AS, Paulsson M, Wadström T. J Clin Microbiol 1988; 26: 1549-1554.
16　Paulsson M, Ljungh Å, Wadström T. J Clin Microbiol 1992; 30: 2006-2012.
17　Paulsson M, Kober M, Freij-Larsson C, Stollenwerk M, et al. Biomaterials 1993;(in press).
18　Paulsson M, Gouda I, Larm O, Ljungh Å. J Biomed Mat Res 1993; (in press).
19　Paulsson M, Petersson A-C, Ljungh, Å. J Med Microbiol 1993; 38: 96-102.
20　Paulsson M, Liang OD, Ascencio F, Wadström T. Zbl Bakt 1992; 277, 54-64.
21　Liang OD, Maccarana M, Flock JI, Paulsson M, et al. Biochim. Biophys. Acta 1993; (in press).
22　Ljungh Å, Kinsman O, Wadström T. In: Jeljaszewicz J, White A, eds. Handbook of Staphylococci. Chicago: Mosby Publisher, 1993; (in press).
23　Liang OD, Ascencio F, Fransson L-Å, Wadström T. Infect Immun 1992; 60,899-906.
24　Steele JG, Johnson G, Norris WD, Underwood PA. Biomaterials 1991; 12:531-539.
25　Steele JG, Dalton BA, Johnson G, Underwood PA. J Biomed Mat Res 1993; 27:927-940.
26　Wesslén B, Kober M, Freij-Larsson C, Ljungh, Å et al. Biomaterials 1993; (in press).

27 Wadström T, Erdei J, Paulsson M, Ljungh Å. In: Wadström T, Eliasson I, Holder IA,Ljungh Å, eds. Pathogenesis of Wound and Biomaterial-associated infections London: Springer Verlag, 1990; 339-34.
28 Vaudaux P, Pittet D, Haeberli A, Lerch PG, *et al.* J Infect Dis 1993; 167: 633-641.
29 Dunne WM, Burd EM. J Appl Bacteriol 1993; 74: 411-416.
30 Valentin-Weigand P, Timmis KN, Chhatwal GS. J Med Microbiol 1993; 38, 90-95.
31 Blaser MJ. J Infect Dis 1990; 161: 626-633.
32 Trust TJ, Doig P, Emödy L, Kienle Z, *et al.* Infect Immun 1991; 59: 4398-4404.
33 Valkonen KH, Ringnér M, Ljungh Å, Wadström T. FEMS Immunol Microbiol 1993; 7: 29-38.
34 Wadström T, Ascencio F, Ljungh Å, Lelwala-Guruge J, *et al.* Europ J Gastroenterol 1993; 11: (in press).
35 Ringnér M, Paulsson M, Wadström T. FEMS Microbiol. Immunol. 1992; 105: 219-224.
36 Lelwala-Guruge J, Ascencio F, Kreger AS, Ljungh Å, Wadström T. Zbl Bakt 1993; 566-572.
37 Jakab E, Paulsson M, Ascencio F, Ljungh Å. APMIS 1993; 101: 187-193.
38 Dahlbäck B Podack ER. Biochem 1985; 24: 2368-2374.
39 Rozalska B, Wadström T. Scand J Immunol 1993; 37: 575-580.

Biology of Vitronectins and their Receptors
K.T. Preissner, S. Rosenblatt, C. Kost, J. Wegerhoff and D.F. Mosher, editors

Bone sialoprotein binding to *Staphylococcus aureus.*

Cecilia Rydén, M.D.,Ph.D., Dept. of Medical and Physiological Chemistry and Dept. of Infectious Diseases, University of Uppsala, Sweden. **Alia Yacoub,** M.Sci., Dept. of Medical and Physiological Chemistry, University of Uppsala, Sweden.

INTRODUCTION

Bone sialoprotein (BSP) is a well characterized glycoprotein specific for bone tissue (1-4), most abundant in the osteoid. BSP binds tightly to hydroxyapatite and contains an Arg-Gly-Asp-sequence responsible for eukaryotic cell binding (5). Infection of bone tissue and joints are mainly caused by *Staphylococcus aureus* (6-10), which reach the tissue either by heamatogenous spread or directly, inoculated through a wound or foreign material such as a prosthesis. Heamatogenously spread staphylococcal infection of bone is most common among young individuals (9,10) who have a high synthesis and turnover of bone tissue. Stahylococci are known to interact with extracellular matrix components which may contribute to their pathogenicity. *Staphylococcus aureus* cells isolated from patients suffering from osteomyelitis and/or septic arthritis bind significantly more BSP compared to staphylococci isolated from other tissues (11). The interaction between staphylococcal cells and BSP has been characterized (12) and fulfills criteria of a specific interaction. Staphylococci bind to the aminoterminal part of the BSP molecule (12), at a site separate from the eukaryotic cell binding RGD-sequence located in the carboxyterminus (4).

Bone matrix composition

Bone is composed of a mineralized scaffold of hydroxyapatite and an organic matrix consisting to 90% of collagen type I with additional glycoproteins and proteoglycans (2). Plasma proteins also add to the proteins found in small amounts in bone (13). One of the most abundant of the non-collagenous proteins of bone is BSP (1). BSP is found only in bone matrix and dentin and has been characterized in detail (3). The gene encoding the BSP molecule has been cloned and sequenced (3). Vitamin D_3 and corticosteroids influence the *in vitro* synthesis of BSP (14). The characterization of other glycoproteins, some of which may be specific for bone tissue as well, is now being undertaken. Several glycoproteins found in bone are found in other tissues as well, such as osteopontin (2,15), thrombospondin (16,17) and fibronectin which might be plasma-derived (13).

Staphylococcal interactions with bone components other than BSP

Staphylococcus aureus, as well as coagulase negative strains of staphylococci interact with several of the extracellular matrix components found in bone tissue. The interaction between staphylococcal cells and fibronectin has been well characterized (18-22) including cloning of two genes encoding two fibronectin binding proteins from S. *aureus* strain 8325-4 (23). The interaction with fibronectin seems common among different clinical isolates of staphylococci (24). Staphylococci also bind collagens (25,26), including collagen type I, the most abundant protein of bone tissue. Strains isolated from patients suffering from septic arthritis bind to bovine nasal cartilage (27), containing collagen type II. A staphylococcal gene encoding a collagen binding protein has been identified (28). Staphylococci bind thrombospondin (29) which can be found in bone tissue, but the significance of that interaction for staphylococcal infection of bone has yet not been revieled.

Interaction between bone sialoprotein and staphylococcal cells

In earlier studies we found that clinical isolates of staphylococci bound BSP (11), and that the interaction was selective such that isolates from patients with osteomyelitis and/or septic arthritis bound significantly more BSP than isolates from other staphylococcal infections such as endocarditis and wound infections. The interaction between staphylococci and BSP was characterized (12), and fulfilled criteria of a specific type of binding including time dependence, saturability, reversibility and ablility to be inhibited only by unlabeled BSP itself and not by other matrix proteins or serum proteins (12). BSP contains clusters of O-glycosidically linked oligosaccharides as well as one or two N-linked oligosaccharides (1), that could be of potential importance for the binding. Purified N-linked oligosaccharides from BSP did not inhibit the binding of BSP to staphylococcal cells, neither did a rat chondrosarcoma proteoglycan that contains clusters of O-linked oligosaccharides similar to those of BSP (12). The main binding of BSP to staphylococci thus seems to be due to interaction with the core protein of the BSP molecule. Since the gene encoding BSP has been cloned (4), fusion protein could be produced and expressed in E. coli, and fusionprotein containing the BSP core protein inhibited the binding as well (12) in contrast to fusion protein containing the osteopontin core protein (12).

The staphylococcal binding domain of BSP

Cleavage of the BSP molecule by CNBr results in two large peptides, as the molecule contains only one methionine residue, located at position 151 (3). These peptides were used as potential inhibitors of the binding of ^{125}I-labelled BSP to staphylococcal cells, and the aminoterminal fragment inhibited the binding whereas the carboxyterminal fragment including the RGD-sequence did not inhibit the binding (12). Deletions of cloned cDNA (30) encoding BSP, priorily fused with the cDNA encoding the C terminus of a 26 kDa glutathione S-transferase encoded by *Schistosoma japonicum*, were performed. The plasmid expression vectors constructed this way were inserted into *Escherichia coli* and soluble fusion proteins containing different amino acid sequences were produced (Rydén C. and Oldberg Å. Unpublished data), which also have been used as potential inhibitors for the BSP binding. By this method it could also be shown that the main binding site for staphylococci in the BSP molecule is located in the aminoterminal part of the BSP molecule. Work is now being performed to identify the staphylococcal binding domain of BSP, and synthetic peptides will be used for confirmation of the data (Rydén and Oldberg, unpublished).

Identification of a BSP binding protein in *S. aureus* strain O24

We could show that lysate of staphylococcal cells inhibited the binding of ^{125}I-labelled BSP to both the same and to other strains of *S. aureus* cells (12). Initially lysates were prepared by Lysostaphin treatment of staphylococcal cells (12,19), a method utilising a specific enzyme produced by *Staphylococcus staphylolyticus* (or the recombinant protein, Lysostaphin) which cleaves pentaglycin bridges in the rigid staphylococcal cell wall. Such lysate had to be heat treated to assure that any remaining Lysostaphin activity was abolished. We had shown that BSP binding occurs also to heat inactivated staphylococci, but heat treatment of the lysate meant loss of a substantial amount of total protein extracted from the cell walls. After heat inactivation of the lysate, protein was precipitated by ammoniumsulphate, dialysed and the BSP binding protein could then be purified by affinity chromatography on BSP coupled to Sepharose. The eluted material inhibited the binding of BSP to staphylococcal cells. When analysed by gelelectrophoresis the eluted material was shown to contain several proteins of varying size, which could be due to cleavage of one larger protein either by the Lysostaphin treatment, or by intracellular staphylococcal enzymes released during the purification procedure (Yacoub et al. Submitted manuscript).

To avoid the rupture of staphylococcal cells while isolating the BSP binding component(s) we developed a method based on caotropic release of surface molecules without lysing the bacteria. Staphylococcal cells were subjected to treatment with LiCl which resulted in a lysate containing less protein than the lysate after Lysostaphin treatment, but still inhibiting the binding of BSP to staphylococci. The bacteria did not disrupt during the procedure, and when lysate was subjected to affinity chromatography on BSP Sepharose a BSP binding protein was identified. The characterization of this protein is currently undertaken by our group.

The BSP binding protein as a targeting factor in staphylococcal bone infection

The BSP binding protein of staphylococci may play a role in the infectious processes occuring in bone tissue and joints. Since not all staphylococci tested have shown to bind BSP, and there seems to be a selectivity for BSP binding in staphylococcal strains regardless of coagulase production in favour of those isolated from bone and joint infection (11,31), it will of course be highly interesting to identify the gene encoding this staphylococcal protein. Strains of staphylococci could then be tested for presence of the gene, and this could be correlated to infectivity of bone tissue. Even if the gene is present, the expression of the gene product may be influenced by external factors leading to variation in BSP binding capacity and maybe in pathogenicity. The ability to bind BSP may correlate with the probability of staphylococci localising to bone tissue, and initialising an infectious process at this site.

Other staphylococcal proteins of potential interest in bone infection

The staphylococci isolated from osteomyelitic and arthritic infections show a varying pattern of binding capacity for extracellular proteins. In addition to BSP binding capacity some strains bind collagen type I and II, and since collagen type I is abundant in bone matrix and collagen type II in joint cartilage these interactions could be of potential importance for the maintenence of bacteria once they reach these tissues. These collagens are not tissue specific (32,33) and thus could hardly be important for the specific targeting of bacteria to these sites, but the abundance of these molecules may make them very important in keeping bacteria stay in the tissue. The collagen-staphylococcal interaction shows quite low affinity (27) but yet this interaction could be of importance in the pathogenicity of staphylococci. Staphylococci also bind fibronectin with very high affinity (19), the binding is virtually irreversible in solution *in vitro* , although on immobilized fibronectin staphylococcal adhesion could be reversed (20). The fibronectin interaction may play a more general role in the infectious process of staphylococci. Some staphylococcal cells show vitronectin binding (34,35), and thrombospondin binding has also been shown to be specific for some staphylococcal strain (29). Such interactions may add to the potential pathogenicity of staphylococci, not only in bone and joint infection.

Identification of a staphylococcal protein with broad affinity for ECM components

Stahylococcal cell lysate contains many different proteins, fewer if treatment with LiCl has been performed to isolate proteins than if Lysostaphin has been used for preparing cell wall proteins. Lysates prepared from staphylococcal cells cultured in Luria broth and subjected to LiCl treatment, were subjected to gel electrophoresis (36). After separation by SDS-PAGE, staphylococcal proteins were transferred to nitrocellulose and incubated with radiolabelled BSP. Some bands could be visualised after exposing the sheet on film, and two bands of apparent M_rs of 72 000 and 60 000 respectively, appeared when different staphylococcal strains were used (36). These bands were not identical to the affinity purified BSP binding protein discussed above, which was isolated from staphylococcal cells grown in tryptic soy broth. The two proteins could be purified by cation- exchange chromatography, in contrast to the BSP binding protein described above. The purified proteins were subjected to Western ligand blotting and then showed to bind several proteins, some with significantly higher affinity than was shown for BSP (36). The 72 kDa protein as well as the 60 kDa protein did bind fibronectin, fibrinogen, vitronectin, thrombospondin and collagen. When purified 60 kDa protein was used as a potential inhibitor of binding of radiolabelled ligands, it was found that the protein rather augmented binding of ligand to staphylococci, especially striking in the case of vitronectin and thrombospondin, where binding increased about 15 and 13-fold respectively (36). The 60 kDa protein did bind several ECM glycoproteins and furthermore was shown to be able to agglutinate sheep red blood cells, an interaction that could be inhibited by L-fucose, D-mannose and melibiose, but not by lactose, N-acetyl- D -glucoaseamine or N-acetyl-D-galactoseamine. These results suggest that the 60 kDa protein is an agglutinating lectin recognising carbohydrate structures present in matrix glycoproteins. Such interactions may be of general importance in the pathogenicity of staphylococcal cells. Surface proteins in other bacteria such as *Treponema denticola* (37) and *Streptococcus pyogenes* (38) also have been shown to have broad ligand-binding capacities.

Experimental septic arthritis in mice

A model to study *S.aureus* induced septic arthritis in mice has been developed (39). When live *S. aureus* cells of certain strains are being injected intravenously into mice, the animals develop arthritis within a few days (39). The strains causing arthritis have been tested for binding capacity of ECM components, and have all been shown to bind BSP, whereas binding of collagen type I and II, fibronectin and vitronectin has varied. Expression of the accessory gene regulator (*agr*) is of importance for the bacterial virulence in this mouse model (40), and strains mutated in the *agr*-locus show lower BSP binding than the wild parental strain (40). It is not known if the expression of the BSP binding protein is regulated by the *agr/hld*-locus. The influence of other gene-products on the virulence of staphylococci in the murine model is currently being studied.

CONCLUSION

Bone sialoprotein (BSP) is a glycoprotein specific for bone tissue and dentin. It is found in high amounts in the area of newly synthesized bone- the osteoid- and binds tightly to hydroxyapatite. Bone sialoprotein binds to eukaryotic cells *in vitro*, a binding dependent on an Arg-Gly-Asp- (RGD) sequence located in the carboxyterminal part of the molecule. Staphylococci, which are the most common cause of bacterial arthritis and osteomyelitis in the so called developed part of the world, interact with BSP. The interaction between staphylococci and BSP is selective in the sense that clinical isolates of *S. aureus* as well as coagulase negative staphylococci from patients with osteomyelitis and/or septic arthritis have a higher affinity for BSP than staphylococcal isolates from other types of infection. The interaction has been characterized, and a staphylococcal protein with specific affinity for BSP has been purified. It has also been possible to determine the binding domain in the BSP- molecule, which is located in the amino-terminal part of the molecule distinct from the eukaryotic cell-binding RGD- sequence. In an experimental model for septic arthritis the staphylococcal strains giving rise to infection of the joints all had the capacity to bind BSP, whereas binding of collagen, fibronectin and vitronectin varied between the strains tested so far. The specific interaction between staphylococci and BSP may contribute to the localization of these bacteria to bone tissue, and may add to the pathogenicity of the staphylococci.

Acknowledgement

The study was supported by grants from the Swedish Medical Research Counsil, The Swedish Medical Association, King Gustaf V:s 80-year fund and Sätra Brunns research fund. This work has been performed in collaboration with Kristofer Rubin, Uppsala University, Dick Heinegård, Åke Oldberg and Mikael Wendel, University of Lund, Andrzej Tarkowski, Tomas Bremell and

Arturo Abdelnour, University of Gothenburg, Joseph Patti, Magnus Höök and Mary Homonylo McGavin, Institute of Bioscience and Technology, Houston, Texas, whom we want to acknowledge.

References

1. Franzén A. and Heinegård D. Isolation and characterization of two sialoproteins present only in bone calcified matrix. 1985. Biochem.J. 232:715-724.

2. Heinegård D. and Oldberg Å. Structure and biology of cartilage and bone matrix non-collagenous macromolecules. 1989. FASEB J. 3:2042-2051.

3. Oldberg Å., Franzén A. and Heinegård D. The primary structure of a cell-binding BSP. 1988. J.Biol.Chem. 263:19430-19432.

4. Oldberg Å., Franzén A. and Heinegård D. Cloning and sequence analysis of rat bone sialoprotein (osteopontin) cDNA reveals an Arg-Gly-Asp cell binding sequence. 1986. Proc.Natl.Acad.Sci. 83:8819-8823.

5. Oldberg Å., Franzén A., Heinegård D., Pierschbacher M. and Rouslahti E. Identification of a BSP receptor in osteosarcoma cells. 1988. J.Biol.Chem. 263:19433-19436.

6. Gentry L.O. Overview of osteomyelitis. 1987. Orthop.Rev. 16:255-258.

7. Dubei L., Krasinski K. and Hernanz-Schulman M. Osteomyelitis secondary to trauma or infected contiguous soft tissue. 1988. Pediatr.Inf.Dis.J. 7:26-34.

8. Goldenberg D.L. Infectious arthritis complicating rheumatoid arthritis and other chronic rheumatic disorders. 1989. Arthritis Rheum. 32:496-502.

9. Gillespie W.J. Epidemiology in bone and join infection. 1990. Inf.Dis.Clin.North Am. 4:361-376.

10. Espersen F., Frimodt-Möller N., Thamdrup-Rosdahl V., Skinhöj P. and Bentzon M.W. Changing pattern of bone and joint infections due to *Staphylococcus aureus*: Study of cases of bacteremia in Denmark, 1959-1988. 1991. Rev.Inf.Dis. 13:347-358.

11. Rydén C., Maxe I., Franzén A., Ljungh Å., Heinegård D. and Rubin K. Selective binding of bone matrix sialoprotein to *Staphylococcus aureus* isolated from patients with osteomyelitis. 1987. Lancet ii:8557:515.

12. Rydén C., Yacoub A., Oldberg Å., Heinegård D., Ljungh Å. and Rubin K. Specific binding of bone sialoprotein to *Staphylococcus aureus* isolated from patients with osteomyelitis. 1989. Eur.J.Biochem. 184:331-336.

13. Triffitt J.T. The special proteins of bone tissue. 1987. J.Clin.Sci. 72:399.

14. Oldberg Å., Jirskog-Hed B., Axelsson S. and Heinegård D. Regulation of bone sialoprotein mRNA by steroid hormones. 1989. J.Cell.Biol. 109:3183-3186.

15. Hultenby K., Reinholt F.P., Oldberg Å. and Heinegård D. Ultrastructural immunolocalization of osteopontin in metaphyseal and cortical bone. 1991. Matrix 11:206-213.

16. Miller R.R. and McDevitt C.A. Thrombospondin is present in articular cartilage and is synthesized by articular chondrocytes. 1988. Biochem.Biophys.Res.Commun. 153:708-714.

17. Mosher D.F. Physiology of thrombospondin. In "Annual review of medicine" ed.41.Palo Alto, Calif. pp 85-97.

18. Kuusela P. Fibronectin binds to *Staphylococcus aureus*. Nature (London) 1978. 276:718-720.

19.   Rydén C., Rubin K., Speziale P., Höök M., Lindberg M. and Wadström T. Fibronectin receptors from *Staphylococcus aureus*. 1983. J.Biol.Chem. 258:3396-3401.

20.   Maxe I., Rydén C., Wadström T. and Rubin K. Specific attachment of *Staphylococcus aureus* to immobilized fibronectin. 1986. Infect.Immun. 54:695-704.

21.   Proctor R., Mosher D.F. and Olbrantz P.J. Fibronectin binds to *Staphylococcus aureus*.. 1982. J.Biol.Chem. 257:14788-14794.

22.   Fröman G., Switalski L.M., Speziale P. and Höök M. Isolation and characterization of a fibronectin receptor from *Staphylococcus aureus*.. 1987 J.Biol.Chem. 262:6564-6571.

23.   Jönsson K., Signäs C., Muller H-P. and Lindberg M. Two different genes encode fibronectin binding proteins in *Staphylococcus aureus*. 1991. Eur.J.Biochem. 202:1041-1048.

24.   Switalski L.M., Rydén C., Rubin K., Ljungh Å., Höök M. and Wadström T. Binding of fibronectin to *Staphylococcus* strains. 1983. Infect.Immun. 42:628-633.

25.   Holderbaum D., Spech R.A. and Erhart L.A. Specific binding of collagen to *Staphylococcus aureus*. 1985. Coll.Rel.Res. 5:261-276.

26.   Switalski L.M., Speziale P. and Höök M. Isolation and characterization of a putative collagen receptor from *Staphylococcus aureus* strain Cowan I. 1989. J.Biol.Chem. 264:21080-21086.

27.   Switalski L.M., Patti J.M., Butcher W., Gristina A.G., Speziale P. and Höök M. A collagen receptor on *Staphylococcus aureus* strains isolated from patients with septic arthritis mediates adhesion to cartilage. Mol.Microbiol. 7:99-107.

28.   Patti J.M., Jonsson H., Guss B., Switalski L.M., Wiberg K., Lindberg M. and Höök M. Molecular characterization and expression of a gene encoding a *Staphylococcus aureus* collagen adhesin. 1992. J.Biol.Chem. 267:4766-4772.

29.   Herrman M., Suchard S.J., Boxer L.A., Waldvogel F.A. and Lew P.D. Thrombospondin binds to *Staphylococcus aureus* and promotes staphylococcal adherence to surfaces. 1991. Infect.Immun. 59:279-288.

30.   Sambrook J., Fritsch E.F. and Maniatis T. Molecular cloning. 1989. Cold Spring Harbor Laboratory Press, New York.

31.   Rydén C., Yacoub A., Hirsch G., Wendel M., Oldberg Å., and Ljungh Å. Binding of bone sialoprotein to *Staphylococcus epidermidis* isolated from a patient with chronic recurrent multifocal osteomyelitis. 1990. J.Inf.Dis. 161:814-815.

32.   Burgeson R.E. New collagens, new concepts. 1988. Ann.Rev.Cell Biol. 4:551-577.

33.   Martin G., Timpl R., Kuller P.K. and Kuhn K. The genetically distinct collagens. 1985. TIBS. 285-287.

34.   Cchatwal G.S., Preissner K.T., Muller-Berghaus G. and Blobel H. Specific binding of the human S protein (Vitronectin) to streptococci, *Staphylococcus aureus* and *Escherichia coli*. 1987. Infect.Immun. 55: 1878-1833.

35.   Paulsson M., Olin D.L., Ascencio F. and Wadström T. Vitronectin-binding surface proteins of *Staphylococcus aureus* . 1992. Zbl.Bakt. 277:54-64.

36.   McGavin Homonylo M., Krajewska-Pietrasik D., Rydén C. and Höök M. Identification of a *Staphylococcus aureus* extracellular matrix-binding protein with broad specificity. 1993. Infect. Immun. 61:2479-2485.

37.   Haapsalo M., Muller H-P., Uitto V-J., Keung Leung W. and Mc Bride B. Characterization, cloning, and binding properties of the major 53-kDa *Treponema denticola* surface antigen. 1992. Infect.Immun. 60:2058-2065.

38. Pancholi V. and Fischetti V.A. A major surface protein on group A streptococci is a glyceraldehyde-3-phosphate-dehydrogenase with multiple binding activity. 1992. J.Exp.Med. 176:415-426.

39. Bremell T., Lange S., Yacoub A., Rydén C. and Tarkowski A. Experimental *Staphylococcus aureus* arthritis in mice. 1991. Infect.Immun. 59:2615-2623.

40. Abdelnour A., Arvidson S., Bremell T., Rydén C. and Tarkowski A. The accessory gene regulator (agr) controls the virulence of *Staphylococcus aureus* in a murine arthritis model. 1993. Infect.Immun. In press.

Biology of Vitronectins and their Receptors
K.T. Preissner, S. Rosenblatt, C. Kost, J. Wegerhoff and D.F. Mosher, editors

273

Adhesive reactions between immobilized platelets and Staphylococcus aureus

Mathias Herrmann[a,1], Ralph M. Albrecht[b], Deane F. Mosher[c,d], Richard A. Proctor[a,d]

Departments of Medical Microbiology and Immunology[a], Animal Health and Biomedical Sciences[b], Biomolecular Chemistry[c], and Medicine[d], University of Wisconsin, Madison, Wisconsin 53706.

SUMMARY

Interaction of S. aureus with specific surface-bound host ligands is a pivotal step in staphylococcal infections such as endocarditis and device infection, however, the ligands involved in Staphylococcus aureus interaction with adherent platelets remain incompletely understood. We studied adhesion of S. aureus to surfaces with adherent platelets in the presence of plasma and purified plasma proteins and found that adherent platelets greatly promote adhesion of S. aureus in the presence of plasma proteins (>150-fold increase over adhesion to uncoated surfaces). Thus, plasma proteins may serve as bridging ligands between specific binding sites on the platelet surface and on staphylococci. Furthermore, platelet pretreatment with anti-GPIIb/IIIa monoclonal antibodies or inhibitors of platelet activation decreased plasma-enhanced adhesion suggesting a role of platelet activation in S. aureus adhesion. Finally, plasma-enhanced adhesion was sensitive to thrombin antagonists, proteinase inhibitors, heparin or antifibrinogen antibodies indicating that fibrinogen/fibrin is necessary for bridging between adherent platelets and S. aureus. Our findings suggest a role of activated platelets adherent to foreign surfaces in the pathogenesis of certain types of invasive infections due to S. aureus.

INTRODUCTION

S. aureus is associated with endocarditis and device infection (1), characterized by exposure of thrombogenic surfaces to blood through damage of the endothelial cell lining or insertion of foreign material. This results to a varying extent in plasma protein (2) and platelet deposition (3) on the surfaces and in platelet activation and thrombus formation (4). One reason for the frequent isolation of S. aureus from device associated infections is the ability of this species to adhere to catheters by recognizing specific host proteins deposited on the material. In addition, S. aureus interacts avidly with platelets in suspension resulting in platelet aggregation and activation (5) and it has been suggested that adherent platelets play a major role in microbial adhesion to an immobilized fibrin matrix (6,7). More recently, it has been shown using flow cytometry that both viridans streptococci (8) and S. aureus (9) bind to platelets even in the absence of plasma proteins. However, these studies have been performed testing adhesion to complex fibrin-plasma protein-platelet matrices, thus they do

[1] Present address: Institut für Medizinische Mikrobiologie, Westfälische Wilhelms-Universität, 48129 Münster, Germany

not allow to describe the specific role of the individual components involved in this interaction, i.e. platelets and purified plasma proteins.

We recently described the S. aureus adhesion promoting characteristics of thrombospondin, a large multifunctional platelet released glycoprotein (10). Here we present results describing more closely the role of platelets and plasma proteins in the promotion of S. aureus adhesion to surfaces.

METHODS

Fibronectin, fibrinogen and vitronectin were purified as previously described (11). Whole human plasma was prepared from ACD (citric acid, 5 mM; sodium citrate, 10 mM, dextrose, 15 mM) anticoagulated blood of healthy volunteers. In some experiments, whole plasma was treated according to either of the following conditions: i) by dialysing against 250 volumes of PBS using Spectra/Por tubings (Spectrum, Los Angeles, CA)(m.w. cutoff 12.000 - 14.000) overnight at 4° C, ii) by filtration using 0.2 micron GS filters (Millipore, Bedford, MA), iii) by heating at 56° C for 30 min in a water bath, iv) by clotting after supplementation of Ca^{2+} at 37° C and centrifugation at 2000 x g. S. aureus strain Cowan 1 was used for most experiments. Selectively mutagenized strains of parent strain Newman were generated using transposon mutagenesis or allelic replacement (12,13). The monoclonal antibodies (mAbs) used were the following: AP 1, directed against the platelet membrane glycoprotein GPIb (14), AP 2 against the GPIIb/IIIa complex (15), 10E5 and 7E3 (IgG_1), against GPIIb/IIIa (16), antibody 6D1 (IgG_1) against GPIb (17), 6F1 (IgG_1) against GPIa/IIa (18), LM 609 against vitronectin receptor (19), OKM 5 directed against GPIV (20)), TM 60 against GPIb (IgG_{2a}) (21). Purified mAbs were used in a working concentration of 20 ug/ml. Final working dilutions of antibodies AP1 and AP2 (stock solution in ascites fluid) were 1:50. Anti-human fibrinogen antibodies, anti human fibrinogen-IgG F(ab')$_2$ fragments, and rabbit preimmune IgG were used at 100 ug/ml, dilutions of anti-thrombospondin antiserum and anti-fibronectin antiserum was 1:50.

For each experiment, bacteria were grown from a fresh overnight culture in Mueller-Hinton broth and radiolabelled as previously described (22). After incubation, bacteria were washed twice and resuspended at $1x10^8$ CFU/ml in Tyrode's buffer. Platelets were purified as described (23). ACD-anticoagulated blood was drawn and platelet rich plasma was prepared by centrifugation. Platelets were separated from plasma proteins by passage through a A-50 column (Biorad, Richmond, CA). The column was equilibrated with a calcium-free Tyrodes buffer pH 7.3 (136 nM NaCl, 2.7 nM KCl, 3.3 mM NaH_2PO_4, 10.5 mM HEPES salt, 4.45 mM HEPES free acid, 2 mM $MgCl_2$, 1 g/l dextrose, 1 g/l BSA). The platelet number was determined and adjusted to $1.5x10^8$ platelets/ml.

Polymethylmethacrylate (PMMA), a material widely used for implants, was used as the solid phase in the S. aureus adhesion assay. The previously described assay (22) was modified and performed as follows: PMMA coverslips (0.8 x 0.8 mm), were cleaned and sterilized. The platelet suspension, 100 ul containing 150,000 platelets/ul, were overlaid onto PMMA coverslips for 15 min at 37°C. The fluid was removed and immediately replaced by 100 ul Tyrode's buffer. For the adhesion assay, a solution containing radiolabelled S. aureus ($4 x 10^6$CFU/ml) in Tyrode's buffer supplemented with Ca^{2+} (1mM), dextrose (0.1%)(total volume: 0.5 mL) and human albumin (0.5%) to block nonspecific adhesion was incubated with the coverslips at 37°C in a shaker bath for 60 min. After incubation, coverslips were washed twice in Tyrode's buffer and cpm of the coverslips were determined.

RESULTS

Adhesion of S. aureus to plasma- and platelet-coated surfaces. Control experiments using coverslips without platelets and incubated with S. aureus in Tyrode's buffer containing Ca^{2+} and albumin confirmed previous observations using this adhesion assay (10,22,24), specifically that adhesion to albumin-coated PMMA was very low (0.2 % ± 0.1 %, mean adhesion [adherent / inoculated S. aureus] ± SE) (Figure 1) revealed a surface with a virtual absence of microorganisms (Figure 2A; bar, 3 μm).

Subsequent experiments were performed by adding human plasma (1%) to the incubation medium, and S. aureus adhesion to plasma-coated coverslips was 1.0 % ± 0.3 % (mean ± SE). When platelets were allowed to adhere to coverslips, staphylococcal adhesion was increased 28-fold when compared to albumin coated PMMA (5.6 % ± 4.4 %, mean adhesion ± SE). SEM micrographs of platelet coated surfaces revealed that microorganisms adhered both to the platelet membranes and to the substrate (Figure 2B; bar, 3 μm). Addition of whole human plasma (1 % vol/vol) to the assay containing S. aureus and platelet-coated coverslips increased adhesion 151-fold when compared to albumin-coated PMMA (30.2 % ± 5.3 % vs. 0.2 %, n = 14, P = .0081, Mann–Whitney test); Examination of parallel coverslips revealed attachment of large amounts of staphylococci preferentially adhering to the platelets (Figure 2C, bar, 2 μm, and D, bar, 3 μm). These findings suggest that S. aureus preferentially interact with binding sites on platelets via plasma proteins which may serve as bridging ligands between S. aureus and activated platelet receptors (25).

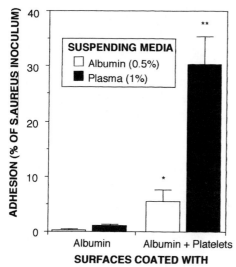

Figure 1. Adhesion of S.aureus to PMMA. *, P <.05; **, P < .01.
Reprinted from (27) with permission.

Determination of specific platelet characteristics associated with promotion of S. aureus adhesion. First, we determined S. aureus adhesion to adherent platelets as a function of platelet activation. If platelets were purified and allowed to adhere to coverslips in the presence of the adenylate cyclase stimulant PGE_1, subsequent S. aureus adhesion was ≈80% inhibited. Adhesion could be restored by pretreatment of adherent platelets with recombinant human α-thrombin. Next, we pretreated platelets with a panel of monoclonal antibodies directed against specific platelet glycoproteins or against platelet proteins putatively expressed on the platelet surface. As shown in Table 1, monoclonal antibodies directed against the platelet membrane glycoprotein GP IIb/IIIa were highly active in inhibiting S. aureus adhesion. Subsequent electronmicroscopic studies and results obtained with PMMA-coverslips precoated with poly-L-lysine revealed that this effect was likely due to decreased activation and spreading of adherent platelets upon contact with the substrate. Thus, platelets

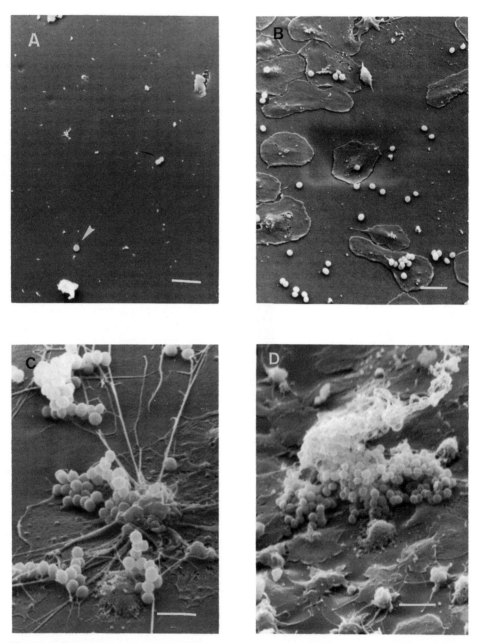

Figure 2. Electron micrographs of <u>S. aureus</u> adherent to uncoated or to platelet-coated PMMA in the presence or absence of plasma. For details please refer to text. Reprinted with permission from (27).

have to become activated upon contact with PMMA (26), and to spread on the surface in order to promote plasma-enhanced staphylococcal adhesion.

Table 1. Effect of Antibody Treatment of Platelets and Platelet-coated Polymethylmethacrylate (PMMA) on Staphylococcus aureus Adhesion.

Antigen	Antibody	Uncoated PMMA		PLL-coated PMMA	
		Adhesion ± SEM (% of control)	n	Adhesion ± SEM (% of control)	n
GPIIb/IIIa	10E5 (20 ug/ml)	21.4 ± 14.0*	6	39.7 ± 11.7*	8
GPIIb/IIIa	7E3 (20 ug/ml)	30.8 ± 14.0*	6	7.2 ± 1.5*	3
GPIIb/IIIa	AP2 (1:50)	43.6 ± 13.9*	6	11.8 ± 2.2*	3
GPIa	AP1 (1:50)	123.6 ± 12.2	4		
GPIa	TM60 (20 ug/ml)	89.0 ± 35.8	4		
GPIb	6D1 (20 ug/ml)	100.1 ± 11.9	2		
GPIa/IIa	6F1 (20 ug/ml)	121.4 ± 11.1	2		
GPIV	OKM5 (20 ug/ml)	91.5 ± 16.2	4		
VN-R	LM609 (20 ug/ml)	77.9 ± 22.0	4		
FG	IgG (100 ug/ml)	101.7 ± 5.4	2		
FG	IgG-F(ab')₂ (100 ug/ml)	86.6 ± 1.9	2	99.6 ± 37.0	5
TSP	Serum (1:50)	106.4 ± 3.3	2		
FN	Serum (1:50)	136.1 ± 5.8	2		
n/a†	IgG1 (20 ug/ml)	116.3 ± 15.0	2		
n/a	IgG2a (20 ug/ml)	91.1 ± 24.2	6		
n/a	IgG (100 ug/ml)	101.7 ± 5.4	2		
n/a	Preimmune serum (1:50)	104.1 ± 12.6	2		

Reprinted from (27) with permission. *, $P < .01$

Delineation of the adhesion-promoting factors contained in plasma. Whole plasma was pretreated using a variety of methods (Figure 3A) or media were supplemented with purified plasma proteins instead of whole plasma (Figure 3B). The factor(s) contained in plasma and are heatlabile, lost upon clotting and sensitive to treatment with thrombininhibitors (PPACK), proteinase-inhibitors (PMSF) and Ca²⁺-free conditions. Furthermore, if purified fibrinogen was used instead of whole plasma, S. aureus adhesion was not promoted (except for thrombin-treated platelets), however, plasma pretreatment with anti-fibrinogen antibodies abolished increased adhesion (data not shown). These findings suggest that fibrinogen is necessary but not sufficient for enhanced S. aureus adhesion, and that other factors contained in plasma, must be present for maximal enhancement of staphylococcal adhesion.

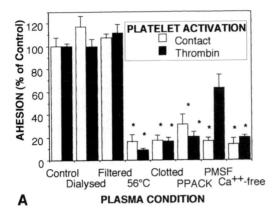

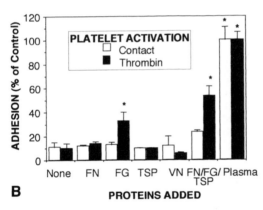

Figure 3. Adhesion of S.aureus to adherent platelets as a function of plasma components. *, P < .01. Reprinted from (27) with permission.

DISCUSSION

Adherence of platelets occurs rapidly after contact of blood with blood (28) through specific mechanisms involving platelet membrane glycoproteins. Upon adherence, platelets become activated resulting in release of α-granule proteins, i.e. thrombospondin, fibrinogen, β-thromboglobulin and PF4 (29), activation of the GPIIb/IIIa receptor on the platelet surface, a rise of intracellular Ca^{2+}-levels (26) and irreversible spreading on the surface. Platelet-released thrombospondin and fibrinogen readily adsorbs to the substrate (30). S. aureus possess binding sites both for fibrinogen (31) and thrombospondin (10), however, receptors for fibrinogen may be saturated. S. aureus recognizes surface-bound thrombospondin resulting in surface colonization and subsequent device infection.

In addition to platelet-released and surface-deposited proteins, adherent platelets themselves may provide binding sites for staphylococci either due to direct interaction of the microorganisms with receptors on the platelet surface or via plasma proteins bridging between platelets and staphylococci. Our data showing a modest increase in adhesion of S. aureus to platelet-coated substrates in the absence of plasma proteins but an extensive increase in the presence of these proteins suggest the latter as the more effective mechanism for S. aureus to bind to platelets.

We attempted to delineate the molecular mechanisms in this interaction on all three levels involved: platelets, plasma, and S. aureus. Platelets have to be activated either by contact with the substrate or by thrombin stimulation, because our experiments with PGE_1-inhibited platelets yielded inhibited staphylococcal adhesion. PGE_1-treatment reduces intracellular Ca^{2+}-mobilization upon contact with PMMA and inhibits platelet activation and aggregation (26,32) and may interfere with fibrinogen binding to GP IIb/IIIa. Thrombin-activation could readily overcome PGE_1-inhibition. Additional evidence for the importance of platelet activation in the promotion of S. aureus adhesion was obtained in our experiments using monoclonal antibodies directed against several platelet membrane glycoproteins: Adhesion was largely inhibited if platelets were pretreated with anti-GPIIb/IIIa antibodies, whereas it was unaffected using antibodies directed against other platelet membrane glycoproteins or plasma proteins. This was supposedly a result of decreased platelet activation as suggested by the microscopic examinations revealing adherent platelets in discoid or early pseudopodial stades of activation.

Candidate factor(s) contained in plasma and interacting with S. aureus and activated platelets include fibrinogen as a necessary cofactor, because antifibrinogen antibodies blocked plasma-enhanced adhesion. However, purified fibrinogen alone was not sufficient in promoting adhesion, and a clumping-factor negative mutant adhered to a similar extent to the platelet-coated coverslips when compared to the parent strain. In addition, PPACK, a specific thrombin-antagonist, was highly active in inhibiting plasma-enhanced adhesion suggesting a role of thrombin and subsequent fibrin polymerization in this system. Other cofactors such as fibronectin-fibrin crosslinking factor XIII or other extracellular matrix proteins may contribute to the formation of a stable interaction between S. aureus and platelets. These findings concur with those of other authors who described fibrinogen as a cofactor necessary for platelet aggregation in suspension after contact with S. aureus (25,33) or streptococci (34) but suggested the presence of additional, unidentified plasma components necessary to elicit maximal aggregation (34).

Among binding sites on S. aureus interacting with surface-adherent platelets, we could identify staphylococcal coagulase, an enzyme forming complexes with prothrombin and resulting in fibrin formation (35)(data not shown). Staphylococcal coagulase is different from the fibrinogen-binding clumping factor (13), and in fact, because our clumping-factor-negative mutant adhered to a similar extent compared to the parent strain, yet adhesion was greatly reduced upon trypsin treatment of the staphylococci, we suspect additional binding sites on S. aureus which may act in concert in the formation of stable complexes of S. aureus, plasma proteins and platelets.

In conclusion, our results demonstrate that both adherent purified platelets and purified platelet proteins promote adhesion of S. aureus to foreign surfaces. Because adhesion is a prerequisite event for colonization and subsequent infection, these findings provide further understanding of the early pathogenic mechanisms of infections at sites of platelet activation and blood clot formation such as implants and may ultimately contribute to the development of strategies to prevent invasive infections due to S. aureus.

ACKNOWLEDGMENTS

This work was supported by the American Heart Association of Wisconsin (grant 92-GS-39) and the Graduate School and Department of Medicine Resarch Funds of the University of Wisconsin. M. Herrmann is a recipient of a fellowship from the Deutsche Forschungsgemeinschaft, FRG. The authors wish to thank D. McDevitt and T. Foster for providing them with selectively mutagenized strains, and B. Coller, N. Yamamoto, P. Rao, D. Cheresh, and T. Kunicki for monoclonal antibodies

REFERENCES

1. Waldvogel, F.A. In: Mandell, G.L., Douglas, R.G.,Jr. and Bennett, J.E. eds. Principles and Practice of Infectious Diseases, New York: Churchill Livingstone, 1990, p. 1489-1510.
2. Vaudaux, P.E., Yasuda, H., Velazco, M.I., et al. J.Biomat.Appl. 1990; 5: 134-153.
3. Engbers, G.H.M., Dost, L., Hennink, W.E., Aarts, P.A.M.M., Sixma, J.J. and Feijen, J. J.Biomed.Mater.Res. 1987; 21: 613-627.

4. Stillman, R.M., Soliman, F., Garcia, L. and Sawyer, P.N. Arch.Surg. 1977; 112: 1497-1499.
5. Clawson, C.C., Rao, G.H.R. and White, J.G. Am.J.Pathol. 1975; 81: 411-419.
6. Scheld, W.M., Valone, J.A. and Sande, M.A. J.Clin.Invest. 1978; 61: 1394-1404.
7. Chugh, T.D., Burns, G.J., Shuhaiber, H.J. and Bahr, G.M. Infect.Immun. 1990; 58: 315-319.
8. Sullam, P.M., Payan, D.G., Dazin, P.F. and Valone, J.A. Infect.Immun. 1990; 58: 3802-3806.
9. Yeaman, M.R., Sullam, P.M., Dazin, P.F., Norman, D.C. and Bayer, A.S. J.Infect.Dis. 1992; 166: 65-73.
10. Herrmann, M., Suchard, S.J., Boxer, L.A., Waldvogel, F.A. and Lew, P.D. Infect.Immun. 1991; 59: 279-288.
11. Murphy-Ullrich, J.E. and Mosher, D.F. Blood 1985; 66: 1098-1104.
12. Kuypers, J.M. and Proctor, R.A. Infect.Immun. 1989; 57: 2306-2312.
13. McDevitt, D., Vaudaux, P.E. and Foster, T.J. Infect.Immun. 1992; 60: 1514-1523.
14. Montgomery, R.R., Kunicki, T.J., Taves, C., Pidard, D. and Corcoran, M. J.Clin.Invest. 1983; 71: 385-389.
15. Pidard, D., Montgomery, R.R., Bennett, J.S. and Kunicki, T.J. J.Biol.Chem. 1983; 258: 12582-12586.
16. Coller, B.S. J.Clin.Invest. 1985; 76: 101-108.
17. Coller, B.S., Peerschke, E.I., Scudder, L.E. and Sullivan, C.A. Blood 1983; 61: 99-110.
18. Coller, B.S., Beer, J.H., Scudder, L.E. and Steinberg, M.H. Blood 1989; 74: 182-192.
19. Cheresh, D.A. and Spiro, R.C. J.Biol.Chem. 1987; 262: 17703-17711.
20. Asch, A.S., Barnwell, J., Silverstein, R.L. and Nachman, R.L. J.Clin.Invest. 1987; 79: 1054-1061.
21. Yamamoto, N., Greco, N.J., Barnard, M.R., et al. Blood 1991; 77: 1740-1748.
22. Vaudaux, P.E., Waldvogel, F.A., Morgenthaler, J.J. and Nydegger, U.E. Infect.Immun. 1984; 45: 768-774.
23. Albrecht, R.M., Simmons, S.R. and Mosher, D.F. In: Matsuda, M., Iwanaga, S., Takada, A. and Henschen, A. eds. Fibrinogen 4. Current basic and clinical aspects, Elsevier, 1990, p. 87-92.
24. Herrmann, M., Vaudaux, P.E., Pittet, D., et al. J.Infect.Dis. 1988; 158: 693-701.
25. Clawson, C.C., White, J.G. and Herzberg, M.C. Am.J.Hematol. 1980; 9: 43-53.
26. Ware, J.A., Kang, J., DeCenzo, M.T., et al. Blood 1991; 78: 1713-1721.
27. Herrmann, M., Lai, Q.J., Albrecht, R.M., Mosher, D.F. and Proctor, R.A. J.Infect.Dis. 1993; 167: 312-322.
28. Salzmann, E.W. and Merrill, E.W. In: Colman, R.W., Hirsh, J., Marder, V.J. and Salzman, E.W. eds. Hemostasis and Thrombosis: Basic Principles and Clinical Practice, Philadelphia: Lippincott, 1987, p. 1335-1347.
29. Mosher, D.F., Pesciotta, D.M., Loftus, J.C. and Albrecht, R.M. In: George, J.N., Nurden, A.T. and Phillips, D.R. eds. Platelet Membrane Glycoproteins, New York: Plenum, 1985, p. 171-191.
30. Tuszynski, G.P. and Kowalska, M.A. J.Clin.Invest. 1991; 87: 1387-1394.
31. Allington, M.J. Br.J.Hematol. 1967; 13: 550-567.
32. Kerins, D.M., Murray, R. and FitzGerald, G.A. In: Coller, B.S. ed. Progress in Hemostasis and Thrombosis. Vol. 10, Philadelphia: W.B.Saunders, 1991, p. 307-337.
33. Clawson, C.C. and White, J.G. Am.J.Pathol. 1980; 98: 197-205.
34. Sullam, P.M., Valone, J.A. and Mills, J. Infect.Immun. 1987; 55: 1743-1750.
35. Kawabata, S., Morita, T., Iwanaga, S. and Igarashi, H. J.Biochem. 1985; 98: 1603-1614.

Vitronectin and pericellular proteolysis

Biology of Vitronectins and their Receptors
K.T. Preissner, S. Rosenblatt, C. Kost, J. Wegerhoff and D.F. Mosher, editors

PLASMINOGEN ACTIVATION:
A MULTIFUNCTIONAL PROTEOLYTIC CASCADE

André-Pascal Sappino and Jean-Dominique Vassalli

Departments of Medicine and Morphology

University of Geneva Medical School, 1211 Geneva 4, Switzerland

Plasminogen activators (PAs) are serine proteases of restricted specificity. Two PAs have been identified in mammals: urokinase-type (uPA) and tissue-type (tPA) plasminogen activators. Products of distinct genes, they differ one from the other by their non-catalytic domains. Funtionally, PAs catalyze the conversion of a common substrate plasminogen into plasmin, a protease of broad specificity. Therefore, PAs induce the formation of a proteolytic enzyme capable of degrading not only fibrin but most extracellular constituents, directly or indirectly, via for instance the activation of inactive metalloproteases (For review see 1 and 2).

PAs are secreted as single chain molecules. The single chain uPA (pro-uPA) has no or little enzymatic activity, while the single and the two-chain forms of tPA appear to be catalytically equivalent. Plasmin itself catalyzes the conversion of pro-uPA into the active two-chain uPA form, illustrating the autoamplification potential of the PA/plasmin system. In vitro biochemical evidence has shown that secreted PAs can interact with a variety of extracellular components that are thought to help maintain a tight temporo-spatial control of plasmin-catalyzed proteolysis. Numerous cells express specific cell-surface binding sites for uPA and for plasminogen that are considered to favor plasmin formation and accumulation in the immediate pericellular environment. Although not yet fully characterized, cell-surface binding sites for tPA may exert the same effect. In addition, natural inhibitors (PAIs), belonging to the family of arginin-serine-protease inhibitors (= serpins) such as PAI-1, PAI-2 and protease-nexine-1 can damp PAs catalytic activity. PAIs, which have distinct distributions in extracellular compartments, form stable complexes with PAs, though with different kinetics, underscoring the variety of modulations susceptible to control plasmin production. Finally, the expression of the genes encoding all members of the PA/plasmin system is influenced by a whole variety of growth factors and hormones (For review see 1 and 2).

PLASMINOGEN ACTIVATION IN PHYSIOLOGY AND PATHOLOGY

In vitro studies have shown that numerous cell phenotypes can synthesize and secrete PAs and PAIs. Since uPA-producing cells are often endowed with invasive properties, it has been postulated that the enzyme plays an essential role in ECM degradation. Recent investigations have confirmed that invasive cells can produce in vivo uPA (3); however, other studies have revealed that PAs are also expressed by cell phenotypes devoid of invasive and migrating properties (4, 5), suggesting that both enzymes could fulfill a larger spectrum of functions than generally suspected. Available evidence indicates that the PA/plasmin system may exert at least three types of function.

1.Plasminogen activators and extracellular matrix remodelling

A series of recent in vivo observations have documented the expression of the uPA gene by physiological and pathological cell phenotypes, such as murine trophoblastic cells which produce uPA during a restricted period of time corresponding to their invasion of the uterine wall (3). Similarly, other cell types endowed with invasive properties produce PAs, such as hematopoietic cells colonizing the bursa of Fabricius (6), neural crest cells (7), and neurons of the cerebellar granule layer (8). The participation of PAs to inflammatory and tissue repair processes is well established. Indeed, monocytes/macrophages (9), polymorphonuclear leukocytes (10), regenerating keratinocytes (11), and endothelial cells during angiogenesis (12), all produce uPA, supporting a role for the enzyme in tissue migration to sites of inflammation and in ECM remodelling that accompanies such reactions. A large body of evidence suggests a role for PAs in cancer cell invasiveness (For review see 1). Experimental models have shown that uPA can promote tissue invasion and metastasis formation (13, 14). In addition, multiple animal and human tumors contain large amounts of uPA, and correlations between uPA expression and metastatic potential have been found in some tumors (15).

2. Plasminogen activators and the maintenance of tubular fluidity

tPA is known to be produced by endothelial cells and is generally thought to be released in the blood stream where it would act to prevent inappropriate coagulation (1, 2). Recent studies have shown that PAs can be synthesized by a variety of epithelial cells lining tubular structures, a situation which is reminiscent of the function attributed to tPA in the vascular space. The analysis of the murine kidney has shown that uPA and tPA are synthesized by epithelial cells lining distinct portions of kidney tubules and that both enzymes are secreted in urine (4). A similar function can be envisaged in the vas deferens and in seminal vesicles (16), and in the mammary gland (17), since uPA is produced by epithelial cells in all these organs.

3. Plasminogen activators and peptidic hormones

PAs have also been observed in tissues devoid of tubular structures and that are not the sites of major ECM remodeling, such as endocrine glands (18, 19). These findings have led to the suggestion that PAs may participate in the conversion of pro-hormones into active molecules. Although this hypothesis has not been formally confirmed, an analogous concept can be derived from more recent observations such those describing the expression of tPA in neurons of the central nervous system of the adult mouse (5). In that context, plasmin could influence the biological activity of peptides acting like hormones, by converting, for instance, inactive precursors of growth factors or by allowing their release from the ECM (20, 21, 22). These data suggest that PAs may exert a larger repertoire of effects on cell growth and differentiation than those generally ascribed to them.

In conclusion, the PA/plasmin system is one of the most extensively studied enzymatic pathways that catalyze extracellular proteolysis. We now know that this system is expressed in a wide variety of conditions and that it contributes to modulate interactions between cellular and extracellular compartments. The exploration of physiological conditions involving plasminogen activation should provide insights into our understanding of their potential participation in pathological processes.

References

1. Danø, K., P.A. Andreasen, J. Grøndhal-Hansen, P. Kristensen, L. S. Nielsen, and L. Skriver. 1985. Adv. Cancer Res. 44: 139-266

2. Vassalli, J.-D., A.-P. Sappino, and D. Belin. 1991. J. Clin. Invest. 88: 1067-1072

3. Sappino A.-P., J. Huarte, D. Belin, and J.-D. Vassalli. 1989. J. Cell Biol. 109: 2471-2479

4. Sappino, A.-P., J. Huarte, J.-D. Vassalli, and D. Belin. 1991. J. Clin. Invest. 87: 962-970

5. Sappino, A.-P., R. Madani, J. Huarte, D. Belin, J.-Z. Kiss, A. Wohlwend, and J.-D. Vassalli. 1993. J. Clin. Invest. In press.

6. Valinsky, J.E., E. Reich, and N. Le Douarin. 1981. Cell 25: 471-476

7. Valinsky, J.E., and N. Le Douarin. 1985. EMBO J. 4: 1403-1406

8. Soreq, H., and R. Miskin. 1983. Develop. Brain Res. 11: 149-158

9. Unkeless, J.C., S. Gordon, and E. Reich. 1974. J. Exp. Med. 139: 834-850

10. Heiple, J.M., and L. Ossowski. 1986. J. Exp. Med. 164: 826-840

11. Grøndahl-Hansen, J., L.R. Lund, E. Ralfkiaer, V. Ottevanger, and K. Danø. 1988. 1988. J. Invest. Dermatol. 90: 790-795

12. Pepper, M. S., R. Montesano, J.-D. Vassalli, and L. Orci. 1987. J. Cell Biol. 105: 2535-2541

13. Ossowski, L., and Reich E. 1983. Cell 35: 611-619

14. Ossowski, L. 1988. J. Cell Biol. 107: 2437-2445

15. Sappino, A.-P., D. Belin, J. Huarte, S. Hirschel-Scholz, J.-H. Saurat, and J.-D. Vassalli. 1991. J. Clin. Invest. 88: 1073-1079

16. Huarte, J., D. Belin, D. Bosco, A.-P. Sappino, and J.-D. Vassalli. 1987. J. Cell Biol. 104: 1281-1289

17. Busso, N., J. Huarte, J.-D. Vassalli, A.-P. Sappino, and D. Belin. 1989. J. Biol. Chem. 264: 7455-7457

18. Kristensen, P., L.S. Nielsen, J. Grøndahl-Hansen, P. B. Andreasen, L.I. Larsson, and K. Danø. 1985. J. Cell. Biol. 101: 305-311

19. Kristensen, P., D.M. Hougaard, L.S. Nielsen, and K. Danø. 1986. Histochemistry. 85: 431-436

20. Pepinsky, R.B., L.K. Sinclair, E.P. Chow, R.J. Mattaliano, T.F. Manganaro, P.K. Donahoe, and R.L. Cate. 1988. J. Biol. Chem. 263: 18961-18964

21. Sato, Y., and D. Rifkin. 1989. J. Cell Biol. 109: 309-315

22. Rifkin, D.B., D Moscatelli, J. Bizik, N. Quarto, F. Blei, P. Dennis, R. Flaumenhaft, and P. Mignatti. 1990. Cell Diff. Dev. 32: 313-318

Biology of Vitronectins and their Receptors
K.T. Preissner, S. Rosenblatt, C. Kost, J. Wegerhoff and D.F. Mosher, editors

Variants of the receptor for urokinase-type plasminogen activator

G. Høyer-Hansen, H. Solberg, E. Rønne, N. Behrendt and K. Danø

Finsen Laboratory, Rigshospitalet, Strandboulevarden 49, DK-2100 Copenhagen Ø, Denmark

Introduction

A central molecule in pericellular proteolysis is the receptor (uPAR) for urokinase-type plasminogen activator (uPA). uPA activates the extracellular zymogen plasminogen into the broad specificity protease plasmin, which is involved in degradation of extracellular matrix proteins in a variety of biological processes including cancer invasion (1). The activity of uPA is regulated by two fast-acting plasminogen activator inhibitors, PAI-1 and PAI-2, which bind to uPA both in solution and when uPA is bound to uPAR (2, 3). uPAR binds both active uPA and its inactive zymogen pro-uPA with high affinity (K_d 0.1-1nM) (4). Pro-uPA can be converted to uPA by plasmin while bound to uPAR and receptor-bound uPA can activate plasminogen. A function of uPAR is to localize uPA at the cell surface and in some cell types to cell-cell and cell-substratum contact sites (5, 6). In addition, concomitant binding of pro-uPA to uPAR and plasminogen to as yet unidentified molecules on the cell surface strongly increases the rate of plasmin formation (7-9).

Structure of the uPA receptor

uPAR is a single polypeptide chain highly glycosylated protein (10), which is COOH-terminally processed and attached to the plasma membrane by a glycosyl-phosphatidyl inositol lipid anchor (11). Analysis of the amino acid sequence of uPAR, deduced from the cDNA sequence (12), shows that it contains three repeats, each consisting of approximately 90 amino acid residues and characterized by a unique pattern of cysteine residues (13). The NH_2-terminal domain 1 of uPAR was identified as the ligand-binding domain by cross-linking to radio-labelled ligand (13). Domain 1 contains 8 cysteines, whereas domains 2 and 3 contain 10 cysteines each. The positions of the disulfide bonds in domain 1 were recently determined to be: Cys^3-Cys^{24}, Cys^6-Cys^{12}, Cys^{17}-Cys^{45} and Cys^{71}-Cys^{76} (14). Domain 1 is liberated by limited proteolysis with chymotrypsin, which cleaves uPAR between Tyr^{87} and Ser^{88}, located in the short peptide sequence (amino acids 78-94), the hinge region, connecting domains 1 and 2. uPAR is involved in internalization and degradation of complexes between uPA and PAI-1 (15, 16) or PAI-2 (17) and inhibition of uPA-PAI-1 binding to uPAR by preincubation with a monoclonal antibody to the ligand-binding domain of uPAR prevents uPA-PAI-1 internalization (16).

The murine uPAR (muPAR) has also been isolated and characterized (18) and its full length cDNA has been cloned (19). muPAR shares many characteristic features with human uPAR (huPAR). It carries a high amount of heterogenous carbohydrates, the molecular mass of the deglycosylated muPAR is very similar to that of huPAR and it is attached to the plasma membrane by a glycosyl phosphatidyl inositol anchor. The muPAR cDNA shows 60% identity to huPAR and conservation of all cysteine residues (19). However, using ligand-blot analysis the binding between uPA and uPAR showed species selectivity; human uPA did not bind muPAR and vice versa (18).

Monoclonal antibodies to uPAR

We have raised monoclonal antibodies to human uPAR purified from phorbol 12-myristate 13-acetate (PMA)-stimulated U937 cells (9). Ten of these have been characterized (R1-R10) and found to react with different epitopes within the huPAR molecule (9, 20). One of the antibodies, R3, reacting with the ligand-binding domain 1, totally inhibits the binding of uPA to U937 cells and completely abolish the potentiation of plasmin generation observed upon incubation of pro-uPA and plasminogen with U937 cells (9). The monoclonal antibodies reacting with domains 2 or 3, do not inhibit uPA binding or cell surface plasmin generation. They react with both free and occupied huPAR and has been used for quantitation of huPAR by ELISA (20), for immunohistochemistry (21) and for identification of huPAR variants (9, 20, 22).

uPAR glycosylation variants

The electrophoretic mobility of uPAR from different cell lines varies depending on the degree of glycosylation (10, 18). huPAR contains five potential sites for N-linked glycosylation (12), one in domain 1 at Asn^{52}, which affect the affinity between uPAR and uPA, and two in each of the other two domains which are important for the cellular transport of uPAR (23). muPAR has two potential glycosylation sites in each of the domains 1 and 2 and three in domain 3 (19). In addition, intracellular glycosylation variants of uPAR have been identified in PMA-stimulated U937 cells (10) as well as in some mouse cell lines (18). These variants contain high mannose carbohydrates and has been identified as immature precursor forms (18, 23).

Soluble uPAR variants

Leukocytes from patients with paroxysmal nocturnal hemoglobinuria (PNH) secrete huPAR due to a deficiency in their ability to synthesize or attach a functional glycolipid anchor, and a soluble form of huPAR is found in the plasma from patients, whereas no huPAR has been detected in plasma from healthy persons (24). Recently, a soluble huPAR variant was identified in ascites fluid and plasma from patients with ovarian cancer. This form of soluble huPAR is probably not secreted but liberated from tumor cells by cleavage with proteases or lipases (25).

Alternatively spliced uPAR mRNA variants

Variants of uPAR mRNA produced by alternative splicing have been identified and characterized in both human and mouse cell lines and tissues (19, 26). These variants are named uPAR 2. The uPAR 2 mRNA variants are formed by different alternative splicing mechanisms occuring at different sites in the two species; huPAR 2 is alternatively spliced in domain 3 and muPAR 2 in domain 2. Recombinant huPAR 2 protein has an apparent molecular mass of 31,000 and the corresponding mouse protein 24,000, whereas the apparent molecular mass of the deglycosylated uPAR 1 from both species is 35,000. The deduced uPAR 2 proteins both include the uPA binding domain, but lack the glycolipid anchor attachment sites, suggesting that they are secreted uPA binding proteins. They may serve a physiological fuction as inhibitors of cell surface proteolysis by inhibition of the binding of uPA to uPAR 1. However, even though mRNA for uPAR 2 has been detected in several cell lines and also in tissues by *in situ* hybridization, expression of the corresponding proteins have untill now not been demonstrated under natural conditions.

Cleavage of uPAR

By Western blotting analysis of extracts from U937 cells with the monoclonal antibody R3, reacting with an epitope on domain 1, we found one immunoreactive band with an apparent molecular mass of 55,000-60,000, corresponding to that of cross-linkable huPAR. In contrast, Western blotting with the R2 and R4 antibodies, recognizing domain 2 + 3, gave two bands, one with the same electrophoretic mobility as above and the other corresponding to an apparent molecular mass of around 40,000 (22). After deglycosylation with peptide:N-glycosidase F (PNGase F) the molecular mass of cross-linkable huPAR is 35,000 (10), whereas deglycosylation with PNGase F of the U937 cell extracts and Western blot analysis with polyclonal antibodies revealed two immunoreactive proteins with apparent molecular masses of 35,000 and 27,000. It was concluded that the protein band reacting with R2 and R4 but not with R3 represents a huPAR variant lacking the ligand-binding domain 1. It was termed uPAR(2+3) and is, as full-length uPAR, removed from cells by trypsin treatment, demonstrating that it is located at the cell surface (22). Cultivation of PMA-stimulated U937 cells with an anticatalytic antibody to uPA decreases the amount of uPAR(2+3) concomitant with an increase in full-length huPAR, demonstrating that uPAR(2+3) is formed from huPAR by a cleavage caused by uPA, either directly or through the activation of plasminogen (22).

Experiments with purified huPAR shows that it can be converted by uPA into uPAR(2+3) as illustrated by Figure 1. Some cleavage occurs at 1 nM uPA for 20 hours at 37 °C and at 100 nM uPA almost all huPAR is converted to uPAR(2+3). The cleavage is a slow process in vitro. Only small amounts are cleaved by 10 nM uPA after 1 hour. After 7 hours an appreciable amount of uPAR(2+3) was found. The effect of uPA on purified huPAR is inhibited by a monoclonal anticatalytic uPA antibody (22).

uPA binds to uPAR via its growth factor domain comprising amino acids 12-32 (27). This domain is absent from low molecular weight (LMW)-uPA, but nevertheless huPAR is cleaved by LMW-uPA. Thus, binding of uPA to huPAR through the growth factor domain is not required for huPAR-cleavage. Also plasmin can cleave purified huPAR into uPAR(2+3). This cleavage is inhibited by aprotinin, whereas the uPA cleavage is not (22).

The removal of the uPA-binding domain from uPAR abolishes the localizing and

292

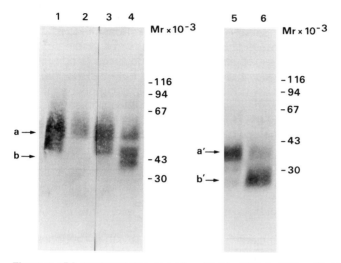

Figure 1. uPA catalysed cleavage of purified huPAR. huPAR purified from PMA-stimulated U937 cells was incubated for 20 h at 37 °C with (lanes 2, 4, 6) or without (lanes 1, 3, 5) 100 nM uPA. After this the samples were divided in three parts; two parts were processed immediately for SDS-PAGE under non-reducing conditions and Western blotting with the monoclonal R3 (lanes 1 and 2) and R4 (lanes 3 and 4) antibodies. The third part was deglycosylated by incubation with 0.5 units of PNGase F for 20 h and analysed by SDS-PAGE under reducing conditions and Western blotting with polyclonal antibodies to reduced huPAR (lanes 5 and 6). Band <u>a</u> is full-lenght huPAR and band <u>b</u> is uPAR(2+3). <u>a</u>´ and <u>b</u>´ represents the corresponding deglycosylated forms. Electrophoretic mobilities of standard proteins are indicated. From ref.22.

enhancing role of uPAR in plasmin generation. On the surface of some cell types unoccupied uPAR has a half-life of more than 4 hours (17). Even though the cleavage of huPAR by uPA and plasmin is slow, it may represent a physiologically important negative feed-back regulation of cell surface plasminogen activation, that could be particularly important when uPA inhibitors are absent. This assumption is supported by the fact that approximately 50% of huPAR on cultured U937 cells is present as uPAR(2+3) (22).

Another serine protease, thrombin, cleaves its receptor (28). A putative cleavage site in the thrombin receptor was deduced from sequence similarity with the site cleaved when thrombin activates the zymogen protein C. Peptides mimicking the proposed new amino terminus after thrombin cleavage were shown to activate the thrombin receptor, which has a well documented mitogenic effect (28). Several studies have suggested that receptor-binding of uPA also has a mitogenic effect, which in some cell types have been shown to be dependant on catalytic activity (29-32). uPA-catalyzed cleavage of huPAR could be involved in such a mitogenic effect.

uPAR(2+3) is present in several neoplastic cell lines. The cleaved huPAR variant is not unique for the monocyte-like cell line U937. By Western blot analysis it has also been detected in the following cell lines: HeLa cervix carcinoma cells, HEp-2 larynx carcinoma cells and 8387 fibrosarcoma cells (unpublished results). Another fibrosarcoma cell line, HT1080, also contains the cleaved form of huPAR (R. Mazzieri; personal communication).

The uPA cleavage site. The consensus sequence for cleavage by serine endoproteases is a basic residue followed by two hydrophobic amino acids. The uPA activation cleavage site in human plasminogen is after Arg^{560} (Gly^{559}-Arg^{560}-Val^{561}-Val^{562}) (33, 34). uPA also activates hepatocyte growth factor/scatter factor by cleavage after Arg^{494}, the sequence around the cleavage site being Leu^{493}-Arg^{494}-Val^{495}-Val^{496} (35). Fibronectin is cleaved by uPA at two sites, after Arg^{259} (Val^{258}-Arg^{259}-Ala^{260}-Ala^{261}) and Arg^{2299} (Gln^{2298}-Arg^{2299}-Thr^{2300}-Asn^{2301}), the latter site being very atypical (36). A sequence, similar to that containing the uPA activation cleavage site in plasminogen, is present in the hinge region (residues 78-94) connecting domains 1 and 2 of huPAR (13): Gly^{82}-Arg^{83}-Ala^{84}-Val^{85}. A putative site for uPA cleavage is therefore between Arg^{83} and Ala^{84}.

Plasmin has a broader substrate specificity, hydrolyzing peptide bonds with arginine or lysine residues on their carboxyl side. The huPAR hinge region contains three potential plasmin cleavage sites, located after Arg^{83}, Arg^{89} and Arg^{91}. Sequencing of uPAR(2+3) purified from U937 cells showed two amino-terminal sequences indicating that cleavage had occurred after Arg^{83} and Arg^{89} (unpublished result), in agreement with the cleavage being catalysed by either plasmin alone or by both plasmin and uPA.

REFERENCES

1   Danø K, Andreasen PA, Grøndahl-Hansen J, Kristensen P, Nielsen LS, Skriver L. Adv Cancer Res 1985; 44: 139-266.
2   Cubellis MV, Andreasen P, Ragno P, Mayer M, Danø K, Blasi F. Proc Natl Acad Sci USA 1989; 86: 4828-4832.
3   Ellis V, Wun T-C, Behrendt N, Rønne E, Danø K. J Biol Chem 1990; 265: 9904-9908.
4   Vassalli J-D, Baccino D, Belin D. J Cell Biol 1985; 100: 86-92
5   Pöllänen J, Hedman K, Nielsen LS, Danø K, Vaheri A. J Cell Biol 1988; 106: 87-95.
6   Myöhänen HT, Stephens RW, Hedman K, Tapiovaara H, Rønne E, Høyer-Hansen G, Danø K, Vaheri A. J Histochem Cytochem 1993; in press.
7   Stephens RW, Pöllänen J, Tapiovaara H, Leung K-C, Sim PS, Salonen E-M, Rønne E, Behrendt N, Danø K, Vaheri A. J Cell Biol 1989; 108: 1987-1995.
8   Ellis V, Behrendt, Danø K. J Biol Chem 1991; 266: 12752-12758.
9   Rønne E, Behrendt N, Ellis V, Ploug M, Danø K, Høyer-Hansen G. FEBS Lett 1991; 288: 233-236.
10   Behrendt N, Rønne E, Ploug M, Petri T, Løber D, Nielsen LS, Schleuning W-D, Blasi F, Appella E, Danø K. J Biol Chem 1990; 265: 6453-6460.
11   Ploug M, Rønne E, Behrendt N, Jensen AL, Blasi F, Danø K. J Biol Chem 1991; 266: 1926-1933.

12   Roldan AL, Cubellis MV, Masucci MT, Behrendt N, Lund LR, Danø K, Appella E, Blasi F. EMBO J 1990; 9:467-474.

13   Behrendt N, Ploug M, Patthy L, Houen G, Blasi F, Danø K. J Biol Chem 1991; 266: 7842-7847.

14   Ploug M, Kjalke M, Rønne E, Weidle U, Høyer-Hansen G, Danø K. J Biol Chem 1993; 268: in press.

15   Cubellis MV, Wun T-C, Blasi F. EMBO J 1990; 9: 1079-1085.

16   Olson D, Pöllänen J, Høyer-Hansen G, Rønne E, Sakaguchi K, Wun T-C, Appella E, Danø K, Blasi F. J Biol Chem 1992; 267: 9129-9133.

17   Estreicher A, Mühlhauser J, Carpentier J-L, Orci L, Vassalli J-D. 1990; 111: 783-792.

18   Solberg H, Løber D, Eriksen J, Ploug M, Rønne E, Behrendt N, Danø K, Høyer-Hansen G. Eur J Biochem 1992; 205:451-458.

19   Kristensen P, Eriksen J, Blasi F, Danø K. J Cell Biol 1991; 115: 1763-1771.

20   Rønne E, Behrendt N, Ploug M, Nielsen HJ, Wöllisch E, Weidle U, Danø K, Høyer-Hansen G. J Immunol Meth 1993; in press.

21   Pyke C, Græm N, Ralfkiær E, Rønne E, Høyer-Hansen G, Brünner N,Danø K. Cancer Res 1993; 53: 1911-1915.

22   Høyer-Hansen G, Rønne E, Solberg H, Behrendt N, Ploug M, Lund LR, Ellis V, Danø K. J Biol Chem 1992; 267: 18224-18229.

23   Møller LB, Pöllänen J, Rønne E, Pedersen N, Blasi F. J Biol Chem 1992; 268: 11152-11159.

24   Ploug M, Eriksen J, Plesner T, Hansen NE, Danø K. Eur J Biochem 1992; 208: 397-404.

25   Pedersen N, Schmitt M, Rønne E, Nicoletti MI, Høyer-Hansen G, Conese M, Giavazzi R, Danø K, Kuhn W, Jänicke F, Blasi F. J Clin Invest 1993; in press

26   Pyke C, Eriksen J, Solberg H, Nielsen BS, Kristensen P, Lund LR, Danø K. FEBS Lett 1993; 326: 397-404

27   Appella E, Robinson EA, Ullrich SJ, Stoppelli P, Corti A, Cassani G, Blasi F. J Biol Chem 1987; 262: 4437-4440.

28   Vu T-K H, Hung DT, Wheaton VI, Coughlin SR. Cell 1991; 64:1057-1068.

29   Baron-Van Evercooren A, Leprince P, Rogister B, Lefebvre PP, Delree P, Selak I, Moonen G. Dev Brain Res 1987; 36: 101-108.

30   Cohen SD, Israel E, Speiss-Meier B, Wainberg MA. J Immmunol 1981; 126: 1415-1420.

31   Kirchheimer JC, Christ G, Binder BR. Eur J Biochem 1989; 181: 103-107.

32   Kirchheimer JC, Wojta J, Christ G, Binder BR. Proc Natl Acad Sci USA 1989; 86: 5424-5428

33   Sottrup-Jensen L, Zajdel M, Claeys H, Petersen TE, Magnusson S. Proc Natl Acad Sci USA 1975; 72: 2577- 2581.

34   Summaria L, Arzadon L, Bernabe P, Robbins KC. J Biol Chem 1975; 250: 3988-3995.

35   Naldini L, Tamagnone L, Vigna E, Sachs M, Hartmann G, Birchmeier W, Daikuhara Y, Tsubouchi H, Blasi F, Comoglio PM. EMBO J 1992; 11:4825-4833.

36   Gold LI, Rostagno A, Frangione B, Passalaris T. J Cell Biochem 1992; 50:441-452.

Biology of Vitronectins and their Receptors
K.T. Preissner, S. Rosenblatt, C. Kost, J. Wegerhoff and D.F. Mosher, editors

Vitronectin and plasmin(ogen) in lesional skin of the bullous pemphigoid: colocalisation suggests binding interactions

M.D. Kramer[a], H.M. Gissler[a], B. Weidenthaler-Barth[a], K.T. Preissner[b]

[a]University Institute for Immunology, Immunopathology, Im Neuenheimer Feld 305
D-69120 Heidelberg, Germany

[b]Max-Planck-Institut, Hemostasis Research Unit, Kerckhoff-Klinik, Sprudelhof 11,
D-61231 Bad Nauheim, Germany

SUMMARY

The bullous pemphigoid is a subepidermal autoimmune bullous dermatosis characterized by autoantibodies specific for hemidesmosomal antigen(s). Deposition of autoantibodies at the epidermal basement membrane zone is accompanied by complement activation and subsequent formation of terminal complement complex C5b-9. *In vitro* studies have revealed that the adhesive protein vitronectin functions as complement inhibitor and binds nascent C5b-9. Moreover, the binding interaction of vitronectin with plasmin(ogen) can regulate plasmin formation and activity. In skin blister fluids of the bullous pemphigoid between 20 and 50 % of the normal plasma concentration of vitronectin was present. Vitronectin was partially degraded, as revealed by immuno-blot analysis. Analysis of the topographical distribution of plasmin(ogen) and vitronectin in lesional skin of the bullous pemphigoid by using specific antibodies, revealed the co-localisation of the two components at the base of the blister roof or at the dermo-epidermal junction.

INTRODUCTION

The bullous pemphigoid (BP) is an autoimmune subepidermal bullous dermatosis characterized by deposition of autoantibodies at the epidermal basement membrane zone [1]. Proteolytic enzymes, including plasmin, have been suggested as one element involved in the pathology, in particular in the dermo-epidermal dyshesion, observed in the BP [2,3]. The autoantibodies are directed against the "bullous pemphigoid antigens", which form part of the hemidesmosome [4]. The hemidesmosome connects the basal keratinocytes with the underlying epidermal basement membrane [5]. As demonstrated by immunohistochemistry the deposition of autoantibodies is associated with complement activation, and subsequent deposition of the complement membrane attack complex C5b-9 at the basement membrane zone [6].

Vitronectin (synonymously termed "complement S-Protein" [7]) is an extracellular glycoprotein present in plasma, interstitial fluid and particularly associated with the extracellular matrix of the blood vessels and the dermis [8]. By incorporation into the nascent attack complex of complement, vitronectin inactivates its lytic function and may serve to limit innocent bystander cytolysis [9,10]. Based on the co-localization of vitronectin with C5b-9 in lesional skin of the bullous pemphigoid it has been suggested that vitronectin may regulate the lytic activity of the membrane attack complex under inflammatory conditions [6,11,12].

Vitronectin has been described to function as regulatory co-factor yet in another ubiquitous system, the plasminogen activator system. The binding of plasmin(ogen) [13], and the plasminogen activator inhibitor-1 [14], to vitronectin may regulate the local activity of the plasminogen activator system [15]. Upregulation of the plasminogen activator system and local generation of plasmin is thought to mediate the local degradation of extracellular matrix glycoproteins [16,17].

Plasminogen is present in plasma and the interstitial fluid. Cellular plasminogen activators (two types of which are known: urokinase-type PA (uPA) and tissue-type PA (tPA)) catalyse the hydrolytic cleavage of a single peptide bond within the plasminogen molecule [16,17]. The plasminogen activators are regulated by plasminogen activator inhibitors (PAI-1 and PAI-2) [18], of which PAI-1 is bound and thus stabilized by vitronectin [14,19]. Recent evidence has suggested that the plasminogen activator system is also involved in lesional skin of the BP: (i) biochemical [20] and molecularbiological [21] studies revealed PAs (primarily tPA) in lesional epidermis, (ii) immunohistological studies provided evidence for a conspicuous deposition of plasmin(ogen) in lesional epidermis [3], and (iii) functional as well as immunochemical studies on blister fluids provided evidence for plasmin generation in lesional skin of the BP [2].

Evidence for the binding interactions of vitronectin with components of the plasminogen activator system has so-far exclusively been derived from *in vitro* studies. We asked whether such binding interactions might also occur under pathological conditions *in vivo*. Consequently, we have immunohistologically analysed the lesional skin of bullous pemphigoid by using monoclonal antibodies specific for complement components, vitronectin, PAI-1 and plasmin(ogen). Moreover, we have analysed the vitronectin present in skin blister fluids by immuno-blotting. Our findings indicated that binding interactions between plasmin(ogen) and vitronectin could occur in lesional skin of the BP *in vivo*.

MATERIALS AND METHODS

Patients and skin biopsies

Skin biopsies were taken from lesional skin of BP patients. All biopsies displayed the characteristic histopathological features of BP: subepidermal blisters and deposition of immunoglobulin and C3 at the epidermal basement membrane zone. As a control, biopsies of normal skin were taken from individuals with no signs and history of disseminated skin disease. All specimens were embedded in OCT-Tissue-Tek II (Miles Laboratories, Naperville, IL), snap frozen in melting isopentane and stored at -70 °C. Five μm cryostat sections were prepared, fixed in acetone, air dried, and used for immunofluorescence staining.

Antibodies used for immunofluorescence

The following antibodies were used for immunohistology. Non-labelled antibodies: rabbit anti-human C3 and anti-human C5 (Dakopatts, Hamburg, FRG), rabbit anti-human C4 and anti-human C9 (Behringwerke, Marburg, FRG); goat anti-human PAI-1 (no. 395 G; American Diagnostica, Greenwich CT, USA); murine monoclonal antibodies against plasmin(ogen) that recognize distinct epitopes of the plasmin(ogen) molecule [3]: HD-PG 2, HD-PG 6, HD-PG 12, and HD-PG 13; murine monoclonal antibody against a neo-epitope of C9, generated during formation of the C5b-9 complex ("anti-C9neo"; Dakopatts); murine monoclonal anti-human vitronectin antibody (clone VN7; [22]); and rabbit anti-human vitronectin IgG [22]. Fluorochrome-labelled antibodies: Fluoresceine-isothiocyanate-(FITC)-conjugated goat anti-human IgG, IgA, or IgM, as well as FITC-conjugated swine anti-rabbit IgG, FITC-conjugated goat anti-mouse IgG, FITC-conjugated rabbit anti-goat IgG (Dianova, Hamburg); FITC-conjugated anti-human fibrinogen (Behring-Werke, Marburg); phycoerythrin-(PE)-conjugated donkey anti-rabbit IgG (Immuno Research, USA), 5-([4,6-dichlorotriazin-2yl]amino)fluorescein (DTAF)-conjugated goat anti-mouse IgG (Immuno Research).

Immunofluorescence

For indirect immunofluorescence staining, the frozen skin sections were incubated with the primary antibody for 30 min at room temperature, washed in phosphate-buffered saline (PBS) for 30 min and then overlaid for 30 min with the respective fluorochrome-labeled secondary antibody. For direct immunofluorescence all FITC-conjugated antibodies were incubated for 30 min at room temperature. Double immunofluorescence was performed using a rabbit anti-vitronectin antibody followed by phycoerythrin-conjugated donkey anti-rabbit IgG and subsequent incubation with a mouse anti-plasmin(ogen) antibody (HD-PG 12 [3]) followed by a DTAF-conjugated goat anti-mouse antibody. Each antibody was incubated for 30 min at room temperature. Controls for the performance of the immunostaining procedures included set-ups where the primary antibody or the secondary (anti-mouse IgG, anti-goat IgG) antibody were omitted.

After washing of stained sections in PBS for 30 min they were mounted in PBS-glycerol (1:2; v:v) and examined using a fluorescence microscope (Zeiss, Oberkochen, FRG) equipped with a vertical illuminator. Photomicrographs were taken with an integrated camera and an Ektachrome 400 colour slide film (Kodak, Hemel Hempstead, UK).

Characterization of vitronectin in blister fluids

Vitronectin was quantified in blister fluids by using a sensitive enzyme-linked immunosorbent assay (ELISA) outlined elsewhere [23]; normal human plasma served as 100 % control. Calibration curves were obtained by using dilutions of normal human plasma (data not shown). In blister fluids the concentration of vitronectin corresponded to ≈ 20 - 50 % of the plasma concentrations. To explore the vitronectin present in blister fluids in more detail, blister fluids were separated by SDS-polyacrylamide gel elctrophoresis (10 % gels), transferred onto nitrocellulose, and incubated with polyclonal anti-vitronectin antibodies.

RESULTS

Immunoglobulin, complement components and vitronectin in lesional skin of the bullous pemphigoid

Subepidermal blistering was observed in the biopsies analysed. The immunopathological features required for diagnosis of BP included deposition of immunoglobulin, as well as of C3, C4 and fibrin at the basement membrane zone (data not shown).

Further immunofluorescence findings on a representative biopsy are shown in Fig. 1. Staining for C9neo antigen is shown in Fig. 1.A. Staining for vitronectin was localised to the epidermal basement membrane zone at the site of immunoglobulin and complement deposition and (Fig. 1.B) and at the base of the blister roof (Fig. 1.C). Normal skin, analysed in parallel did neither display blistering, nor deposits of immunoglobulin, C9neo, or vitronectin at the basement membrane zone.

PAI-1 and plasmin(ogen) in the lesional skin of bullous pemphigoid

We have explored the deposition of the putative vitronectin ligands plasmin(ogen) and PAI-1 in lesional skin of the BP by using specific antibodies. In all biopsies tested, staining for PAI-1 was negative; to control for reactivity of the antibodies used, PAI-1-expressing cultured cells were stained, which yielded positive results (data not shown). Staining was observed with the anti-plasmin(ogen) antibodies. Staining for plasmin(ogen) was pronounced at the basal epidermal cell layers (Fig. 1.G) with a particularly strong staining of the blister roof (Fig. 1.E). The staining pattern was basically similar to that seen previously in lesional skin of the BP by using the alkaline phosphatase anti-alkaline phosphatase technique [3].

Based on these findings, double-staining of vitronectin and plasmin(ogen) in lesional skin of the BP was performed. As shown in Fig. 1.D-G stainings for vitronectin and for plasmin(ogen) were colocalised, in particular at the base of the blister roof as well as at the epidermo-dermal junction, where staining for vitronectin was not observed in normal skin.

298

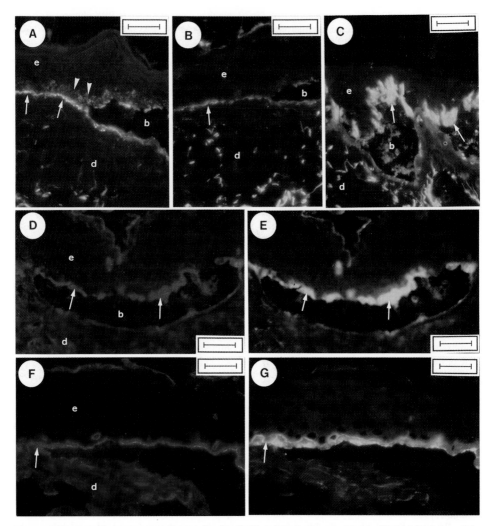

Figure 1. Immunofluorescence staining of lesional skin of the bullous pemphigoid. (A) Staining of the C9neo antigen. Continuous granular deposits (arrows) along the basement membrane zone starting out from the blister edge. Speckled staining of basal keratinocytes (arrowheads). (B) Staining of vitronectin (arrow) along the dermo-epidermal junction. (C) Staining of vitronectin (arrows) at the base of the blister roof. (D-G) Double-immunofluorescence staining for vitronectin and plasmin(ogen) in lesional skin of the bullous pemphigoid. Sections were double-stained for vitronectin (D,F; using a rabbit anti-vitronectin IgG and phycoerythrin-labelled anti-rabbit IgG antibody) and for plasmin(ogen) (E,G; using the monoclonal anti-plamin(ogen) antibody "HD-PG 12" and a DTAF-labelled anti-mouse IgG antibody). Vitronectin was colocalised with plasmin(ogen) in lesional skin of the bullous pemphigoid at the base of the blister roof (D,E; arrows) and at the dermo-epidermal junction (F,G; arows). e = epidermis, b = blister cavity, d = dermis. Scale bars A = 150 μm; B = 180 μm; C = 200 μm; D - G = 140 μm.

Vitronectin in blister fluids

Vitronectin present in skin blister fluids of the BP was analysed by ELISA (data not shown) and immuno-blotting (Fig. 2). Vitronectin was present in blister fluids at concentrations that corresponded to 20 - 50 % of that in normal plasma (data not shown). Vitronectin was largely fragmented to products of ≈ 61 kDa (compare Fig. 2; lanes 2-6).

The detection of mostly fragmented vitronectin (Fig. 2) is compatible with a proteolysis-favouring situation of plasmin-modified extracellular matrix *in vitro* [25]. The presence of degraded vitronectin in blister fluids (Fig. 2; lanes 2-6), together with the previous finding that plasmin is generated in lesional skin of the BP [2], as well as the the colocalization of nitronectin and plasmin(ogen) shown here, suggests that both components may be directly involved in the transient expression of a proteolysis-rich microenvironment in this skin disease.

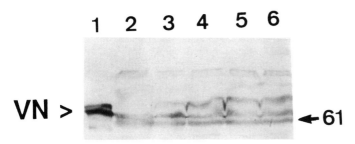

Figure 2. Characterization of vitronectin in skin blister fluids by immuno-blotting. Isolated plasma vitronectin (lane 1; arrow head), as well as blister fluids aspirated from bullous pemphigoid skin blisters (lanes 2-6) were separated by SDS-PAGE using 10 % gels, and transferred onto nitrocellulose. Nitrocellulose was reacted with polyclonal anti-vitronectin antibodies. Arrow (right hand side) indicates a ≈ 61 kDa fragment of vitronectin that is not observed in normal plasma or purified vitronectin.

DISCUSSION

In the present study we have analysed the distribution of immunoglobulins, complement components, vitronectin, plasmin(ogen), and PAI-1 in lesional skin of the bullous pemphigoid. Moreover, vitronectin was quantified in skin blister fluid by ELISA and the fragmentation pattern of vitronectin was analysed by immuno-blotting. Emphasis was put on clarifying the relative topographical distribution of vitronectin and its putative ligands PAI-1 and plasmin(ogen). Vitronectin was found to be colocalized with plasmin(ogen), but not with PAI-1 suggesting that binding interactions with plasmin(ogen) would prevail in the lesional skin of the bullous pemphigoid. This finding was further corroborated by the two-color immunofluorescence analysis that revealed colocalization of vitronectin and plasmin(ogen) at the dermo-epidermal junction and at the bottom of the blister roof in lesional skin of the bullous pemphigoid.

Fragmented vitronectin (61 kDa and smaller polypeptides) was detectable in appreciable quantities in blister fluids. The functional implications of plasmin(ogen)/vitronectin binding interactions are not entirely understood, but the present data as well as previous *in vitro* results [24] are in accordance with the concept that partially degraded vitronectin has lost its PAI-1 binding capacity but gained a specific plasmin(ogen) binding site, such that a switch from anti- into pro-fibrinolytic cofactor properties of vitronectin is expected. Upon partial degradation of vitronectin, a microenvironment may be generated that locally and transiently favors proteolytic activity [24]. Our present data favour this hypothesis in inflammatory areas of skin, since under normal conditions vitronectin-specific immuno-staining is exclusively confined to dermal extracellular matrix in connection with elastic fibres [8], and no co-staining with plasminogen is seen.

In previous studies the expression of plasminogen activators, in particular of tPA, in lesional epidermis of the bullous pemphigoid was demonstrated [20]. This allows the contention that keratinocyte-derived (t)PA may interact with plasminogen enriched in the lesional epidermis to yield active plasmin. Presently, immuno-staining experiments with specific anti-PA antibodies as well as zymography on frozen sections are being carried out to corroborate the presence of tPA in lesional epidermis and to identify the causal proteolytic events observed in lesional skin.

The monoclonal antibody to a neoantigen of the C9 [25] was used to identify the deposition of the membrane attack complex at the basement membrane zone *in vivo* in the bullous pemphigoid (Fig. 1.A). Colocalization of C5b-9 with vitronectin supports previous findings [summarized in: 26], that had already been taken to suggest the deposition of possibly non-lytic vitronectin-C5b-9 complex in the skin. Vitronectin may however also be associated in a non-stoichiometric manner with the lytic complex [27]. The extent to which the non-lytic vitronectin-C5b-9 complex may contribute to the local accumulation of plasmin(ogen) is not known and is presently explored in our laboratory.

Our present findings have identified aspects of an *in vivo* interaction which involves the plasminogen activator system, complement activation products including the terminal complement complex (C5b-9), [28] as well as the multifunctional co-factor vitronectin. A vitronectin-C5b-9-dependent pathway promoting plasminogen activation may occur during dermo-epidermal dyshesion and would focus enzyme activity to the site where dermo-epidermal cohesion is disturbed: the hemidesmsosomal structures at their extracellular binding sites in lesional skin. Moreover, partially proteolyzed vitronectin lacks its heparin-binding site, which constitutes the functional domain for complement inhibitory activity [29] and thereby allows formation of the membrane attack complex to proceed.

ACKNOWLEDGEMENTS

The authors wish to thank Sabine Wentrup, Ilka Neumann, Antje Heidtmann, and Brigitta Richter for expert technical assistance. We also acknowledge Dr. Klaus Rother for for continuous generous support. The work has in part been supported by the Deutsche Forschungsgemeinschaft, Bonn-Bad Godesberg (grants Kr 931/3-1, and Pr 327/1-1).

REFERENCES

1 Farmer ER. J Cutan Pathol 1985; 12: 316-321
2 Kramer MD, Reinartz J. J Clin Invest 1993; in press
3 Gissler HM, Simon MM, Kramer MD. Br J Dermatol 1992; 127: 272-277
4 Stanley JR, et al. J Clin Invest 1988; 82: 1864-1870
5 Uitto J, Christiano AM. J Clin Invest 1992; 90: 687-692
6 Dahl MV et al. J Invest Dermatol 1984; 82: 132-135
7 Preissner KT. Annu Rev Cell Biol 1991; 7: 275-310
8 Dahlbäck K, Lofberg H, Dahlbäck B. Acta Dermatol Venereol 1986; 66: 461-467
9 Podack ER, Preissner KT, Müller-Eberhard HJ. Acta Pathol Microbiol Immunol Scand 1984; 92 (Suppl. 284): 89-96
10 Dahlbäck B, Podack ER. Biochemistry 1985; 24: 2368-2374
11 Dahlbäck K, Löfberg H, Dahlbäck B. Br J Dermatol 1989; 120: 725-733

12 Hintner et al. Br J Dermatol 1990; 123:39-47
13 Preissner KT et al. J Biol Chem 1990; 265: 18490-18498
14 Declerck PJ et al. J Biol Chem 1988; 263: 15454-15461
15 Preissner KT, Jenne D. Thromb Haemost 1991; 66: 189-194
16 Mayer M. Clin Biochem 1990; 23: 197-211
17 Saksela O. Biochim Biophys Acta 1985; 823: 35-65
18 Andreasen PA et al. Mol Cell Endocrinol 1990; 68: 1-19
19 Preissner KT. Biochem Biophys Res Commun 1990; 168: 966-971
20 Jensen PJ et al. J Invest Dermatol 1988; 90: 777-782
21 Baird J et al. J Invest Dermatol 1990; 95: 548-552
22 Preissner KT, Wassmuth R, Müller-Berghaus G. Biochem J 1985; 231:349-355.
23 Preissner KT et al. Blood 1989; 74: 1989-1996
24 Kost C et al. J Biol Chem 1992; 267: 12098-12105
25 Mollness TE et al. Scand J Immunol 1985; 22: 197-201
26 Hintner H, Breathnach SM, Dahlbäck K, Dahlbäck B, Fritsch P. Hautarzt 1991; 42: 16-22
27 Bhakdi et al. Clin Exp Immunol 1988; 74: 459-464
28 Kramer MD, Hänsch GMH, Rother KO. Behring Communications 1992; 91: 145-156
29 Tschopp J, Masson D, Schafer S, Peitsch M, Preissner KT. Biochemistry 1988; 27: 4103-4109

Biology of Vitronectins and their Receptors
K.T. Preissner, S. Rosenblatt, C. Kost, J. Wegerhoff and D.F. Mosher, editors

303

The Interaction of Plasminogen Activator Inhibitor-1 with Vitronectin.

D.A. Lawrence[†], M.C. Naski[*¶], M.B. Berkenpas[*], S. Palaniappan[*], T.J. Podor[#], D.F. Mosher[§],, & D. Ginsburg[†*]. [†]Department. of Internal Medicine, University of Michigan and [*]Howard Hughes Medical Institute, 4520 MSRB I, Ann Arbor, MI., 48109-0650, U.S.A. [#]Department of Pathology, McMaster University, Hamilton Civic Hospitals Research Centre, Hamilton, Ontario., Canada. [§]Departments. of Medicine and Biomolecular Chemistry, University of Wisconsin, Madison, WI, U.S.A. [¶]Current address, Department. of Pathology, Washington University, St. Louis, MO., U.S.A.

I. INTRODUCTION

PAI-1 is a member of the serine protease inhibitor, or SERPIN, super family and is thought to be one of the principal regulators of the PA-system. It is a single chain glycoprotein with a molecular weight of 50 kDa [1] and is the most efficient inhibitor known of the single- and two-chain forms of tPA and of uPA, with second order rate constants ranging between 0.45 and 2.7 X 10^7 M^{-1} s^{-1} [2]. PAI-1 is also reported to inhibit plasmin and trypsin with second order rate constants of about 10^6 $M^{-1}s^{-1}$ [3], and less efficiently inhibits thrombin and activated protein C (10^3 to 10^5 M^{-1} s^{-1}) [4,5]. PAI-1 is present in plasma at very low concentrations, ranging from 0 to 60 ng/ml, with an average level of about 20 ng/ml (0.5 nM) [6,7]. PAI-1 is present in platelets and many other tissues and is produced by many cells in culture [8]. In vivo the primary extravascular source of PAI-1 appears to be vascular smooth muscle cells [9]. However, in response to endotoxemia or other pathological conditions, endothelial cells become a major site of PAI-1 synthesis [10,11]. In plasma PAI-1 is present as a complex with vitronectin (Vn) or S protein [12,13]. PAI-1 is also associated with Vn in the extracellular matrix in culture, and may be involved in maintaining the integrity of the cell substratum in vivo [14-17].

PAI-1 exists in at least two conformations, active and latent [18,19]. The active form spontaneously converts to the latent form with a half-life of about 1 hour at 37° C at neutral or slightly alkaline pH [2,3,20]. The latent form can be converted into the active form by treatment with denaturants, negatively charged phospholipids or Vn, though the latter reaction is very slow [18,21,22]. In one report, analysis of latent PAI-1 infused into rabbits suggests that latent PAI-1 may also be physiologically reactivated in vivo, though the exact mechanism remains unknown [23]. This reversible interconvertion between the active and latent structures, is a unique feature of PAI-1 which distinguishes it from other SERPINs.

Vn-bound PAI-1 in solution is approximately twice as stable as unbound PAI-1 [12] and on extracellular matrix the half-life is reported to be greater than 24 hours [24]. Given the high concentration of Vn in plasma (200-400 mg/ml [25]) and the reported clearance half-life of PAI-1 in plasma of only 6-7 minutes [23], the conversion from active to latent PAI-1 seen in cell culture or with purified preparations may be an artifact of these in vitro systems. Consistent with this hypothesis, very little latent PAI-1 can be demonstrated in normal, fresh plasma. However, most of the PAI-1 found in platelets appears to be latent [6,7], though this point is controversial [4,26]. Of note, platelets contain Vn [27], which could potentially function to reactivate latent platelet PAI-1 [22] but this interaction has not yet been reported.

The interaction of PAI-1 with Vn has generated considerable debate. In one study only active PAI-1 was shown to bind Vn [28]. However in another study no apparent difference in the binding was seen between active and latent PAI-1 [29]. The reported dissociation constant is also controversial, with one group reporting a Kd of 0.3 nM for active PAI-1 and Vn [30], while another reports a major dissociation constant of 55-190 nM with a second, low capacity but high affinity binding site (Kd <0.1 nM) [31].

Vitronectin is purified from plasma using both denaturing [32] and nondenaturing [33] methods. For convenience we will refer to protein purified using denaturing conditions as urea-treated vitronectin and the protein purified using nondenaturing conditions as native vitronectin. Urea-treated vitronectin differs from native vitronectin in that urea treatment causes the formation of disulfide bonded multimers as well as a conformational change [34,35]. The evidence for different conformations of vitronectin is based on the altered reactivity of native and urea-treated vitronectin with certain antibodies. In this study we address several questions; i) can the apparently contradictory results showing either an acceleration of the rate of inactivation of thrombin by PAI-1 in the presence of Vn or lack thereof be accounted for by the choice of either native or urea-treated Vn? and if so ii) what differences might exist between the properties of native and urea-treated Vn to distinguish the ability of one form to accelerate the inactivation of thrombin by PAI-1?, and iii) what are the kinetic parameters for the reaction of Vn-bound PAI-1 with a-thrombin and how do these parameters account for the accelerated rate? We show that native Vn accelerates the rate of inactivation of α-thrombin by PAI-1 whereas urea-treated Vn does not. We provide evidence that urea-treated Vn does not interact with α-thrombin as does native Vn and suggest that this may account for the inability of urea-treated Vn to accelerate the reaction between α-thrombin and PAI-1. In addition, we show that i) the reaction of α-thrombin with the native Vn:PAI-1 complex is governed by a second order rate constant 270-fold greater than the second order rate constant in the absence of Vn and ii) the acceleration is largely accounted for by an increase in the affinity (as reflected by a more than 25-fold decrease in value of the dissociation constant) of α-thrombin for PAI-1 bound to Vn relative to that for unbound PAI-1.

II. RESULTS AND DISCUSSION

To determine whether Vn can accelerate the inactivation of α-thrombin by PAI-1 we compared the influence of both urea-treated and native Vn on the first order rate constant for the inactivation of α-thrombin by PAI-1. The experiments in Figure 1 demonstrate that native Vn can accelerate the rate of inactivation of α-thrombin by PAI-1. The apparent first order rate constant approaches a maximum (which is approximately 50-fold greater than the first order rate constant in the absence of Vn) when the concentration of PAI-1 (0.34 μM) and native Vn are nearly equal. This observation suggests that PAI-1 and native Vn form a tight complex such that at concentrations of Vn less than the concentration of PAI-1 all of the available Vn is complexed with PAI-1. At Vn concentrations well in excess of the PAI-1 concentration the accelerated rate of inactivation of α-thrombin is slightly diminished.

Surprisingly, increasing the concentration of PAI-1 to 1.9 μM in the presence of excess native Vn (4 μM) resulted in no further increase in the rate of inactivation of α-thrombin (k_{app}=8.4 x 10^{-3} s^{-1}). Complementary experiments performed with urea-treated Vn (Figure 1) resulted in no more than a 1.15-fold increase in the apparent first order rate constant for the reaction of PAI-1 with α-thrombin. Our data are consistent with PAI-1 binding very tightly with Vn. The data of Figure 1 show that the rate of inactivation increases in proportion to the concentration of Vn until [Vn] ≅ [PAI-1], suggesting that essentially all the available Vn is complexed with PAI-1 (when [Vn] < [PAI-1]). These data suggest that the concentrations of PAI-1 and Vn used in these experiments are significantly greater than the dissociation constant for the native Vn:PAI-1 complex, and are thus consistent with a value for the dissociation constant that is much less than 130 nM (the lowest concentration of Vn in these reactions).

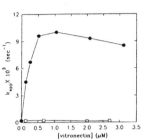

Figure 1. Dependence of the first order rate constant for the inactivation of α-thrombin by PAI-1 on the concentration of native (circles) or urea-treated (squares) Vn. The kinetics of inactivation of α-thrombin (1.58 pM) by PAI-1 (0.34 μM) were measured at 25°C in 0.1 M HEPES, 0.1 M NaCl, 0.1 mg/ml BSA, 0.1%, PEG, pH 7.4 at the indicated concentrations of native or urea-treated Vn.

In general the inactivation of a serine protease by a SERPIN (like PAI-1) behaves according to a second order rate law. This occurs because of the relatively weak initial interaction that governs the encounter complex between SERPIN and cognate protease prior to the reaction of the SERPIN with the active site of the protease and formation of the inactive terminal complex. This is demonstrated by the large value of the dissociation constant generally observed for SERPIN:protease encounter complexes. Experimentally the reaction between protease and inhibitor is often measured under conditions where the concentration of inhibitor is much greater than the concentration of protease. Using these conditions the reaction behaves according to a pseudo first order rate law, whereby the reaction rate is directly proportional to the inhibitor concentration. Our observation that increasing the concentration of PAI-1 did not result in a proportional increase in the first order rate constant (in the presence of a concentration of native Vn which exceeds that of PAI-1) indicated that the reaction of α-thrombin with the Vn:PAI-1 complex did not behave according to a second order rate law under all conditions. This suggested that in the presence of Vn, the dissociation constant for the protease:protease inhibitor complex is substantially less than in the absence of Vn.

We therefore undertook an examination of the dependence of the apparent first order rate constant for the reaction of α-thrombin with PAI-1 in the presence of native Vn on the concentration of the PAI-1:Vn complex (Figure 2A). We performed these experiments with a concentration of native Vn that was greater than the concentration of PAI-1. Given that the concentration of PAI-1 and Vn are much greater than the dissociation constant for the Vn:PAI-1 complex [30], then we can assume that all of the PAI-1 is bound to Vn and that the concentration of the Vn:PAI-1 complex is equal to the concentration of PAI-1. The data in Figure 2A show a saturable dependence of the rate of α-thrombin inactivation on the concentration of the Vn:PAI-1 complex. From these data the dissociation constant (K_d) for the interaction of α-thrombin with the Vn:PAI-1 complex is calculated to be 75 ± 9 nM and a value for the maximal inhibition rate constant (kI) equal to $(1.23 \pm 0.05) \times 10^{-2}$ sec^{-1}. In contrast to the data in Figure 2A, Figure 2B shows the results obtained for the inactivation of α-thrombin by PAI-1 in the absence of Vn. These data show a linear dependence of the first order rate constant (k_{app}) on the PAI-1 concentration, indicating that this reaction behaves according to a second order rate law and that the value of K_d for the interaction of α-thrombin with uncomplexed PAI-1 is much greater than 2 μM. The slope of the line in Figure 2B is equal to the second order rate constant (k_I/K_d = 610 M^{-1}s^{-1}) for the reaction of α-thrombin with PAI-1. Comparison of the second order rate constant for the inactivation of α-thrombin by PAI-1 in the absence of Vn to the corresponding second order rate constant (k_I/K_d) which governs the reaction of α-thrombin with the Vn:PAI-1 complex at low concentrations of the complex, reveals a 270-fold increase in the presence of Vn. Interestingly, the second order rate constant in the presence of Vn is 10-20-fold greater than the second order rate constant for α-thrombin reacting with its

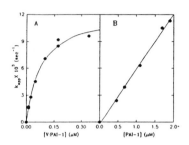

Figure 2. Dependence of the first order rate constant for the inactivation of α-thrombin on the concentration of the Vn:PAI-1 complex or uncomplexed PAI-1. The kinetics of inactivation of α-thrombin (1.58 pM) by PAI-1 (0-1.9 μM) in the presence (A) or absence (B) of native Vn (0.52 μM) were measured at 25°C in 0.1 M HEPES, 0.1 M NaCl, 0.1 mg/ml BSA, 0.1% PEG, pH 7.4.

principle inhibitor, antithrombin III. The magnitude of the acceleration observed here in the presence of Vn is similar to that previously reported [5]. α-Thrombin has a much greater affinity for the native Vn:PAI-1 complex ($K_d \cong 75$ nM) compared to that for uncomplexed PAI-1 ($K_d \gg 1.9$ μM). This suggests that the acceleration is largely due to a reduction in the dissociation constant of α-thrombin for PAI-1 when PAI-1 is bound to native Vn. The maximal inhibition rate constant for the inactivation of α-thrombin by the Vn:PAI-1 complex is equal to 0.0123 sec^{-1}. This value is much slower than the maximal inhibition rate constant (5 sec^{-1}) for the inactivation of thrombin by antithrombin III [36-38]. This observation is supportive of our conclusion that the acceleration of the inactivation of α-thrombin by PAI-1 in the presence of Vn is largely due to a reduction in the dissociation constant.

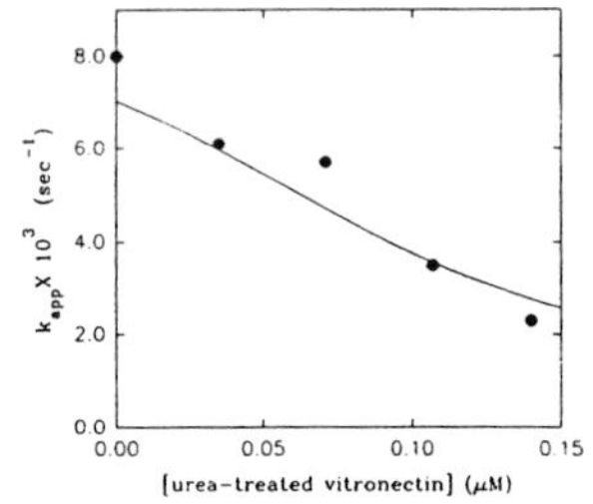

Figure 3. Dependence of the first order rate constant for the inactivation of α-thrombin on the concentration of urea-treated Vn in the presence of native Vn. The kinetics of inactivation of α-thrombin (1.58 pM) by PAI-1 (0.1 μM) in the presence of 0.26 μM native Vn and 0 - 0.14 μM urea-treated Vn were measured at 25°C in 0.1 M HEPES, 0.1 M NaCl, 0.1 mg/ml BSA, 0.1% PEG, pH 7.4.

As described earlier, urea-treated Vn did not accelerate the reaction of PAI-1 with α-thrombin. A simple explanation for this finding is that PAI-1 does not bind to urea-treated Vn under the conditions of these experiments. To determine whether PAI-1 binds to urea-treated Vn, we measured the effect of urea-treated Vn on the accelerated rate of inactivation of α-thrombin at a constant concentration of native Vn (Figure 3). We reasoned that if PAI-1 binds to urea-treated Vn, then native and urea-treated Vn should compete for the binding of the available PAI-1.

The presence of increasing amounts of urea-treated Vn in the reaction mixture should then result in progressively more PAI-1 bound to urea-treated Vn and therefore a reduction in the accelerated rate of inactivation. The data of Figure 3 show a reduction in the accelerated rate for the inactivation of α-thrombin by PAI-1 when urea-treated Vn is present in the reaction mixture. These data demonstrate that urea-treated Vn binds to PAI-1 and acts as an effective antagonist of the accelerated reaction between PAI-1 and α-thrombin. Furthermore, analysis of these data suggests that PAI-1 binds urea-treated Vn with an affinity approximately 10-fold greater than that for native Vn [39]. These findings demonstrated that formation of the Vn:PAI-1 complex was not sufficient for acceleration of the inactivation of α-thrombin by PAI-1.

The competition experiments show that both urea-treated and native Vn bind tightly to PAI-1. The complex of PAI-1 with urea-treated Vn, however, was unable to react with thrombin at an accelerated rate. This suggested that perhaps an additional interaction with Vn was required in order for the Vn:PAI-1 complex to react with α-thrombin at an accelerated rate. We therefore examined the interaction of both forms of Vn with thrombin in the absence of PAI-1. In these experiments we measured the influence of Vn on the α-thrombin-catalyzed hydrolysis of tosGPRamc (Figure 4). The results obtained for native Vn show a modest saturable inhibition of the rate of α-thrombin-catalyzed hydrolysis of tosGPRamc. These results are consistent with native Vn behaving as a noncompetitive inhibitor ($K_d=350 \pm 40$ nM) of the action of α-thrombin on tosGPRamc, suggesting that native Vn interacts with α-thrombin at a site distinct from the active site. In contrast, urea-treated Vn showed no interaction with α-thrombin as reflected by the absence of an effect of urea-treated Vn on the action of α-thrombin on tosGPRamc (Figure 4). Previous investigators have shown that native Vn binds to thrombin, as evidenced by complex formation with agarose gel electrophoresis [40]. The results reported here show that native Vn interacts with α-thrombin in a saturable manner to effect a modest inhibition of the action of α-thrombin on a fluorogenic substrate. The non-zero limiting rate of substrate hydrolysis at saturating concentrations of Vn indicates that native Vn does not bind to the active site of thrombin as does a substrate, but rather that native Vn binds to α-thrombin at a site distinct from the active site. Our observations that native Vn accelerates the inactivation of α-thrombin by PAI-1 and binds to α-thrombin whereas urea-treated Vn has neither of these properties yet does bind tightly to PAI-1, suggest that the interaction of Vn with α-thrombin is an important determinant of the competence of Vn to accelerate the reaction between PAI-1 and α-thrombin.

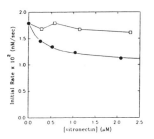

Figure 4. Dependence of the initial rate of α-thrombin-catalyzed hydrolysis of tosGPRamc on the concentration of Vn. Initial rates were performed at 25°C in 0.1 M HEPES 0.1 M NaCl, 0.1 mg/ml BSA, 0.1% PEG, pH 7.4 at the indicated concentrations of native (circles) or urea-treated (squares) Vn.

Our finding that native Vn binds to α-thrombin in the absence of PAI-1 suggests that perhaps native Vn behaves simply as a template for the assembly of α-thrombin and PAI-1, analogous to the role of heparin as a catalyst of the reaction between antithrombin III and thrombin [41,42]. During heparin-catalyzed inactivation of α-thrombin by antithrombin III both thrombin and antithrombin III bind to heparin. The acceleration results from the approximation of the two proteins on the template, heparin. A template mechanism for the acceleration of protease-protease inhibitor reactions predicts that at concentrations of the template which exceed the concentrations of both protease and protease inhibitor as well as the values of the dissociation constants for the protease:template and inhibitor:template complexes there should be a reduction in the magnitude of the acceleration. The rate reduction results from the formation of two independent binary complexes of protease:template and inhibitor:template which react at a slower rate than if both protease and protease inhibitor assemble on the same template molecule. The data of Figure 1 demonstrate that at concentrations of native Vn greater than the concentrations of PAI-1, α-thrombin, and the dissociation constant for the Vn:α-thrombin complex, there is a diminution of the accelerated rate. This is the expected result if Vn behaves as a template. However, the magnitude of the diminution is not as great as predicted if native Vn behaves solely as a template. Assuming that PAI-1 is completely bound to native Vn and that the free, uncomplexed native Vn binds to α-thrombin and behaves as a competitive inhibitor of the reaction between native Vn:PAI-1 and α-thrombin, then we predict an apparent first order rate constant for the inactivation of α-thrombin equal to $4.0 \times 10^{-3} \text{sec}^{-1}$ at the greatest concentration of native Vn in Figure 1. This value is approximately half of the experimentally determined value of $8.5 \times 10^{-3} \text{sec}^{-1}$. This observation suggests that in addition to a template mechanism, other alterations in either PAI-1, Vn, or α-thrombin occur which contribute to the accelerated reactivity of PAI-1 and α-thrombin in the presence of native Vn. The inactive SERPIN-protease complex, thrombin:ATIII, binds to and induces a conformational change in Vn [34,35]. Perhaps the binding of PAI-1 or α-thrombin to native Vn results in a conformational change in Vn that is important for the acceleration of the inactivation of α-thrombin by PAI-1. Other possibilities include alterations in the reactivity of PAI-1 or α-thrombin when bound to Vn. The data of Figure 4 show that a perturbation occurs at the active site of α-thrombin when Vn binds to a nonactive site domain of α-thrombin. Further experimentation is required to determine whether this perturbation contributes to the accelerated reaction between PAI-1 and α-thrombin.

The observation that native Vn accelerates the reaction of PAI-1 with α-thrombin whereas urea-treated Vn does not, suggests that the capacity of Vn to accelerate this reaction is sensitive to its conformation. Similarly, it appears that substrate-adsorbed Vn shows a variable capacity to accelerate the inactivation of α-thrombin by PAI-1 [5,43,44]. This too may be a consequence of alternate conformations of substrate-adsorbed Vn as evidenced by infrared spectroscopy, elutability with SDS, thrombogenicity [45] and reactivity with a conformationally specific antibody [46].

Acknowledgments We thank Suzann Labun for assistance in manuscript preparation. Parts of this manuscript, including figures 1-4, have been adapted from [39] and [8]. This work was supported by National Institutes of Health Grant 1-RO1-HL39137 (DG) and 1-PO1-HL29586 (DFM); DG is a Howard Hughes Medical Institute Investigator. T.J.P. is a Research Scholar of the H.S.F.O. and is supported by grant A1729 from the Heart and Stroke Foundation of Ontario.

III. REFERENCES

1. van Mourik, J. A., Lawrence, D. A. & Loskutoff, D. J. *J. Biol. Chem.* **259**, 14914-14921 (1984).

2. Lawrence, D., Strandberg, L., Grundström, T. & Ny, T. *Eur. J. Biochem.* **186**, 523-533 (1989).

3. Hekman, C. M. & Loskutoff, D. J. *Biochem.* **27**, 2911-2918 (1988).

4. Fay, W. P. & Owen, W. G. *Biochem.* **28**, 5773-5778 (1989).

5. Ehrlich, H. J., Gebbink, R. K., Keijer, J., Linders, M., Preissner, K. T. & Pannekoek, H. *J. Biol. Chem.* **265**, 13029-13035 (1990).

6. Declerck, P. J., Verstreken, M., Kruithof, E. K. O., Juhan-Vague, I. & Collen, D. *Blood* **71**, 220-225 (1988).

7. Booth, N. A., Simpson, A. J., Croll, A., Bennett, B. & MacGregor, I. R. *Br. J Haematol.* **70**, 327-333 (1988).

8. Lawrence, D. A. & Ginsburg, D. *Molecular Biology of Thrombosis and Hemostasis* (in press)

9. Loskutoff, D. J. *Fibrinolysis* **5**, 197-206 (1991).

10. Pyke, C., Kristensen, P., Ralfkiaer, E., Eriksen, J. & Dano, K. *Cancer Res.* **51**, 4067-4071 (1991).

11. Keeton, M., Eguchi, Y., Sawdey, M., Ahn, C. & Loskutoff, D. J. *Am. J. Pathol.* **142**, 59-70 (1993).

12. Declerck, P. J., De Mol, M., Alessi, M. C., et al. *J. Biol. Chem.* **263**, 15454-15461 (1988).

13. Wiman, B., Almquist, Å., Sigurdardottir, O. & Lindahl, T. *FEBS Lett.* **242**, 125-128 (1988).

14. Levin, E. G. & Santell, L. *J. Cell. Biol.* **105**, 2543-2549 (1987).

15. Pöllänen, J., Saksela, O., Salonen, E. M., et al. *J. Cell. Biol.* **104**, 1085-1096 (1987).

16. Knudsen, B. S. & Nachman, R. L. *J. Biol. Chem.* **263**, 9476-9481 (1988).

17. Schleef, R. R., Loskutoff, D. J. & Podor, T. J. *J. Cell. Biol.* **113**, 1413-1423 (1991).

18. Hekman, C. M. & Loskutoff, D. J. *J. Biol. Chem.* **260**, 11581-11587 (1985).

19. Levin, E. G. & Santell, L. *Blood* **70**, 1090-1098 (1987).

20. Lindahl, T. L., Sigurdardóttir, O. & Wiman, B. *Thromb. Haemost.* **62**, 748-751 (1989).

21. Lambers, J. W., Cammenga, M., Konig, B. W., Mertens, K., Pannekoek, H. & van Mourik, J. A. *J. Biol. Chem.* **262**, 17492-17496 (1987).

22. Wun, T. -C., Palmier, M. O., Siegel, N. R. & Smith, C. E. *J. Biol. Chem.* **264**, 7862-7868 (1989).

23. Vaughan, D. E., Declerck, P. J., Van Houtte, E., De Mol, M. & Collen, D. *Circ. Res.* **67**, 1281-1286 (1990).

24. Mimuro, J., Schleef, R. R. & Loskutoff, D. J. *Blood* **70**, 721-728 (1987).

25. Tomasini, B. R. & Mosher, D. F. *Prog. Hemost. Thromb.* **10**, 269-305 (1991).

26. Lang, I. M., Marsh, J. J., Moser, K. M. & Schleef, R. R. *Blood* **80**, 2269-2274 (1992).

27. Preissner, K. T., Holzhüter, S., Justus, C. & Muller-Berghaus, G. *Blood* **74**, 1989-1996 (1989).

28. Sigurdardóttir, O. & Wiman, B. *Biochim. Biophys. Acta* **1035**, 56-61 (1990).

29. Kost, C., Stüber, W., Ehrlich, H. J., Pannekoek, H. & Preissner, K. T. *J. Biol. Chem.* **267**, 12098-12105 (1992).

30. Seiffert, D. & Loskutoff, D. J. *Biochim. Biophys. Acta* **1078**, 23-30 (1991).

31. Salonen, E. -M., Vaheri, A., Pollanen, J., et al. *J. Biol. Chem.* **264**, 6339-6343 (1989).

32. Yatohgo, T., Izumi, M., Kashiwagi, H. & Hayashi, M. *Cell Struct. Funct* **13**, 281-292 (1988).

33. Dahlbäck, B. & Podack, E. R. *Biochem.* **24**, 2368-2374 (1985).

34. Tomasini, B. R. & Mosher, D. F. *Blood* **72**, 903-912 (1988).

35. Tomasini, B. R., Owen, M. C., Fenton, J. W.,II & Mosher, D. F. *Biochem.* **28**, 7617-7623 (1989).

36. Olson, S. T. & Shore, J. D. *J. Biol. Chem.* **261**, 13151-13159 (1986).

37. Olson, S. T. & Shore, J. D. *J. Biol. Chem.* **257**, 14891-14895 (1982).

38. Olson, S. T. *J. Biol. Chem.* **263**, 1698-1708 (1988).

39. Naski, M. C., Lawrence, D. A., Mosher, D. F., Podor, T. J. & Ginsburg, D. *J. Biol. Chem.* **268**, 12367-12372 (1993).

40. Podack, E. R., Dahlbäck, B. & Griffin, J. H. *J. Biol. Chem.* **261**, 7387-7392 (1986).

41. Olson, S. T. & Björk, I. *J. Biol. Chem.* **266**, 6353-6364 (1992).

42. Olson, S. T., Björk, I., Sheffer, R., Craig, P. A., Shore, J. D. & Choay, J. *J. Biol. Chem.* **267**, 12528-12538 (1992).

43. Keijer, J., Linders, M., Wegman, J. J., Ehrlich, H. J., Mertens, K. & Pannekoek, H. *Blood* **78**, 1254-1261 (1991).

44. Greengard, J. S., Seiffert, D., Schleef, R. R., Loskutoff, D. J. & Griffin, J. H. *Thromb. Haemost.* **65**, 1277 (1991).(Abstract)

45. Fabrizius-Homan, D. J., Cooper, S. L. & Mosher, D. F. *Thromb. Haemost.* **68**, 194-202 (1992).

46. Bale, M. D., Wohlfahrt, L. A., Mosher, D. F., Tomasini, B. & Sutton, R. C. *Blood* **74**, 2698-2706 (1989).

Biology of Vitronectins and their Receptors
K.T. Preissner, S. Rosenblatt, C. Kost, J. Wegerhoff and D.F. Mosher, editors

Covalent modulation of vitronectin structure for the control of plasminogen activation by PAI-1.

S. Shaltiel, I. Schvartz, Z. Gechtman and T. Kreizman

Department of Chemical Immunology, The Weizmann Institute of Science, Rehovot, Israel, 76100.

INTRODUCTION

Our interest in vitronectin stemmed from an observation made in our laboratory, showing that upon physiological stimulation of blood platelets with thrombin they release into the plasma the enzyme cAMP-dependent protein kinase (PKA) [1], and that in human blood it selectively phosphorylates one plasma protein which we identified as vitronectin [2].

Originally discovered in serum as a cell spreading factor with high affinity for glass surfaces [3], vitronectin is now recognized as a multifunctional regulatory protein [4] and has been implicated in a variety of extracellular processes such as the cell attachment and spreading of normal and of neoplastic cells [5], the function of complement [6,7] and the function of the blood coagulation pathways [6,8,9]. More recently, vitronectin was shown to bind and stabilize the active form of the inhibitor-1 of plasminogen activators (PAI-1) both in circulating blood and in the extracellular matrix [10-14]. This inhibitor has an important role in the control of plasminogen activation, and as such it is involved in the dissolution of hemostatic plugs and the penetration of tissues. Consequently, through its functional interaction with PAI-1, vitronectin is involved not only in fibrinolysis, but also in inflammation, ovulation, tissue remodeling and development, angiogenesis, nerve regeneration, malignancy and tumor cell invasion [15-17]. In view of this diversity of functions, these processes have to be localized and strictly regulated. Therefore, it sounds reasonable to assume that vitronectin may well be functionally modulated for the purpose of control.

THE PHOSPHORYLATION OF VITRONECTIN BY PKA CAN TAKE PLACE IN WHOLE BLOOD

The phosphorylation of plasma vitronectin by platelet-released PKA is quite selective [2,18]: if $[\gamma^{32}P]ATP$ is added to human plasma, no phosphorylation of vitronectin will occur, provided, of course, that appropriate measures are taken to secure that no accidental activation of platelets will occur during the isolation of the plasma. However, the phosphorylation of vitronectin which occurs in human serum or in platelet-containing plasma after platelet activation can be reproduced by addition of the pure catalytic subunit of PKA to plasma devoid of platelets. The selectivity of this phosphorylation is especially noteworthy in view of the multitude of proteins present in plasma which are not phosphorylated by this kinase [2,18].

The phosphorylation of vitronectin by PKA in human serum occurs even if no exogenous cAMP is added [2], showing that this enzymatic modification can take place in whole blood, at least *ex vivo*. If, in addition to $[\gamma^{32}P]ATP$ (used to reveal the phosphorylation of vitronectin) we also add exogenous cAMP, the phosphorylation of vitronectin is enhanced, illustrating the cAMP-dependency of this reaction. Most importantly, the specific inhibitor of PKA known as PKI [19] inhibits both the basal phosphorylation of vitronectin (obtained without addition of exogenous cAMP) and the additional phosphorylation observed upon adding exogenous cAMP.

Table 1
Comparison of the Km values associated with the phosphorylation of vitronectin by PKA with the concentration of the relevant constituents in blood*.

Constituent	Concentration in blood [Reference]	Km
Vitronectin	3-6 μM [28, 33]	3.0 ± 0.3 μM
ATP	2-4 μM [32] ; ~12 μM** [34]	5.0 ± 0.5 μM
Mg++	~1 mM [35]	0.50 ± 0.05 mM

 * For experimental details see [36,37].
** At the locus of a hemostatic event.

If we take into account that cAMP occurs in blood and that its level fluctuates under normal and pathological conditions [20-23], that intracellular cAMP is released to the cell exterior [24,25], and that cAMP has been implicated in hemostasis and thrombosis [26,27], then it is reasonable to assume that from the point of view of its cAMP-dependency, the phosphorylation of vitronectin by PKA is physiologically feasible.

Table I shows that the apparent Km of PKA for vitronectin, ATP and Mg++ will allow for vitronectin phosphorylation in blood. However, since vitronectin is secreted by stimulated platelets [28] and deposited in the vascular wall [29-31], and since both PKA [1,2] and ATP [32] are released by physiologically stimulated platelets, then the phosphorylation of vitronectin may certainly take place at the locus of a hemostatic event, where the concentrations of all the constituents necessary for the reaction are (under the appropriate circumstances) above their respective Km. In fact, the phosphorylation of vitronectin may thus be localized, to ensure that the control of clot formation (hemostasis) and subsequent clot dissolution (fibrinolysis) will be confined to the site of a hemostatic event and to that site only.

S^{378} - THE TARGET SITE OF PKA IN VITRONECTIN

Work done in several laboratories has shown that in the liver, or in cultured Hep G2 cells, vitronectin exists as a single chain glycoprotein with an apparent molecular mass of ~75 kDa [38,39]. But the ultimate protein isolated from pooled plasma contains two forms of this protein: a single chain polypeptide (V$_{75}$) and a nicked polypeptide(V$_{65+10)}$ in which the two chains are linked by an interchain disulfide bridge (Fig. 1) [6,40] The cleavage site for the conversion of V$_{75}$ into V$_{65+10}$ is most likely between R^{379} and A^{380} [6,40,41] although the particular enzyme or locus where this specific cleavage takes place in vivo has not been established yet. Until recently it was assumed that there are "no major functional differences" between V$_{75}$ and V$_{65+10}$ [42]. However, in the course of our work we found that these two forms are distinctly different in their properties, since they exhibit a different affinity for immobilized heparin, and behave differently as substrates for PKA [43,44].

The phosphorylation of vitronectin by PKA was found to be targeted to one site (S$^{378)}$) [2,44] at the C-terminal edge of the cluster of basic amino acids (residues 348-379) in vitronectin. This serine residue is located within a typical PKA phosphorylation consensus sequence (RRPSR, residues 375-379) downstream from two adjacent glycosaminoglycan recognition consensus sequences (XBBXBX and XBBBXXBX (Fig.2), where B is a basic amino acid residue and X — a hydrophilic residue [45]). These two heparin binding motifs are located between residues 347 and 359: AKKQRFRHRNRKG (Fig. 2).

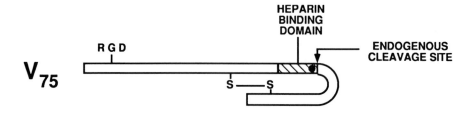

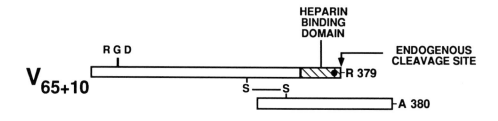

Fig. 1. The two forms of vitronectin (V$_{75}$ and V$_{65+10}$).

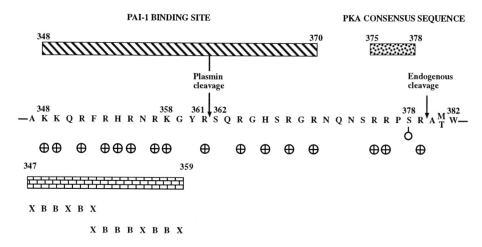

Fig. 2. Identification of the binding sites for heparin and PAI-1 within the cluster of basic amino acids in vitronectin.

THE PHOSPHORYLATION OF S[378] IS CONFORMATION-DEPENDENT

One of the unexpected findings related to this phosphorylation was that PKA differentiates between the intact (V75) form of vitronectin and the nicked (V65+10) form of this protein. Under physiological conditions (pH 7.5), PKA phosphorylates V75 but not V65+10, in spite of the fact that the large chain of vitronectin (V65) contains the full consensus sequence for PKA phosphorylation (RRPSR) and the fact that a synthetic heptapeptide containing that sequence, as well as a partially denatured vitronectin, were found to be phosphorylated by PKA [45]. These facts led us to the conclusion that the PKA phosphorylation site (S[378]) is cryptic or "buried" in V65+10 (Fig. 3) [43,44]. In view of the fact that heparin has been shown to induce significant conformational changes in vitronectin [46,47], we investigated the effect of heparin and other glycosaminoglycans on the accessibility of S[378] to phosphorylation by PKA [43,44].

Heparin, as well as heparan sulfate, were found to have a pronounced effect on the phosphorylation of V65+10, enhancing its relative phosphorylation and shifting its pH dependency so as to allow its phosphorylation at physiological pH [43,44]. It should be noted that this heparin dependency of the PKA phosphorylation is characteristic of V65+10 and does not occur with other substrates of PKA [44]. For example, heparin enhances the rate of the PKA phosphorylation of V65+10 4.7 fold, while it inhibits the phosphorylation of histone H2B and has essentially no effect on the phosphorylation of the synthetic peptide LRRASLG (Kemptide). This interaction between a functional site of vitronectin and its phosphorylation site supports our working hypothesis that the phosphorylation of vitronectin may be important in a physiological modulation of its structure and function. For example, this heparin dependency could be instrumental in allowing the phosphorylation of those V65+10 molecules which presumably interact with GAGs in the subendothelium or on the surface of cells.

THE PLASMIN CLEAVAGE OF VITRONECTIN ATTENUATES ITS BINDING OF PAI-1

While studying the V75 → V65+10 conversion, we observed [48] that when plasmin acts on a regular vitronectin preparation composed of V75 and V65+10, both the 75 kDa and the 65 kDa chains are cleaved to yield a somewhat smaller chain of about 63 kDa. This was indicated by protein staining and by immunoblotting with polyclonal anti-vitronectin antibodies. To identify the exact plasmin cleavage site, we took advantage of the fact that in vitronectin, PKA phosphorylates exclusively S[378] [2,44] (*i. e.* the penultimate amino acid in the 65 kDa chain), and the fact that this selective phosphorylation also takes

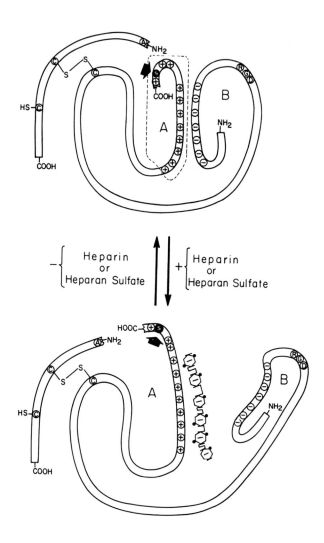

Fig. 3. Schematic representation of the conformation of V65+10 at physiological pH, in the absence and in the presence of heparin or heparan sulfate. Note the electrostatic interactions between the positively charged heparin binding domain (A) and the negatively charged N-terminal domain (B). Also note that in the absence of heparin, S^{378} (arrow) is inaccessible to phosphorylation by PKA (buried), and that upon binding of heparin (or heparan sulfate), S^{378} becomes exposed [43,44].

place in vitronectin peptides which contain S^{378}, provided that they contain the entire consensus sequence of PKA (which, in the case of vitronectin, is RRPSR, i.e. positions 375-379). Indeed, when the plasmin cleavage was monitored by PKA phosphorylation of the cleavage products with $[\gamma\text{-}^{32}P]ATP$, gel electrophoresis and autoradiography, it became clear that the plasmin cleavage occurs not at $R^{379}\text{-}A^{380}$ (as originally postulated) but a site preceding R^{375}, since the single PKA phosphorylation site (S^{378}) was then found in two small-chain cleavage products (~12 kDa and ~2 kDa) which should be obtained from V_{75} and V_{65+10} respectively if the plasmin cleavage site occurs about twenty amino acid residues (~2 kDa) before R^{379}.

To establish the exact site of the plasmin cleavage, we phosphorylated the V_{75} form of vitronectin by carrying out the reaction at pH 7.5 and in the absence of glycosaminoglycans [2,43,44]. The $[^{32}P]$ labeled cleavage product (V_{12}) was identified, cut out and sequenced. Thus, it was established that plasmin cleaves vitronectin at the $R^{361}\text{-}S^{362}$ bond, reducing the molecular mass of the heavy chain of vitronectin to ~63 kDa [48]. This cleavage clearly occurs in V_{75} and in V_{65+10}, both of which yield V_{63} as a single heavy chain product.

Since the plasmin cleavage occurs within the heparin binding domain (positions 348-379) and since this domain may probably be involved in the interaction of vitronectin with PAI-1 [10-14] we explored the effect of the plasmin cleavage of vitronectin on its binding onto immobilized PAI-1. Using the enzyme immunoassay developed by Salonen et al. [13] we found that the plasmin cleavage of vitronectin significantly reduces its ability to bind PAI-1 [48]. This attenuation of binding was observed in four experiments, with PAI-1 activated by the two different procedures described in the literature [13,49] for the conversion of "latent" PAI-1 to its "active" form (Fig. 4) [50]. The extent of the attenuation differed somewhat in the different experiments, but in all four experiments the apparent affinity between vitronectin and PAI-1 was at least 10-fold higher than the affinity between plasmin clipped vitronectin and PAI-1.

As in the case of vitronectin phosphorylation, its plasmin cleavage was also found to be stimulated by heparin and by heparan-sulfate [48]. This stimulation occurred in both the V_{75} and the V_{65+10} forms of vitronectin. However, the stimulation was more pronounced in the case of V_{65+10}, where the rate of cleavage was very slow in the absence of glycosaminoglycans. This result would indicate that glycosaminoglycan-bound vitronectin may well be cleaved by plasmin, forwarding this cleavage to the locus of a hemostatic event.

318

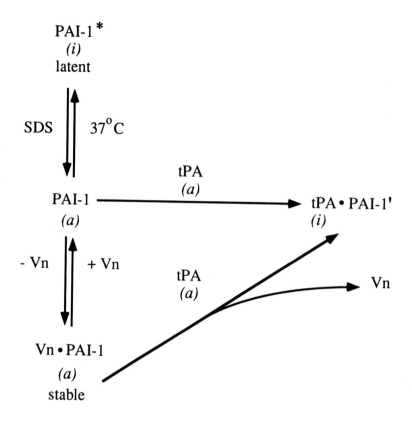

Fig. 4. The inactive ("latent") form (*i*) of PAI-1, its inhibitory "active" form (*a*) and the effect of vitronectin on the stabilization of the active form, which arrests plasminogen activation by inhibiting tPA.

MAPPING THE PAI-1 BINDING SITE IN VITRONECTIN WITH SYNTHETIC PEPTIDES

On the basis of the above mentioned findings regarding the functional consequences of the plasmin cleavage of vitronectin, we set out to map its PAI-1 binding site. For this purpose we synthesized a series of peptides, the structure of which was derived from the amino acid sequence surrounding its plasmin cleavage site (R^{361}-S^{362}) [48]. Each of these synthetic peptides (Fig. 5) was tested by a competition ELISA assay for its ability to inhibit the binding of vitronectin to immobilized activated PAI-1. It was found that the peptides BP1, BP2, BP3 and BP4 which cover the vitronectin sequence between the plasmin cleavage site (R^{361}-S^{362}) and the endogenous cleavage site (R^{379}-A^{380}) inhibit to some extent the binding of vitronectin to PAI-1. This inhibition is mainly due to the N-terminus moiety of BP4, since BP4-1 (S^{362}-R^{370}) was a rather effective inhibitor while BP1 (N^{371}-A^{380}) exhibited a very minor inhibition, if any. It should be noted that BP4-2 (S^{362}-R^{368}) which is only two amino acids shorter than BP4-1, was a less effective inhibitor, suggesting that the G^{369} and/or R^{370} may contain an important biorecognition element. Furthermore, the inhibition capacity of the various peptides could not be attributed merely to their positive charge, since BP4-1 (net charge +4) was found to be more effective than BP3 (net charge +6) or BP4 (net charge +7). The peptide BP5 (K^{358}-R^{370}), which contains all of BP4-1 and a further extension beyond the plasmin cleavage site, was found to be more effective than BP4-1 as an inhibitor. Furthermore, the peptide BP6 (K^{348}-Q^{363}) was found to be the most effective in this series.

The relative inhibitory power of the peptides was assessed by measuring the concentration dependence of the inhibition [51]. It was found that all three of the peptides BP4, BP5 and BP6 achieved a maximal inhibition with increasing concentration. However, while BP4 did so at a concentration of 25 µM and BP5 at a concentration of 100 µM, the peptide BP6 was maximally effective already at a concentration of 2.5 µM (Fig. 2 in [51]). This was further illustrated in an experiment where the relative inhibitory efficacy of these peptides was compared at one concentration of vitronectin (0.5 µg/ml) and with increasing peptide concentrations (Fig. 3 in [51]). In this experiment all three peptides achieved the same level of inhibition (~65%). However, BP6 achieved this level of inhibition at a concentration which was 200- to 500-fold lower than that of BP4 and BP5 [51].

These results indicate that while some high affinity elements of the PAI-1 binding site reside within the K^{348}-R^{357} region, the complete PAI-1 binding site in that region encompasses the K^{348}-R^{370} sequence. This positioning of the PAI-1 binding site can account for our previous finding [48], that upon cleavage of the R^{361}-S^{362} bond by plasmin, the affinity between vitronectin and PAI-1 is attenuated, since this cleavage splits the PAI-1 binding site in two. Furthermore, in the case of the two-chain (endogenously clipped) form of vitronectin, the plasmin cleavage detaches the S^{362}-R^{379} peptide, between the plasmin cleavage site and the endogenous cleavage site, which provides some additional affinity elements for the binding of PAI-1.

A recent study by Preissner, Pannekoek and their coworkers [52] also assigns the PAI-1 binding site in vitronectin within the cluster of basic amino acids, partially overlapping the heparin binding site. Two other reports assign the binding site of this inhibitor to the N-terminus part of vitronectin [53] or to its middle part [54]. To account for this apparent discrepancy, the possibility was raised that vitronectin my possess "multiple PAI-1 binding sites" [54]. However, these different site assignments need not be contradictory. In fact, they may turn out to be complementary, since the interaction of proteins with their ligands (or with other proteins) usually involves amino acid residues which are distal in the linear sequence of the molecule but vicinal in the folded, 3D structure.

320

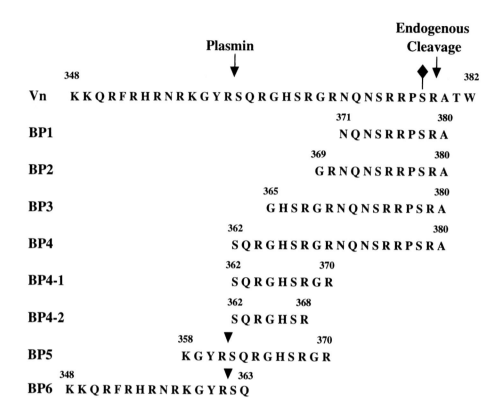

Fig. 5. Structure of synthetic peptides used in mapping the PAI-1 binding site of vitronectin. Note the location of the plasmin cleavage site and the PKA phosphorylation site [51].

CONTROL OF PLASMINOGEN ACTIVATION BY LEASHING AND UNLEASHING PAI-1 FROM VITRONECTIN

The results summarized above suggest that vitronectin acts as a physiological regulatory protein and that it exercises this regulatory function through modulation of its structure (a) by a specific PKA phosphorylation (S^{378}), and (b) by a specific plasmin cleavage (R^{361}-S^{362}). In the case of plasmin cleavage, the covalent modulation was shown to lower the affinity of vitronectin for PAI-1, *i.e.*, to unleash it [48].

The activation of plasminogen by its natural activators (e.g. tPA or uPA) is now considered to be a multipurpose biological tool [15,17]. It is involved

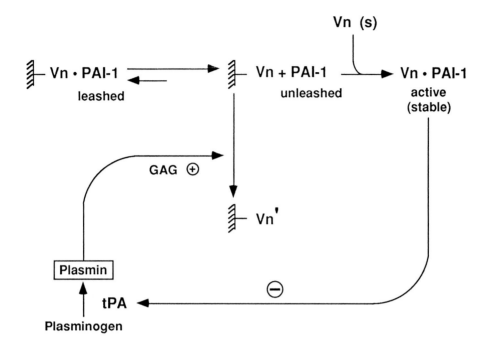

Fig. 6. Proposed mechanism for the control of plasminogen activation by unleashing PAI-1, translocating it from immobilized to mobile vitronectin (Vn) which can then reach and inhibit tPA to prevent or arrest plasmin production. According to this mechanism, at the completion of fibrinolysis, when plasmin is no longer needed, it triggers the arrest of its own production by cleaving Vn and converting the "leashed" PAI-1 into a mobile and active ("unleashed") inhibitor of the plasminogen activator. Localization is achieved by the glycosaminoglycan dependency of this covalent modulation. ⊕ activation, ⊖ inhibition.

not only in the dissolution of blood clots (fibrinolysis) once their plugging function has been completed, but also in paving the way for tissue penetration of nerves (nerve regeneration) and blood vessels (angiogenesis), in cell invasion (metastasis) in inflammation, ovulation, tissue remodeling and development. One would expect *a priori* that excessive or mislocalized plasmin action may have detrimental consequences and therefore that the trigger for the activation of plasminogen and especially for its arrest should be under a strict control by appropriate signaling.

On the basis of our findings we suggested [48] that when plasmin levels become too high, the excess plasmin clips preferentially the vitronectin molecules immobilized in the ECM or on cells at the locus of the hemostatic event. Consequently, the equilibrium between anchored PAI-1 and the detached (mobile) PAI-1 that is bound to soluble vitronectin is displaced, enabling this active and mobile PAI-1 to reach and inhibit the plasminogen activator and arrest plasmin production [48].

Several reports in the literature have described the conformational lability of vitronectin as reflected in its binding to heparin and its recognition by conformation-dependent monoclonal antibodies [46,55-59]. Work done in our laboratory [36,37,43,44,51] shows that his conformational lability may well reflect an intrinsic flexibility designed to detect biological signals and to implement them by means of post-translational modulations, so as to achieve localization and control of vitronectin functions.

ACKNOWLEDGMENTS

This work was supported by grants from the Israel Science Foundation, Jerusalem, from the Minerva Foundation, Munich, Germany, and from the Ministry of Science and Technology, Israel, jointly with the GSF, Munich, Germany. Shmuel Shaltiel is the incumbent of the Kleeman Chair in Biochemistry at the Weizmann Institute of Science. We thank Mrs. Marcia Zelinger for her valuable secretarial assistance.

REFERENCES

1   Korc-Grodzicki B, Tauber-Finkelstein M, Shaltiel S. Proc Natl Acad Sci USA 1988; 85: 7541-7545.
2   Korc-Grodzicki B, Tauber-Finkelstein M, Chain D, Shaltiel S. Biochem Biophys Res Commun 1988; 157: 1131-1138.
3   Holmes R. J Cell Biol 1967; 32: 297-308.
4   Preissner KT. Ann Rev Cell Biol 1991; 7: 275-310.
5   Pytela T, Pierschbacher MD, Ruoslahti E. Proc Natl Acad Sci USA 1985; 82: 5766-5770.
6   Dahlbäck B, Podack ER. Biochemistry 1985; 24: 2368-2374.
7   Podack ER, Kolb WP, Müller-Eberhard HJ. J Immunol 1978; 120: 1841-1848.
8   Preissner KT, Wassmuth R, Müller-Berghaus G. Biochem J 1985; 231: 349-355.
9   Ill CR, Ruoslahti E. J Biol Chem 1985; 260: 15610-15615.
10 Declerck PJ, De Mol M, Alessi MC, Baudner S, Paques EP, Preissner KT, Müller-Berghaus G, Collen D. J Biol Chem 1988; 263: 15454-15461.
11 Wiman B, Almquist A, Sigurdardottir O, Lindahl T. FEBS Lett 1988; 242: 125-128.

12 Mimuro J, Loskutoff DJ. J Biol Chem 1989; 264: 936-939.
13 Salonen EM, Vaheri A, Pöllänen J, Stephens R, Andreasen P, Mayer M, Danø K, Gailit J, Ruoslahti E. J Biol Chem 1989; 264: 6339-6343.
14 Wun T-C, Palmier MO, Siegel NR, Smith CE. J Biol Chem 1989; 264: 7862-7868.
15 Danø K, Andréasen PA, Grondahl-Hansen J, Kristensen P, Nielsen LS, Shriver L. Adv Cancer Res 1985; 44: 139-266.
16 Mignatti P, Tsuboi R, Robbins E, Rifkin DB. J Cell Biol 1989; 108: 671-682.
17 Cajot JF, Bamat J, Bergonzelli GE, Kruithof EKO, Metcalf RL, Testus J, Sordat B. Proc Natl Acad Sci USA 1990; 87: 6939-6943.
18 Korc-Grodzicki B, Chain D, Kreizman T, Shaltiel S. Anal Biochem 1990; 188: 288-294.
19 Walsh DA, Ashby CD, Gonzalez C, Calkins D, Fischer EH, Krebs EG. J Biol Chem 1971; 246: 1977-1985.
20 Broadus AE, Kaminsky NI Hardman JG, Sutherland EW, Liddle GW. J Clin Invest 1970; 49: 2221-2245.
21 Aurbach GD. Adv Cyclic Nucleotde Res 1980; 12: 1-9.
22 Hamet P, Franks DJ, Adnot S, Coquil JF. Adv Cyclic Nucleotide Res 1980; 12: 11-23.
23 Hamet P. Trends Pharmacol Sci 1983; 4: 218-221.
24 Barber R, Butcher RW. Adv Cyclic Nucleotide Res 1983; 15: 119-138.
25 Brunton LL, Heasley LE . Methods Enzymology 1988; 159: 83-93.
26 Steer ML, Salzman EW. Adv Cyclic Nucleotide Res 1980; 12: 71-92.
27 Haslam RJ, Vanderwell M. Methods Enzymology 1989; 169: 455-471.
28 Preissner KT. In: Harenberg O, Heene DL, Stehle G, Schettler G, eds. New Trends in Haemostasis. Berlin, New York: Springer,1990; 123-135.
29 Niculescu F, Rus HG, Vlaicu R. Atherosclerosis 1987; 65: 1-11.
30 Guettier C, Hinglais N, Bruneval P, Kazatchkine M, Bariety J, Camilleri, JP. Virchows Arc A Pathol. Anal Histopathol 1989; 414: 309-313.
31 Reilly JT, Nash JRG. J Clin Pathol 1988; 41: 1269-1272.
32 Gordon JL. Biochem J 1986; 233: 309-319.
33 Shaffer MC, Foley TP, Barnes DW. J Lab Clin Med 1984; 103: 783-791.
34 Ingerman CM, Smith JB, Silver MJ. Thromb Res 1979; 16: 335-344.
35 Friedman PA. In: Isselbacher KJ, Adams RD, Braunwald E, Petersdorf RG, Wilson JD (eds.) Harrison's principles of internal medicine. McGraw-Hill Kogakush, Ltd., 1981; 953-965.
36 Schvartz I, Korc-Grodzicki B, Kreizman T, Shaltiel S. In preparation.
37 Shaltiel S, Schvartz I, Korc-Grodzicki B,Kreizman T. Mol Cell Biol In press.
38 Barnes DW, Reing J. J Cell Physiol 1985; 125: 207-214.
39 Jenne D, Stanley KK. Biochemistry 1987; 26: 6735-6742.
40 Suzuki S, Pierschbacher MD, Hayman EG, Nguyen K, Ohgren Y, Ruoslahti E. J Biol Chem 1984; 259: 15307-15314.

41 Conlan MG, Tomasini BR, Schultz RL, Mosher DF. Blood 1988; 72: 185-190.
42 Preissner KT. Blut 1989; 59: 419-431.
43 Chain D, Korc-Grodzicki B, Kreizman T, Shaltiel S. FEBS Lett 1990; 269: 221-225.
44 Chain D, Korc-Grodzicki B, Kreizman T, Shaltiel S. Biochem J 1991; 274: 387-394.
45 Cardin AD, Weintraub HJR. Arteriosclerosis 1989; 9: 21-31.
46 Tomasini B, Mosher DF. Blood 1988; 72: 903-912.
47 Hildebrand A, Schweigerer L, Teschemacher H. J Biol Chem , 1988; 263: 2436-2441.
48 Chain D, Kreizman T, Shapira H, Shaltiel S. FEBS Lett 1991; 285: 251-256.
49 Katagiri K, Okada K, Hattori H, Yano M. Eur J Biochem 1988; 176: 81-87.
50 Loskutoff DJ, Sawdey M, Mimuro J. in: Progress in Thrombosis and Hemostasis, Coller B (ed.) W. Saunders, Philadelphia 1988; 9: 87-115.
51 Gechtman Z, Sharma R, Kreizman T, Fridkin M, Shaltiel S. FEBS Lett 1993; 315: 293-297.
52 Kost C, Stüber W, Ehrlich HJ, Pannekoek H, Preissner KT. J Biol Chem 1992;267: 12098-12105.
53 Siefert D, Loskutoff DJ, J Biol Chem, 1991; 266: 2824-2830.
54 Mimuro J, Miramatsu S-I, Kurano Y, Ushida Y, Ikadai H, Watahabe S-I, Sakata Y, Biochemistry 1993; 32: 2314-2320.
55 Tomasini BR, Mosher DF. In: Coller B, ed. Progress in Hemostasis and Thrombosis. Philadelphia: W. Saunders,1991; 10: 169-305.
56 Barnes DW, Reing J, Amos B. J Biol Chem 1985; 260: 9117-9122.
57 Hayashi M, Akama T, Kono I, Kashiwagi H. J Biochem 1985; 98: 1135-1138.
58 Preissner KT, Müller-Berghaus G. J Biol Chem 1987; 262: 12247-12253.
59 Tomasini BR, Owen MC, Fenton JW, Mosher DF. Biochemistry 1989; 28: 7617-7623.

Biology of Vitronectins and their Receptors
K.T. Preissner, S. Rosenblatt, C. Kost, J. Wegerhoff and D.F. Mosher, editors

Novel approaches towards PAI-1 interactions

Hans Pannekoek[a,b], Marja van Meijer[a], Anton Jan van Zonneveld[a], David J. Loskutoff[b] and Carlos F. Barbas[b]

[a]Department of Molecular Biology, Central Laboratory of the Netherlands Red Cross Blood Transfusion Service, Plesmanlaan 125, 1066 CX Amsterdam, The Netherlands

[b]Departments of Molecular Biology and Vascular Biology, The Scripps Research Institute, 10666 N. Torrey Pines Road, La Jolla (Cal.), U.S.A.

INTRODUCTION

In the eukaryotic cell, newly synthesized polypeptides are transported from the cytoplasm to the lumen of the endoplasmic reticulum (ER). During the translocation over the ER-membrane, the signal peptide is proteolytically removed and, simultaneously or subsequently, the polypeptide is folded into its proper configuration, which is sustained by correct disulfide bond formation. It has been demonstrated that the translocation of polypeptides over the inner-membrane of a prokaryotic cell (*i.e.* the bacterium *E.coli*) into the periplasm is functionally equivalent to transport of a protein to the lumen of the ER [1-2]. Initially, these findings have been exploited to express heterodimeric Fv and Fab fragments of human and murine immunoglobulins (Ig's) in the periplasm of *E.coli*. Such truncated proteins display similar antigen-binding affinities as their native, intact counterparts [1-2]. These observations emphasize the prominent role of the periplasm in directing the folding of proteins as well as in intra- and intermolecular disulfide bonding. The routing of proteins, involved in the assembly and morphogenesis of filamentous phages, largely resembles that of periplasmic proteins. Both the parental and the newly synthesized gene VIII and gene III coat proteins (cp) of phage M13 are inserted in a polarized mode into the inner-membrane and subsequently encapsulate extruding phage DNA to generate infectious virusparticles. Consequently, peptides or proteins, preceded by an appropriate leader peptide and carboxyl-terminally fused to either cpVIII or cpIII, are routed to the periplasm and subsequently expressed on the surface of phages [3]. This concept has formed the basis for the design of the "combinatorial immunoglobulin (Ig) repertoire cloning technique" with the use of phagemids [4-5]. This procedure allows an analysis of Fab fragments of human or murine Ig repertoires after "immortalization" of Ig mRNA by the "polymerase chain reaction" (PCR) and consecutive insertion of heavy chain cDNAs and light chain cDNAs into a selected phagemid DNA. Specific antigen-antibody interactions can be easily selected by several rounds of "panning" with immobilized antigen. At this point it should be stressed that functional display of proteins on phages permits the simultaneous selection of very large numbers of different molecules (e.g. 10^7), a technical feature that cannot be met by conventional procedures.

RESULTS

In this study, these recent developments have been applied to attempt functional monovalent display of the human plasminogen activator inhibitor 1 (PAI-1) protein (42 kDa) on the surface of the M13-derived phage pComb3 (carbenicillin-resistant) [4]. PAI-1 is an essential regulatory protein of the fibrinolytic system and has been defined as the physiological, fast-acting inhibitor of both tissue-type (t-PA) and urokinase-type (u-PA) plasminogen activator [6]. The mature protein is produced as an inactive (latent) polypeptide after prolonged, cytoplasmic synthesis in transformed *E.coli* cells [7-8]. PAI-1 activity can be recovered upon denaturation with chaotropic agents, followed by renaturation. Here, we used the phagemid pComb3, designed for combinatorial immunoglobulin repertoire cloning, for routing of PAI-1 to the periplasm and subsequent exposure on the surface of filamentous phages. Phage-displayed PAI-1 specifically bound to immobilized polyclonal and monoclonal (mAb) anti-human PAI-1 antibodies. In addition, PAI-1 retains its capacity to form equimolar complexes with its target serine protease t-PA, as well as its ability to inhibit t-PA activity (Fig. 1). Moreover, PAI-1 exposed on the surface of pComb3 phages specifically binds to its carrier protein vitronectin. This property was demonstrated by incubating pComb3/PAI-1 phages with an increasing amount of vitronectin, followed by panning with either an immobilized, anti-vitronectin monoclonal antibody or an anti-t-PA monoclonal antibody. Clearly, binding of pComb3/PAI-1 phages

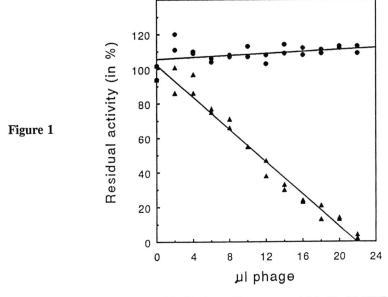

Figure 1

Inhibition of t-PA activity by pComb3/PAI-1 phages. Phages, suspended in 50 mM HEPES (pH 7.4), 100 mM NaCl, 0.01 % (v/v) Tween-80, 1 mM EDTA, were adjusted to 10^{13} cfu/ml and the indicated volumes were incubated for 2 h at 37° C with 0.5 nM t-PA in a volume of 25 μl. Residual amidolytic t-PA activity was determined, after adding 50 μl 4 mM chromogenic substrate S-2288 and 150 μl of the buffer outlined above, by continuous recording of the optical density at 405 nm. Open triangles, pComb3/PAI-1; open circles, pComb3.

was only observed with the anti-vitronectin antibody and not with the anti-t-PA antibody (data not shown).

Furthermore, we explored and manipulated the error-prone property of *Taq*I DNA polymerase during PCR amplification of the full-length PAI-1 cDNA, to generate a large library of predominantly single, random PAI-1 mutants. In addition, a computer simulation program has been devised that converts the number of mutations per codogenic region (*i.e.* PAI-1) into actual mutant proteins (Illustrated by Table 1). The PAI-1-phage mutant library is composed of 46% single, 34% double mutants and 20% wt PAI-1, and can be employed to isolate mutants defective in interactions of PAI-1 with other components. Another application of the phage-display technique is the construction of "epitope libraries" as a highly efficient method to study the structure and function of a protein and to localize interaction sites. In this approach, we assume that domains or sub-domains of a protein are correctly folded when they are routed to the periplasm and, subsequently, displayed on phages by fusion to cpVIII or cpIII. The validity of this strategy was assessed by constructing a phage library that displays random peptide fragments (30 to 60 amino acids) of PAI-1, followed by "panning" against an antibody of which the corresponding epitope had been previously determined. For that purpose, full-length PAI-1 cDNA was randomly digested with pancreatic DNase I and cDNA fragments with a length of 100 to 200 bp were isolated and provided with asymmetric BstX1 linkers. The resulting fragments were inserted in a modified pComb3 vector (pComb3-B) in which we had replaced the XhoI and SpeI sites by BstX1 restriction sites. The phage library was "panned" for several rounds

Table 1

Determination of the percentage of single and multiple PAI-1 mutants at a given number of mutations

Mutations/	Percentage of single or multiple PAI-1 mutants						
1137 bp	1	2	3	4	5	>5	total
1	57.9	0	0	0	0	0	57.9
2	46	34	0	0	0	0	80
3	33.5	40.3	16.7	0	0	0	90.5
4	19.4	37.0	30.7	9.3	0	0	96.4
5	12.3	27.5	33.2	20.7	4.5	0	98.2
6	6.3	17.2	30.5	28.1	14	2.8	98.9
7	3.6	12.9	23.5	27.5	22.6	9.4	99.5
8	2.1	8.4	18.5	27.5	23.7	19.5	99.7
9	0.8	4.4	13.6	23.5	23.6	34.1	100
10	0.5	2.4	8.9	17.5	23.9	46.8	100

A computer simulation program was devised to calculate the number of amino-acid substitutions in a PAI-1 phage library as a result of error-prone DNA synthesis by *Taq*I DNA polymerase. The program incorporates the genetic code and its degeneracy and can be loaded with the codogenic DNA sequence of choice. In this case, the program was loaded with 1137 bp, corresponding with 379 amino-acid residues that constitutes mature PAI-1 [8]. The percentage of single, double, etc. mutants is calculated for a number of iterations (*i.e.* 1000).

against an anti-PAI-1 monoclonal antibody (CLB-2C8) to select phages that harbor a PAI-1 fragment with CLB-2C8 binding properties. Our choice of this particular monoclonal antibody was based on previous work in which we had determined that this antibody blocks the inhibitory activity of PAI-1 and that the epitope for CLB-2C8 is located distant from the reactive center P1 (arg-346), *i.e.* between amino-acid residues 110 and 145 [9]. After two rounds of panning, we found that about 25 % of the selected phages contained inserts that overlap with the region between amino acids 110 and 145 of PAI-1. DNA sequence analysis of a number inserts of pComb3-B/PAI-1 phages allowed us to restrict the corresponding epitope of CLB-2C8 from residue 128 to 145. These results confirm the validity of the "epitope library" and illustrate the selection power of the method.

CONCLUSIONS

In this paper, applications are described demonstrating that proteins, other than the Fab fragments of immunoglobulins, can be functionally expressed on the surface of filamentous phages. Specifically, we have fused full-length cDNA, encoding the human plasminogen activator inhibitor 1 (PAI-1), to the gene III coat protein (cpIII) of phage pComb 3 [4]. Subsequently, display of PAI-1 protein on the surface of pComb3 provided the phage with t-PA-inhibitory and vitronectin binding properties. One of the main advantages of this new technology is the facility to rapidly screen very large numbers of different macro-molecules (*i.e.* a phage library). We have exploited this possibility by constructing mutant libraries. To that end, the error-prone property of *Taq*I DNA polymerase was employed to randomly introduce mutations in PAI-1 cDNA during a PCR. Consequently, a pComb3/PAI-1 phage library was constructed that predominantly harbors single point mutations. To convert the number of mutations into actual mutants, we developed a computer simulation program that incorporates the degeneracy of the genetic code and the appropriate nucleotide sequence. Finally, we illustrated yet another application of the phage-display technique by the construction of an epitope library. Random fragments of PAI-1 (30 to 60 amino acids) were expressed on the surface of a modified pComb3 phage and panned with an anti-PAI-1 monoclonal antibody (CLB-2C8). The location of the epitope (between residues 110 and

145) of CLB-2C8 had been previously determined, using different methods. PAI-1 fragments were selected that specifically bound to the indicated antibody. The corresponding PAI-1 cDNA sequence matched the cDNA sequence, encoding the region between the amino-acid residues 110 and 145 of PAI-1.

REFERENCES

1   Skerra A, Plückthun A. Science 1988; 240: 1038-1041.

2   Better M, Chang, CP, Robinson RR, Horwitz AH. Science 1988; 240: 1041-1043.

3   Smith GP. Gene 1993; 128: 1-2.

4   Barbas CF, Kang AS, Lerner RA, Benkovic SJ. Proc. Natl. Acad. Sci. USA 1991; 88: 7978-7982.

5   Kang AS, Barbas CF, Janda KD, Benkovic SJ, Lerner RA. Proc. Natl. Acad. Sci. USA 1991; 88: 4363-4366.

6   Schleef RR, Loskutoff DJ. Haemostasis 1988; 18: 328-341.

7   Ny T, Sawdey M, Lawrence DA, Millan JL, Loskutoff DJ. Proc. Natl. Acad. Sci. USA 1986; 83: 6776-6780.

8   Pannekoek H, Veerman H, Lambers H, Diergaarde P, Verweij CL, Van Zonneveld AJ, Van Mourik JA. EMBO J. 1986; 5: 2539-2544.

9   Keijer J, Linders M, Van Zonneveld AJ, Ehrlich HJ, De Boer JP, Pannekoek H. Blood 1991; 78: 401-409.

Biology of Vitronectins and their Receptors
K.T. Preissner, S. Rosenblatt, C. Kost, J. Wegerhoff and D.F. Mosher, editors

Clinical relevance of the plasminogen activator system in tumor invasion and metastasis in breast cancer

M. Schmitt[a], F. Jänicke[a], C. Thomssen[a], L. Pache[a], M. Kramer[b], J. Bläser[c], H. Tschesche[c], O. Wilhelm[a], U. Weidle[d] and H. Graeff[a]

[a] Frauenklinik der Technischen Universität München, Germany

[b] Immunologisches Institut der Universität Heidelberg, Germany

[c] Institut für Biochemie der Universität Bielefeld, Germany

[d] Boehringer-Mannheim, Abteilung Molekulare Onkologie, Penzberg, Germany

INTRODUCTION

The capacity of solid tumors to invade the surrounding tissue and to metastasize correlates with the formation and degradation of structural elements in the vicinity of the tumor cells. Substances with both procoagulant activity and fibrinolytic activity, released by tumor cells or stromal cells, are important factors in the formation or degradation of the extracellular matrix (tumor stroma) surrounding the tumor nests. Evidence has accumulated that invasion and metastasis in solid tumors require the action of proteases which promote the dissolution of the tumor stroma and the basement membranes [1-3].

Cancer invasion and metastasis is considered to be a multifactorial process involving a complex interaction of a variety of proteolytic enzymes, growth factors, steroid hormones, oncogenes, cell-cell and cell-substrate adhesion molecules all effecting the control of cell growth and cell adherence [4,5]. In molecular terms, these changes are characterized by modifications in gene expression which involve oncogenes, tumor suppressor genes and other genes which, although they do not transform on their own normal into neoplastic cells, are involved in invasive and metastatic processes which characterize cancer progression [6]. Various types of proteases have been identified in tumor tissues which are believed to play a key role in tumor invasion and metastasis [3,7]. Four different classes of proteases are known to be correlated with the malignant phenotype in various types of solid cancer including cancer of the breast, ovary, lung and cancer of the gastrointestinal tract.

1) Serine proteases (e.g. plasmin, plasminogen activators uPA and tPA)
2) Metalloproteases (e.g. collagenases, gelatinases, stromelysins)
3) Cysteine proteases (e.g. cathepsin B, H, L)
4) Aspartyl proteases (e.g. cathepsin D)

332

Tumor cells and stromal cells, e.g. fibroblasts, synthesize and secrete metallo-proteases, cathepsins and the urokinase-type plasminogen activator (uPA) as inactive proenzymes which may be activated by other proteases **(Fig. 1)**; similar as it is known for complement activation or the blood clotting system.

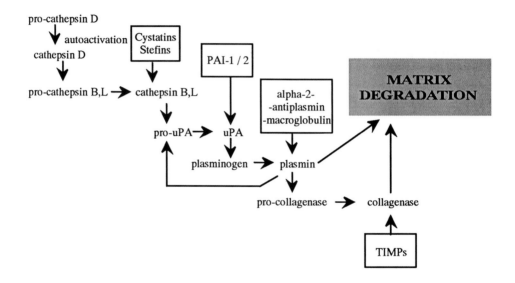

Figure 1

The activation of tumor-associated proteases is a complex pattern involving enzymatically inactive proenzymes, active proteases and inhibitors (boxes) directed to these proteases

uPA, plasmin and the 72 kDa type IV collagenase bind to specific receptors on the tumor cell surface, incorporation of cathepsin B into tumor cell membranes has also been reported [8-11]. Hence the proteolytic activity is focused to the surface of the tumor cell and will thereby promote the dissolution of the tumor stroma and the basement membrane, which may be a prerequisite for tumor invasion and metastasis. Receptor-bound proteases may be inactivated by naturally occuring inhibitors and thus the invasive and metastatic phenotype of cancer cells be modulated. A prominent example in this respect is the binding of the plasminogen activator inhibitor PAI-1 to receptor-bound uPA; this trimeric complex (PAI-1/uPA/uPA-receptor) is subsequently internalized by the tumor cell [12-14]. Evidently, the presence of both uPA and PAI-1 will modulate the proteolytic capacity of tumor cells.

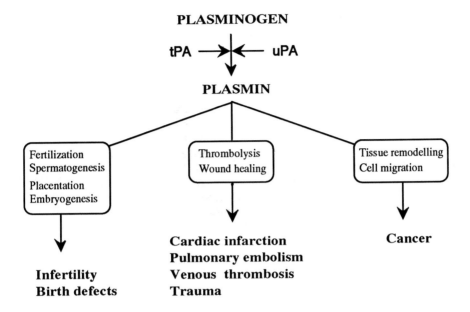

Figure 2

Plasminogen is activated by two different types of plasminogen activators, the tissue-type plasminogen activator (tPA) or by the urokinase-type plasminogen activator (uPA). Under physiological conditions the enzymatic activity of plasmin is the net result of a positive balance of uPA / tPA and their inhibitors PAI-1 / 2 (limited proteolysis). In tumors this balance seems to be deregulated. Excessive high levels of the serine protease uPA (but not tPA) or cathepsins B, D, L, metalloproteases are claimed to be linked to the metastatic potential of solid tumors.

Proteases are not only important pathophysiological factors for tumor invasion and metastasis but also play an important role in trophoblast invasion, embryonic development, tissue remodeling and wound healing under physiological conditions [15] (**Fig. 2**). An impressive number of studies on the biological and/or clinical significance of proteases, their inhibitors and receptors under physiological or pathophysiological conditions have been undertaken in Europe where basic and clinic-oriented research on tissue-associated proteases is strong [1,3]. Proteases are of prognostic relevance in breast cancer, ovarian cancer, lung cancer and gastrointestinal cancer, most detailed information, however, is available for breast cancer in which two different proteases have been reported to be of high prognostic relevance for relapse-free and overall survival: uPA and cathepsin D [16-30].

CLINICAL IMPACT OF TUMOR-ASSOCIATED PROTEASES IN BREAST CANCER PATIENTS

Evidently, in addition to the established histomorphological factors (e.g. tumor size, nodal status, evidence of distant metastasis) biological markers indicative for tumor cell invasion and metastasis are urgently needed by clinicians to devise the correct form of adjuvant therapy to breast cancer patients. A broad panel of prognostic indicators close to 100 describing the proliferative capacity of tumor cells rather than their invasive capacity have been documented in the literature [31]. As most of these "prognostic" factors have only been established in a number of clinical laboratories and are still awaiting multicentric justification one thus should be careful to assess the relative risk for relapse formation or overall survival on the basis of these factors alone. In fact, clinicians in general do not base their therapy decision only on biological factors such as oncogenes, EGF receptor, p53, pS-2, Ki-67, ploidy, or S-phase. In breast cancer it is recommended that these factors should only be considered for therapy decision in conjunction with established histomorphological factors and steroid hormone receptor determinations [32].

Proteases have been shown in recent years by several independent groups to be associated with the malignant phenotype of tumor cells in solid cancer and are known to play a key role in tumor invasion and metastasis. Most of the very recent data about the clinical significance of proteases, however, were collected in patients with breast cancer or gastrointestinal cancer and mainly on two types of proteases, uPA and cathepsin D. Metalloproteases, the cysteine proteases cathepsin B,H,L,S and related inhibitors and receptors were also known to be associated with the invasive capacity of tumor cells but due to lack of specific antibodies, gene probes and highly sensitive immunoassays quantitative determination of these proteolytic factors to assess clinical outcome were not performed [3].

The pattern has changed. Within the past years a huge battery of new antibodies, gene probes and sensitive test kits were made available through researchers or industry [3]. We therefore expect within the next few years a rapid increase in information about the biological and clinical significance of tumor-associated proteases, their receptors and inhibitors in solid cancers.

This is in line with the European programme "Europe against Cancer" sponsored by the Commission of European Communities (CEC). A newly established project "Clinical Relevance of Proteases in Tumor Invasion and Metastasis" comprising of 23 research groups from 11 European countries addresses 5 major topics dealing with proteases, cancer and prognosis:

1) Increase of the knowledge of mechanisms by which proteases facilitate tumor invasion and metastasis in solid tumors, melanoma and leukemia.

2) Standardization of reagents, antibodies, gene probes, kits and assay procedures to determine expression of proteolytic factors at the gene and protein level.

3) Quantitative assessment of proteases, their receptors and inhibitors in human tumor tissue specimens of various origin by ELISA, immunohistochemistry or molecular biological approaches. Correlation of findings with established clinical and histomorphological data.

4) Determination of the prognostic relevance of proteolytic factors for disease-free and overall survival by standardized statistical methods.

5) In which type of cancer do proteolytic factors serve as predictors for poor prognosis; will this knowledge lead to new therapy modalities? What is the impact of proteases in rare types of cancer (e.g. ovarian cancer) and subgroup analysis in more frequent types of cancer (e.g. in node-negative breast cancer).

UPDATE ON PROTEOLYTIC FACTORS AND THEIR PROGNOSTIC POWER IN BREAST CANCER

In the western world, one in eight women will experience breast cancer at some time in her life; every second patient with breast cancer will eventually die of metastatic disease [25]. Patients with no incidence of metastases in their axillary lymph nodes have a good prognosis as about 70 % are cured by locoregional surgery alone. Recommendations for treatment vary considerably and there remains substantial controversy regarding adjuvant therapy in women with node-negative breast cancer [21,25,32]. In 1990 the National Cancer Institute (USA) recommended that all breast cancer patients be considered for adjuvant therapy regardless of their lymph node status. A major concern is that on the basis of this recommendation a majority of women would receive unnecessary systemic adjuvant treatment including potentially toxic chemotherapy. Alternatively, especially in the group of node-negative patients, women could be selected for adjuvant therapy by a combination of histomorphological and biological prognostic factors [21,25,32].

Proteolytic factors are of special interest in search for independent prognostic factors as they play a prominent role in tumor invasion and metastasis [3,22]. Cathepsin D and uPA are potentially important proteases and there have been several clinical investigations published concerning poor prognosis and elevated levels of cathepsin D or uPA in breast cancer tissues [16-30].

uPA and PAI-1

uPA-mediated conversion of plasminogen into plasmin is controled by the inhibitors PAI-1 and PAI-2 **(Fig. 1)** which can inactivate enzymatically active uPA by binding to uPA in its free form in solution or even when uPA is tightly bound to its receptor. During this inhibition process, PAI-1 may be stabilized through the interaction with vitronectin. We and others developed highly sensitive ELISAs for uPA and PAI-1 which are now available commercially through American Diagnostica (Greenwich, CT, USA) and Monozyme (Copenhagen, Denmark) to determine both factors in breast cancer tissue extracts (cytosols or detergent extract). The antibodies selected in these ELISAs detect both enzymatically active and inactive uPA or PAI-1, respectively. The detection of uPA is not impaired even when uPA is complexed with the uPA receptor or PAI-1 / 2. Likewise PAI-1 determination is not impaired even if PAI-1 is complexed with uPA, tPA or vitronectin. The antibodies selected in these ELISAs are also suitable for flow cytometrical or immunohistochemical detection of uPA or PAI-1 in tumor or stromal cells [14,21,24,25,33]. Detection of PAI-2 in tissue extracts is still a problem and matter of debate. At present the ELISAs available for PAI-2 are only tailored to detect free PAI-2. Once a complex is formed between PAI-2 and uPA or tPA, antibodies will not recognize PAI-2 in such a protease-inhibitor complex quantitatively and the PAI-2-ELISA will dramatically underestimate the content of PAI-2 in tumor tissue extracts. An ELISA for the uPA receptor has been made public by Danish researchers of the Finsen Laboratory, Copenhagen, but so far the uPA receptor content in breast cancer tissue extracts has not been published.

Alternatively Del Vecchio et al. measured the tissue concentration of the receptor for uPA in 22 breast carcinomas and 9 benign lesions using *in vitro* quantitative autoradiography applying radiolabeled pro-uPA [33]. They found a 19-fold increase in uPA (ELISA) and a 5 times higher content of uPA receptor in breast cancer tissues compared to benign lesions. The authors therefore concluded that their findings further support the notion that the uPA / uPA receptor system may play a central role in the acquisition of an invasive phenotype and therefore proteolytic factors should be considered as potential indicators for prognosis.

The prognostic value of factors of the plasminogen activator system in breast cancer has indeed been demonstrated before. Due to the extensive and detailed work of German (Jänicke, Schmitt, Graeff), Irish (Duffy), Dutch (Foekens, Klijn) Danish (Grondahl-Hansen, Brünner, Danø) and French (Spyratos, Martin, Oglobine) researchers, uPA and and its inhibitor PAI-1 have been established as markers for the metastatic capacity of breast tumor cells [16-30].

High uPA or PAI-1 values in tumor tissues correlate with increased incidence of relapse or death [16-25,27]. This metastatic capacity reflected by uPA and PAI-1 is independent of the nodal status at diagnosis [19-21,23-25]. Earlier observations by Duffy et al. in 1988 [16] applying enzyme activity measurements already indicated a prognostic relevance of uPA. Independent findings by Jänicke et al. [19] which were later confirmed by Duffy et al. [17], Foekens et al. [25] and Grondahl-Hansen et al. [23] demonstrated that antigen measurements of uPA by ELISA in breast cancer tissue extracts are far superior and determined uPA antigen to be an independent prognostic factor in both node-negative and node-positive patients.

It seems somewhat contradictory that the uPA inhibitor PAI-1 is also of independent value for poor prognosis and that its ranking in multivariate analysis is close in order to that of uPA. Jänicke et al. in 1991 showed for the first time that either high levels of uPA or PAI-1 are associated with shorter relapse-free survival and higher incidence of death in breast cancer patients [20]. One would expect PAI-1 to act protectively by blocking the enzymatic activity of free and receptor-bound uPA. Recent reports, however, documented a specific role of the inhibitors PAI-1 and PAI-2 in uPA receptor clearance. Once enzymatically uPA is bound to tumor cell surface receptors, PAI-1/2 might bind to and inactivate uPA. Within several minutes after binding at physiological temperature this trimeric complex (uPA receptor / uPA / PAI-1) is internalized [12-14]. Moreover, excess release of PAI-1 in tumor tissues which would lead to inactivation of uPA and internalization could be important for reimplantation of circulating tumor cells at distant loci. Generation and growth of metastases should be supported by formation of a new tumor stroma via preventing the uPA-mediated degradation of the extracellular matrix. There exists also a remarkable statistically significant correlation between S-phase and PAI-1 levels and also the inverse correlation of PAI-1 to hormone receptor status indicating a role of PAI-1 in cell proliferation.

In the most recents studies by Jänicke et al. [21] and Foekens et al [24,25] it was shown that the independent prognostic impact of uPA and PAI-1 determined and weighted by multivariate analyses very much depended on the mode of tissue extraction. Foekens et al. [24,25] and Grondahl-Hansen et al. [23] reported that the so-called "cytosol fractions" obtained from breast cancer tissues by mechanical disruption and subsequent ultracentrifugation may be used for uPA or PAI-1 determination. Nevertheless, Jänicke et al. had shown previously that an approximately 100 % increase in yield of uPA, but not of PAI-1, will be obtained if tumor tissues are extracted in the presence of the nonionic detergent Triton X-100 [19,20].

The different modes of extraction have consequences for the relative impact of uPA and PAI-1 for prognosis. Foekens et al. quantified uPA and PAI-1 in close to 700 cytosol fractions and determined PAI-1 to be an an even more powerful prognostic factor than uPA, also in the clinically important subgroup of node-negative patients [24,25]. The underestimated content of uPA in cytosols may be the reason for differing results. Jänicke et al. assessing 247 breast cancer specimens have shown before that in Triton X-100 extracts the prognostic impact of uPA is as high as the lymph node status and that in node-negative patients the impact of uPA is closely followed by that of PAI-1. Since uPA and PAI-1 are independent prognostic factors, node-negative patients could be subdivided further by combining uPA and PAI-1. This may also apply for uPA and PAI-1 determined in the cytosol fractions.

If plasminogen activators convert plasminogen into enzymatically active plasmin, why don't we measure plasmin activity instead of uPA ? Several obvious reasons point against this approach. Plasmin(ogen) is an abundant plasma protein found in high concentrations in normal and also in malignant tissues; it has not been possible to demonstrate a clear difference in plasmin(ogen) content in normal and malignant tissues. In addition, ELISAs which detect plasmin(ogen) in tissues extracts do not distinguish between free and tumor cell receptor-bound plasmin. It is assumed that only receptor-bound plasmin facilitates invasion of tumor cells as free plasmin is rapidly inactivated by plasma inhibitors [3]. Measurement of the enzymatic activity of plasmin in vitro is also not advised. As plasmin is rapidly dissociated from its low affinity receptor it is an easy target for the inhibitors $\alpha2$-antiplasmin and $\alpha2$-macroglobulin as soon as it is released.

Cathepsins B, D, L and metalloproteases

Clinical research on the involvement of the cysteine proteases cathepsins B and L in tumor invasion and metastasis is still in its infancy. Increased expression, membrane association and secretion of cathepsin B and L by tumor cells have been linked to malignancy [11,34]. Incomplete proteolytic processing of cathepsin B is a consistent finding in malignant cells which might be due to 1) mutation in cathepsin B or altered glycosylation/phosphorylation, 2) reduction in the amount or activity of cathepsin B or 3) an increase in the pH of the lysosomal compartment such that processing of cathepsin B is not done under optimal conditions. Tumor cathepsin B does not differ in major biochemical and physical characteristics from normal cathepsin B [34]. Sloane et al. suspect that alterations in trafficking of cathepsin B occur during neoplastic progression of breast epithelium [34]. Vashista et al determined elevated activities of cathepsins B and L in breast cancer tissue [35]. This is in line with the results of Gabrijelcic et al. and Lah et al. who reported that breast carcinoma cytosols contain elevated amounts of cathepsin B and L antigen [36,37]. In the work of Gabrijelcic et al. the amount of cathepsin B antigen in tissue extracts of breast cancer patients correlated with the degree of malignancy within the histological subtypes of invasive ductal carcinomas. A negative correlation of values for cathepsin B with the involvement of regional lymph nodes was found.

In a prospective study we screened 316 breast cancer patients (median follow-up: 40 months) for the content of uPA, PAI-1 and also cathepsin B, D and L antigen in primary breast cancer tumor extracts and found that patients with either high uPA (>3.5 ng / mg protein; p<0.000) or high PAI-1 (>13.7 ng / mg protein; p<0.000) in their primary tumors experienced an increased risk of relapse and death. In this group of patients the aspartyl protease cathepsin D (p=0.007) was also of prognostic value and so were cathepsin B (p=0.005) and cathepsin L (p=0.002). All three types of cathepsins, when elevated, were associated with poor prognosis which confirms the assumption that in addition to uPA lysosomal proteases are involved in tumor invasion and metastasis. (**Fig. 1**).

Cathepsin D has been described by several authors to have prognostic value for breast cancer patients, and similarly to uPA, is considered as a marker of metastasis [27,29,30]. In a large group of 710 patients in spring 1993 Foekens et al. confirmed and extended the reports by Tandon et al. [29], Spyratos et al. [27] and Thorpe et al. [30] in which they demonstrated the prognostic value of cathepsin D. Foekens et al. could also endorse the observation by Spyratos et al. and Thorpe et al. that cathepsin D is of prognostic relevance [26].

Tumor-associated metalloproteases, e.g. collagenase IV, have also been reported to degrade components of the tumor stroma and the basement membrane. Davies et al [38] and Daidone et al. [39] demonstrated a clear relationship between production of type IV collagenases and malignant breast disease. Daidone, however, published that in node-negative patients the expression of collagenase IV was independent of established prognostic factors as well as proliferation markers [39]. In their small group of node-negative patients they did not find any significant impact of collagenase IV on relapse-free or overall-survival.

It has been proposed to classify the various types of tissue metalloproteases more precisely and to use a new nomenclature (**matrix metalloprotease, MMP**). The so far known collagenase IV is now defined of at least two different MMPs, MMP-2 (72 kDalton) and MMP-9 (92 kDalton). Zucker et al. [40,41] determined by ELISA that assessment of MMP-2 and MMP-9 in plasma of breast cancer patients yields different clinically relevant information. MMP-9 is increased in breast cancer patients' plasma [40], MMP-2 is not [41]. In their study, on the other hand, MMP-9 levels in plasma were not significantly increased in patients with metastatic disease as compared to those with nonmetastatic disease.

We determined MMP-8 (leukocyte-type collagenase), MMP-9 and the inhibitor TIMP-1 in tissue extracts of 150 breast cancer patients. MMP-8, MMP-9 and TIMP-1 were elevated in the tumor tissue extracts in comparison to benign tissue controls. By univariate analysis (disease-free survival), we calculated that MMP-8 (p=0.026) and TIMP-1 (p=0.043) to be independent prognostic factors, MMP-9 was less significant (p=0.091). Another metalloprotease, stromelysin-3, is expressed in invasive breast carcinomas in fibroblasts immediately surrounding the neoplastic cells [6,42]. Basset and coworkers suggest that stromelysin-3 may represent a stroma-derived factor necessary for the progression of epithelial malignancies [6,42]. Its prognostic significance has not been demonstrated yet.

340

ACKNOWLEDGEMENTS

Supported by the Deutsche Forschungsgemeinschaft (Klinische Forschergruppe GR28/4-1), the Wilhelm-Sander-Stiftung and the Commission of European Communities (4th Medical Health Research Program). We gratefully acknowledge the support provided by American Diagnostica, Greenwich, CT, USA, and Grünenthal GmbH, Stolberg, Germany.

REFERENCES

1   Dano K, Andreasen PA, Grondahl-Hansen J, Kristensen P, Nielsen LS, Skriver L. Adv Cancer Res 1985; 44:139-266.
2   Markus G. Enzyme 1988; 40:158-172.
3   Schmitt M, Jänicke F, Graeff H. Fibrinolysis 1992; 6 (4):3-26.
4   Brünner N, Dano K. Breast Cancer Res Treatment 1993; 24:173.
5   Clark WH. Br. J. Cancer 1991; 64:631-644
6   Basset P, Wolf C, Chambon P. Breast Cancer Res Treat 1993; 24:185-193.
7   Blasi F. J Surg Oncol Suppl 1993; 3:21-23.
8   Blasi F. Fibrinolysis 1988; 2:73-84.
9   Emonard HP, Remacle AG, Noel AC, Grimaud JA, Stetler-Stevenso WG, Foidart JM. Cancer Res. 52:5845-5848.
10   Burtin P, Fondaneche MC. JNCI 1988; 80:762-765.
11   Lah TT, Buck MR, Honn KV, Crissman JD, Rao NC, Liotta LA, Sloane BF. Clin Exp Metastasis 1989; 7:461-468.
12   Cubellis MV, Wun TC, Blasi F. EMBO J 1990; 9:1079-1085.
13   Estreicher A, Muhlhauser J, Carpentier JL, Orci L, Vassalli JD. J Cell Biol 1990; 111:783-792.
14   Chucholowski N, Schmitt M, Rettenberger P, Schüren E, Moniwa N, Goretzki L, Wilhelm O, Weidle U, Jänicke F, Graeff H. Fibrinolysis 1992; 6, Suppl. 4:95-102.
15   Woessner JF. FASEB J 1991; 5:2145-2154.
16   Duffy MJ, O'Grady P, Devaney D, O'Siorain L, Fennelly JJ, Lijnen HR. Cancer Res 1988; 48:1348-1349.
17   Duffy MJ, Reilley D, O'Sullivan C, O'Higgins N, Fennelly JJ, Andreasen P. Cancer Res 1990; 50: 6827-6829.
18   Jänicke F, Schmitt M, Ulm K, Gössner W, Graeff H. The Lancet 1989; 8670:1049.
19   Jänicke F, Schmitt M, Hafter R, Hollrieder A, Babic R, Ulm K, Gössner W, Graeff H. Fibrinolysis 1990; 4:69-78.
20   Jänicke F, Schmitt M, Graeff H. Sem Thromb Hem 1991; 17:303-312.
21   Jänicke F, Schmitt M, Pache L, Ulm K, Harbeck N, Höfler H, Graeff H. Breast Cancer Res Treatment 1993; 24:195-208.
22   Schmitt M, Jänicke F, Chucholowski N, Goretzki L, Moniwa N, Schüren E, Wilhelm O, Graeff H. In: Molecular Diagnostics of Cancer. Wagener, C and Neumann, S, eds. Springer Verlag Berlin, 1993; 129-149.

341

23 Grondahl-Hansen J, Christensen IJ, Rosenquist C, Brünner N, Mouridsen HT, Dano K, Blichert-Toft M,. Cancer Res 1993; 53:2513-2521.
24 Foekens JA, Schmitt M, van Putten WLJ, Peters HA, Jänicke F, Klijn JGM. J Clin Oncol 1993; in press
25 Foekens JA, Schmitt M, van Putten WLJ, Peters HA, Bontenbal M, Jänicke F, Klijn JMG. Cancer Res 1992; 52:6101-6105.
26 Foekens JA, van Putten WLJ, Portengen H, de Koning HYWCM, Thirion B, Alexia-Figusch J, Klijn JGM. J Clin Oncol 1993; 11:899-908.
27 Spyratos F, Maudelonde T, Brouillet JP, Brunet M, Defrenne A, Andrieu C, Hacene K, Desplaces A, Rouesse J, Rochefort H. The Lancet 1989; 8672:1115-1118.
28 Spyratos F, Martin PM, Hacene K, Romain S, Andrieu C, Ferrero-Pous M, Deytieux S, Le Doussal V, Tubiana-Hulin M, Brunet M. J Natl Cancer Inst 1992; 84:1266-1272.
29 Tandon AT, Clark GM, Chamness GC, Chrigwin JM, McGuire WL. New Engl J Med 1990; 322:297-302.
30 Thorpe SM, Rochefort H, Garcia M, Freiss G, Christensen IJ, Khalaf S, Paolucci F, Pau B, Rasmussen BB, Rose C. Cancer Res 1989; 49:6008-6014.
31 Klijn JG, Berns EM, Bontenbal M, Foekens J. Cancer Treat Rev 19, Suppl B:45-63.
32 McGuire WL, Tandon AK, Allred DC, Chamness GC, Clark GM. JNCI 1990; 82:1006-1015.
33 Del Vecchio S, Stoppelli MP, Carriero MV, Fonti R, Massa O, Li PY, Botti G, Cerra M, D'Aiuto G, Esposito G, Salvatore M. Cancer Res 1993; 53:3198-3206.
34 Sloane BF, Cao L, Sameni M, Rozhin J, Moin K, Ziegler G. Proceed Am Assoc Cancer Res 1993; 34:603.
35 Vashista A, Baker PR, Preece PE, Wood RA, Cushieri A. J Cancer Res Clin Oncol 1989; 115:89-92.
36 Gabrijelcic D, Svetic B, Spaic D, Skrk J, Budihna M, Dolenc I, Popovic T, Cotic V, Turk V. Eur J Clin Chem Clin Biochem 1992; 30:69-74.
37 Lah TT, Kokalj-Kunovar M, Drobnic-Kosoroc M, Babnik J, Golouh R, Vrhovec I, Turk V. Hoppe Seyler's Z Biol Chem 1992; 373:595-604.
38 Davies B, Miles SW, Happerfield LC, Naylor MS, Bobrow LG, Rubens RD, Balkwill FR. Br J Cancer 1993; 67:1126-1131.
39 Daidone MG, Silvestrini R, D'Errico A, Di Fronzo G, Benini E, Mancini AM, Garbisa S, Liotta LA, Grigioni WF. Int J Cancer 1992; 48:529-532.
40 Zucker S, Lysik RM, Zarrabi MH, Stetler-Stevenson W, Liotta LA, Birkedal-Hansen H, Mann W, Furie M. Cancer Epidemiol Biomarkers Prev 1992; 1:475-479.
41 Zucker S, Lysik RM, Zarrabi MH, Moll U. Cancer Res 1993:140-146.
42 Wolf C, Rouyer N, Lutz Y, Adida C, Loriot M, Bellocq JP, Chambon P, Basset P. Proc Natl Acad Sci USA 1993; 90:1843-1847.

Biology of Vitronectins and their Receptors
K.T. Preissner, S. Rosenblatt, C. Kost, J. Wegerhoff and D.F. Mosher, editors

Low density lipoprotein receptor-negative fibroblasts internalize lipoprotein (a) by the low density lipoprotein receptor-related protein/α2-macroglobulin receptor

Angela Beckmann,[a] H. Scharnagl,[a] Birgit Hertwig,[a] R. Siekmeier,[a] W. Schneider,[b] W. Groß[a] and W. März[a]

[a]Gustav Embden-Center of Biological Chemistry and [b]Department of Cardiology, Center of Internal Medicine, J. W. Goethe-University, Theodor Stern-Kai 7, 60590 Frankfurt, FRG

INTRODUCTION

The protein moiety of lipoprotein (a) (Lp(a))[1] is composed of apolipoprotein (apo) B-100 and apo(a) [1, 2]. Apo(a) is homologous to plasminogen [3]. This led to the suggestion that Lp(a) exerts thrombogenic effects by inhibiting fibrinolysis [4].

Plasma levels of Lp(a) are largely determined by the rate of synthesis [5]. The plasma half life of Lp(a) is similar to that of low density lipoproteins (LDL) [5]. However, reports about the role of the LDL receptor in Lp(a) clearance are conflicting. Evidence that Lp(a) can specifically bind to the LDL receptor comes from studies with cultured cells [2, 6-9] and with partially purified receptors [10]. Most investigators, however, agree that the LDL receptor binds Lp(a) with significantly lower affinity than LDL [7, 8], apo(a) presumably 'masking' the apoB-100 receptor binding domain [2, 9]. In humans, up-regulation of LDL receptors by HMG-CoA reductase inhibitors does not reduce Lp(a) [11]. We previously reported that Lp(a) is heterogeneous with respect to particle particle size [12]. We operationally defined two subpopulations of Lp(a) particles, high molecular mass Lp(a), HM_r-Lp(a), and low molecular mass Lp(a), LM_r-Lp(a), and demonstrated that a substantial proportion of HM_r-Lp(a) is taken up into cells by LDL receptor-independent pathways, which also include the LDL receptor related protein/α2-macroglobulin receptor (LRP) [13]. LRP is a multifunctional cell surface receptor [14], which has also been implicated in the clearance of chylomicron remnants [15] and protease-inhibitor complexes [16-20].

To examine in further detail the LDL receptor-independent component of the Lp(a) catabolism, we studied the interaction of Lp(a) size isoforms with homozygous familial hypercholesterolemia fibroblasts.

MATERIALS AND METHODS

Preparation of Lp(a) size isoforms

Lp(a) was purified from the regenerate fluid of a dextran sulfate column based LDL-apheresis system (Kanegafuchi MA 01 - Liposorber LA 15) by sequential ultracentrifugation and gel filtration on Biogel A-15m [21]. The leading and the trailing part of the Lp(a) peak were

---

[1] *Abbreviations:* apo, apolipoprotein; α_2M, α2-macroglobulin; FH, familial hypercholesterolemia; GST, gluthatione-S-transferase; Lp(a), lipoprotein (a); M_r, relative molecular mass; HM_r-Lp(a), high molecular mass Lp(a); LM_r-Lp(a), low molecular mass Lp(a); LRP, low density lipoprotein receptor-related protein/α2-macroglobulin receptor; tPA, tissue-type plasminogen activator; RAP, 39 kDa LRP associated protein; LDL, HDL, low density lipoproteins.

344

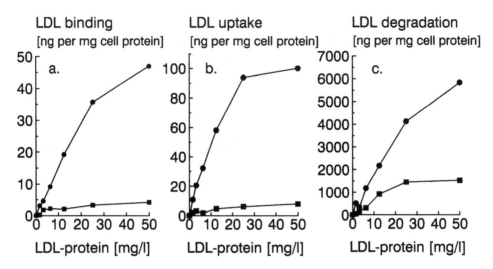

Figure 1. Binding (*a*), uptake (*b*), and degradation (*c*) of LDL in normal and in FH fibroblasts. Normal fibroblasts (*circles*) and fibroblasts from a 16 year old female patient with homozygous FH (*squares*) were grown as described in Methods. Cells were were switched to 10 % (v/v) human lipoprotein deficient serum 40 h before the experiment. Binding (*a*), uptake (*b*) and degradation (*c*) of ^{125}I-LDL were determined as described [24]. Each point is the mean of triplicates; data are adjusted for non-specific binding.

considered high HM$_r$-Lp(a) and LM$_r$-Lp(a), respectively. Both Lp(a) subfractions were free of LDL as determined by agarose gel electrophoresis [22] and intermediate gel immunoelectrophoresis [23].

Cells

Fibroblasts were from skin biopsies obtained from normolipidemic healthy individuals and from a 16 year old female patient with homozygous familial hypercholesterolemia (FH). Cells were grown in 24-well polystyrene plates in RPMI 1640 medium supplemented with 10 % (v/v) fetal calf serum and used at 75 % confluence.

Binding, uptake and degradation of Lp(a)

Binding, uptake (binding *plus* internalization), and degradation of radioactively labelled Lp(a) and α_2-macroglobulin (α_2M) were measured as described [24].

Other methods

α_2M (Sigma) was activated with methylamine according to *Ashcom et al.* [25]. Recombinant tissue-type plasminogen activator (tPA, Thomae) was inactivated with 2 mM phenylmethylsulfonyl fluoride. LDL were prepared by preparative ultracentrifugation (1.019 to 1.050 kg/l). Iodination was performed as described [26]. A recombinant fusion protein consisting of gluthatione-S-transferase sequences (GST) and the 39 kDa LRP/AMR associated protein (RAP) [27] was provided by J. Herz.

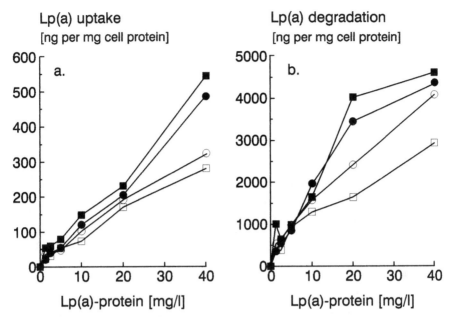

Figure 2. Uptake (*a*) and degradation (*b*) of HM$_r$-Lp(a) and LM$_r$-Lp(a) in normal (*circles*) and FH fibroblasts (*squares*). 40 h before the experiment, cells received medium containing 10 % (v/v) human lipoprotein deficient serum. Uptake and degradation of ^{125}I-HM$_r$-Lp(a) (*solid symbols*) and ^{125}I-LM$_r$-Lp(a) (*open symbols*) were determined as described [24]. No adjustment was made for uptake or degradation in the presence of excess LDL.

RESULTS

The results presented in Figure 1 show that the FH fibroblasts were completely deficient in binding and internalizing LDL, indicating that they lacked functional LDL receptors. We compared the uptake of HMr-Lp(a) and LM$_r$-Lp(a) in normal fibroblasts and in FH fibro-blasts. HM$_r$-Lp(a) was internalized at approximately the same rate in normal fibroblasts and in FH fibroblasts. In contrast, uptake of LM$_r$-Lp(a) was lower in FH fibroblasts than in normal fibroblasts. In both cell types, LM$_r$-Lp(a) was internalized at lower rates than HMr-Lp(a) (Figure 2). In normal fibroblasts, LDL competed more effectively for the degradation of LM$_r$-Lp(a) than for the degradation of HM$_r$-Lp(a) (not shown).

We further analyzed the ability of a$_2$M and tPA to compete with LM$_r$-Lp(a) or HM$_r$-Lp(a) for endocytosis in normal and FH fibroblasts. In both cell types, unlabelled a$_2$M and tPA competed with labelled HM$_r$-Lp(a), but not with labelled LM$_r$-Lp(a) (Figure 3).

To demonstrate that LRP contributed to the uptake of HM$_r$-Lp(a), we examined whether the recombinant GST-RAP fusion protein was active as a competitor. As shown in Figure 4, the GST-RAP inhibited internalization of HM$_r$-Lp(a) and a$_2$M at approximately the same rate. In contrast, the GST-RAP did not influence endocytosis of LM$_r$-Lp(a).

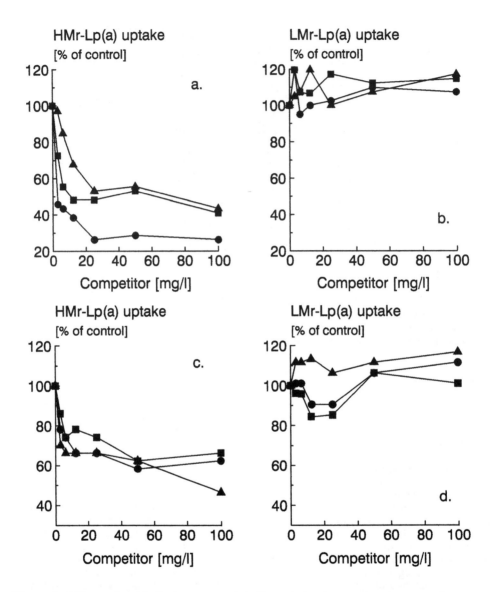

Figure 3. Effects of unlabelled α_2-macroglobulin and tissue-type plasminogen activator on the uptake of ^{125}I-HM$_r$-Lp(a) and ^{125}I-LM$_r$-Lp(a) in normal (*a and b*) and FH fibroblasts (*c and d*). Cells were maintained in medium containing 10 %(v/v) fetal calf serum and incubated with radioactively labelled HM$_r$-Lp(a) (*a and c*) and LM$_r$-Lp(a) (*b and d*) at 6.8 mg/l Lp(a)-protein. Activated α_2-macroglobulin (*squares*), tissue-type plasminogen activator (*circles*) and mixtures of both (*triangles*) were added as unlabelled competitors at the indicated concentrations.

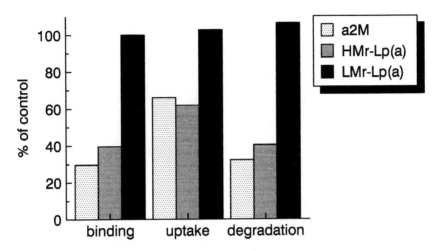

Figure 4. Effect of the 39kDa LRP associated protein (RAP) on the interaction of HM$_r$-Lp(a) and LM$_r$-Lp(a) with FH fibroblasts. Cells were cultured in RPMI 1640 containing 10 % (v/v) fetal calf serum. Binding, uptake, and degradation of ^{125}I-labelled HM$_r$-Lp(a), LM$_r$-Lp(a), and α_2M (at 5 mg/l protein each) were determined as described [24]. The bars represent binding, uptake, and degradation in the presence of 50 mg/l recombinant GST-RAP compared to control incubations without GST-RAP.

DISCUSSION

In this study, we confirm and extend previous observations [13] that heterogeneous size isoforms of Lp(a) differ by their interaction with the LDL receptor and with non-LDL receptor sites of endocytosis and that LRP, apart from its role in the clearance of chylomicron remnants [15] and protease-inhibitor complexes [16-20], contributes to the LDL receptor independent component of Lp(a) catabolism.

We have previously shown that circulating Lp(a) is heterogeneous with respect to size [12]. This size polymorphism is independent of the genetic polymorphism of apo(a) described by Utermann *et al.* [28].

In this work, we used two Lp(a) subfractions of different molecular masses, HM$_r$-Lp(a) and LM$_r$-Lp(a). They were operationally defined as the leading and the trailing part of the Lp(a) peak eluted during gel filtration, which is the last step of the purification procedure.

The structural correlates for the size heterogeneity of Lp(a) are not known so far. Nevertheless, Lp(a) size isoforms appear to differ by their metabolic properties. Using FH fibroblasts lacking functional LDL receptors we were able to study the LDL-receptor independent endocytosis of Lp(a) size isoforms in further detail. HM$_r$-Lp(a) was internalized at virtually the same rate in normal and in FH fibroblasts, suggesting that the uptake of HM$_r$-Lp(a) into fibroblasts proceeds largely independent of LDL receptors. In contrast, LM$_r$-Lp(a) was taken up lower at a lower rate into FH fibroblasts than into normal fibroblasts, indicating that the LDL receptor contributes to LM$_r$-Lp(a) endocytosis. The uptake of HM$_r$-Lp(a), but not of LMr-Lp(a) was inhibited by methylamine-activated a$_2$M, tPA and GST-RAP, indicating that the two Lp(a) subclasses used different routes of internalization. α_2M and tPA are ligands for LRP [16-18], and the GST-RAP fusion protein interferes with the binding of all known ligands to LRP. These results strongly suggest that LRP functions as a receptor for

Lp(a) particles with high molecular masses, whereas low molecular mass isoforms may preferentially be targeted to the LDL receptor.

It is in line with this concept, that the receptor binding domain of apolipoprotein B-100 is more exposed on LM_r-Lp(a) than on HM_r-Lp(a) [12, 13] and that on ligand blots HM_r-Lp(a) appeared to bind stronger to LRP than LM_r-Lp(a) [13].

The precise mechanism by which Lp(a) interacts with LRP is not clear. Some homology exists between kringles 1 and 2 in tPA and the kringles in apo(a) [29]. It is, therefore, tempting to speculate that binding of Lp(a) to LRP depends on apo(a). Beyond this, endocytosis of Lp(a) may require the coordinate formation of multimeric complexes consisting of Lp(a), LRP and other Lp(a) binding proteins (*e.g.* plasminogen receptors or LDL receptors), similar to the model proposed by *Herz et al.* [20] for the uptake of urokinase-type plasminogen activator-plasminogen activator inhibitor 1 complexes by LRP. In many of our experiments, a substantial part of the Lp(a) uptake could not be attributed to the LDL receptor nor to LRP. It is, therefore, not unlikely that further mechanisms, specific or non-specific, contribute to the endocytosis of Lp(a).

Although these issues have to be worked out in future studies our data are clear with respect to the fact that some of the uptake of Lp(a) into cells does not rely on the LDL receptor and that LRP contributes to the LDL receptor-independent component. There are several lines of evidence suggesting that the greater part of the catabolism of Lp(a) does not proceed *via* the LDL receptor: Turnover studies in rats failed to show a relationship between LDL receptor activity and Lp(a) clearance [30, 31]. In a heterozygous FH patient, *Krempler et al.* found that Lp(a) was metabolized faster than LDL [5], and *Knight et al.* reported that the catabolic rate of Lp(a) was normal in FH [32]. Drugs known to stimulate LDL receptors do not lower Lp(a) [11]. In individuals with heterozygous familial defective apoB-100, Lp(a) contained approximately 50 % of the binding-defective apoB-100, compared to approximately 75 % in the LDL fraction from these patients, suggesting that the LDL receptor plays a minor role in Lp(a) clearance [33].

Further progress in our understanding of Lp(a)-cell interactions will probably provide new insights into the way how Lp(a) affects cellular cholesterol homeostasis and causes atherosclerosis.

ACKNOWLEDGMENTS

We thank Dr. Joachim Herz (The Department of Molecular Genetics, University of Texas Southwestern Medical Center, Dallas) for providing recombinant GST-RAP fusion protein, Dr. P. Grützmacher and Yvonne Helmstetter (Department of Internal Medicine II, St. Markus-Hospital, Frankfurt/Main) for providing the skin biopsy from the patient with homozygous FH. The financial support from Bristol Myers/Squibb to W.M. is gratefully acknowledged.

REFERENCES

1. Utermann G. Science 1989; 246: 904-910.
2. Armstrong VW, Walli AK, Seidel D. J Lipid Res 1985; 26: 1314-1323.
3. McLean JW, Tomlinson JE, Kuang W-J, Eaton DL, Chen EY, Fless GM, Scanu AM, Lawn RM. Nature 1987; 330: 132-137.
4. Miles LA, Plow EF. Thromb Haemostas 1990; 63: 331-335.
5. Krempler F, Kostner GM, Bolzano K. J Clin Invest 1980; 65: 1483-1490.
6. Havekes L, Vermeer BJ, Brugman T, Emeis J. FEBS Lett. 1981; 132: 169-173.
7. Floren C-H, Albers JJ, Biermann EL. Biochem Biophys Res Comm 1981; 102: 636-639.
8. Krempler F, Kostner GM, Roscher A, Haslauer F, Bolzano K, Sandhofer F. J Clin Invest 1983; 71: 1431-1441.

9. Armstrong VW, Harrach B, Robenek H, Helmhold M, Walli AK, Seidel D. J. Lipid Res. 1990; 31: 429-441.
10. Steyrer E, Kostner GM. J Lipid Res 1990; 31: 1247-1253.
11. März W, Grützmacher P, Paul D, Siekmeier R, Schoeppe W, Groß W. Clin Pharmacol Ther Toxicol (in press).
12. März W, Siekmeier R, Groß W. Thromb Haemostas 1991; 65: 714.
13. März W, Beckmann A, Scharnagl H, Siekmeier R, et al. FEBS Lett 1993; 325: 271-275.
14. Herz J, Hamann U, Rogne S, Myklebost O, Gausepohl H, Stanley KK. EMBO J 1988; 7: 4119-4127.
15. Kowal RC, Herz J, Weisgraber KH, Mahley RW, Brown MS, Goldstein JL. J Biol Chem 1990; 265: 10771-10779.
16. Strickland DK, Ashcom JD, Williams S, Burgess WH, Migliorini M, Argraves WS. J Biol Chem 1990; 265: 17401-17404.
17. Kristensen T, Moestrup SK, Glieman J, Bendtsen L, Sand O, Sottrup-Jensen L. FEBS Lett 1990; 276: 151-155.
18. Bu G, Williams S, Strickland DK, Schwartz AL. Proc Natl Acad Sci USA 1992; 89: 7426-7431.
19. Orth K, Madison EL, Gething M-J, Sambrook JF, Herz J. Proc Natl Acad Sci USA 1992; 89: 7422-7426.
20. Herz J, Clouthier DE, Hammer RE. Cell 1992; 71: 411-421.
21. Groß E, März W, Siekmeier R, Scharrer I, Groß W. Protein Expression and Purification (in press).
22. Neubeck W, Wieland H, Habenicht A, Müller P, Baggio G, Seidel D. Clin. Chem. 1977; 23: 1296-1300.
23. Gaubatz JW, Cushing GL, Morrisett JD. In: Albers JJ, Segrest JJ, eds. Plasma Lipoproteins. Part B. Characterization, Cell Biology, and Metabolism. Meth Enzymol. Vol 129. Orlando: Academic Press Inc, 1986; 167-186.
24. Goldstein JL, Basu SK, Brown MS. In: Fleischer S, Fleischer B, eds. Biomembranes. Part L. Membrane Biogenesis. Meth Enzymol. Vol 98. Orlando: Academic Press Inc, 1983: 241-260.
25. Ashcom JD, Tiller SE, Dickerson K, Cravens JL, Argraves WS, Strickland DK. J Cell Biol 1990; 110: 1041-1048.
26. Sinn HJ, Schrenk HH, Friedrich EA, Via DP, Dresel HA. Anal Biochem 1988; 170: 186-192.
27. Herz J, Goldstein JL, Strickland DK, Ho YK, Brown MS. J Biol Chem 1991; 266: 21232-21238.
28. Utermann G, Kraft HG, Menzel HJ, Hopferwieser T, Seitz C. Hum Genet 1988; 78: 41-46.
29. Castellino FJ, Urano T, deSerrano VS, Beals JM. In: Scanu AM, ed. Lipoprotein (a). San Diego: Academic Press Inc, 1990: 87-101.
30. Harkes L, Jürgens G, Holasek A, Berkel TJC. FEBS Lett 1988; 227: 27-31.
31. Ye SQ, Keeling J, Stein O, Stein Y, McConathy WJ. Biochim Biophys Acta 1988; 963: 534-540.
32. Knight BL, Perombelon YFN, Soutar AK, Wade DP, Seed M. Atherosclersis 1991; 87: 227.
33. Perombelon YFN, Gallagher JJ, Myant NB, Soutar AK, Knight BL. Atherosclerosis 1992; 92: 203-212.

Biology of Vitronectins and their Receptors
K.T. Preissner, S. Rosenblatt, C. Kost, J. Wegerhoff and D.F. Mosher, editors

BIOLOGICAL PROPERTIES OF HUMAN IMMUNODEFICIENCY VIRUS TYPE-1 TAT PROTEIN: ANGIOGENIC EFFECTS AND ADHESIVE INTERACTIONS OF EXTRACELLULAR TAT

V. Fiorelli, G. Barillari, R.C. Gallo and B. Ensoli

Laboratory of Tumor Cell Biology, National Cancer Institute, National Institutes of Health, Bethesda, MD 20902, USA.

INTRODUCTION

The human immunodeficiency virus type 1 (HIV-1) possesses several regulatory genes. Among them the tat gene encodes an early protein (Tat) that transactivates viral gene expression (1). The Tat protein is encoded by two exons (Fig.1). The product of exon 1 (72 aa) contains a cystein rich region and a basic region. The product of exon 2 (14 aa) includes the tripeptide sequence RGD.

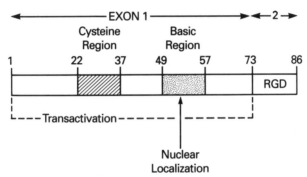

Fig. 1. HIV-1 Tat protein.

Tat possesses other activities on cellular functions. During acute infection of T cells by HIV-1, Tat is released into the extracellular fluid (2,3). In this form Tat promotes the growth of spindle cells of vascular origin derived from Kaposi's sarcoma (KS) lesions of AIDS patients (AIDS-KS cells) (2,3). Normal vascular cells (endothelial and smooth muscle cells), progenitors of AIDS-KS cells (4,5) acquire spindle cell morphology and growth responsiveness to Tat after exposure to inflammatory cytokines present in conditioned media (CM) from activated immune cells (6). These results, and data showing that tat-transgenic mice develop dermal lesions resembling KS (7) have suggested that Tat plays a role in the development of this AIDS-associated angioproliferative disease (8).

RELEASE AND VASCULAR CELL GROWTH PROMOTING ACTIVITY OF HIV-1 TAT PROTEIN

We recently developed a system for the long-term culture of

352

AIDS-KS cells dependent on CM derived from activated CD4+ T cells or HIV-1 infected T cells (9-11). The presence of a KS-growth promoting activity in CM from HIV-1 infected T cells, the absence of HIV-1 sequences in DNA from KS cells (10) and the observation that tat-transgenic mice developing KS-like lesion express tat in the skin but not in the tumor cells (7), indicated that the role of HIV-1 in KS is indirect. In particular, Tat itself might be released by infected cells and promote growth of target cells involved in the formation of KS. To test this hypothesis, we

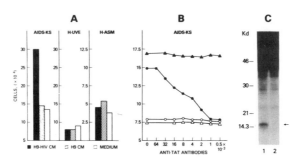

Fig. 2: (A) Growth response of AIDS-KS and normal vascular cells to CM from HIV-1 acutely infected or uninfected H9 cells; (B) block of AIDS-KS cell growth to H9-HIV-1 CM by affinity purified anti-Tat antibodies (●). The antibodies did not block growth induced by HTLV-II CM (▲) or basal cell growth (O) (Δ); (C) radioimmunoprecipitation analysis of Tat in CM from infected (1) or uninfected (2) H9 cells. Reprinted from Ensoli et al. Nature 1990; 345: 84-86.

examined the effect of CM from HIV-1-infected T cells (H9) on the growth of AIDS-KS cells and on normal vascular cells (human umbilical vein endothelial cells, H-UVE; and human aortic smooth muscle cells, H-ASM). H9-HIV-1-derived CM, but not H9 CM, stimulated AIDS-KS cell proliferation but had no effect on H-UVE or H-ASM cell growth (Fig.2A) (2). AIDS-KS cell growth stimulated by H9-HIV-1 CM was specifically inhibited by anti-Tat antibodies (Fig. 2B) and supernatants from H9-HIV-1 infected cells contained Tat (Fig 2C). These results indicated that Tat is released into the extracellular media and possesses a KS growth promoting activity.

In order to assess the contribution of cell death in the release of Tat, Cos-1 cells were transfected with a tat-expressing plasmid (PCV-TAT) or the control vector DNA (PCV-0) and Tat content in cell extracts and supernatants was analyzed by radioimmunoprecipitation assay (RIPA) (Fig. 3). Tat was detected in the cell supernatants in the absence of cell death. In addition, the level of LDH in supernatants from transfected cells was similar to that detected in growth medium and lower than in supernatants from cells containing 1% dead cells (Fig. 3). CM from cells transfected with <u>tat</u> induced AIDS-KS cell proliferation which was inhibited by anti-Tat antibodies (Fig.4) (3).

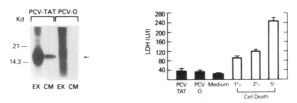

Fig. 3. Cos-1 cells were transfected with PCV-TAT or PCV-0. Cells were monitored for Tat expression by RIPA in cell extracts (EX) and supernatants (CM) from metabolically labeled cells (left panel). LDH was measured in the same supernatants, growth medium, and supernatants containing 1, 2 or 5% dead cells (right panel). Reprinted from Ensoli et al. J Virol 1993; 67: 277-287.

In order to verify the concentration of Tat required for cell growth, serial dilutions of recombinant purified Tat protein were

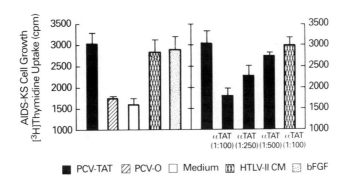

Fig.4. AIDS-KS cell growth ([^{3}H]-Thymidine incorporation) induced by supernatants from Cos-1 cells transfected with PCV-TAT, PCV-0, media alone or known growth inducers of AIDS-KS cells (HTLV-II CM and bFGF). The right panel shows blocking experiments with serial dilutions of anti-Tat antibodies. Reprinted from Ensoli et al. J Virol 1993; 67: 277-287.

added to AIDS-KS cells. Tat stimulated cell growth at concentrations of 0.05 to 50 ng/ml with peak activity between 0.1 and 1 ng/ml (Fig. 5) (3).

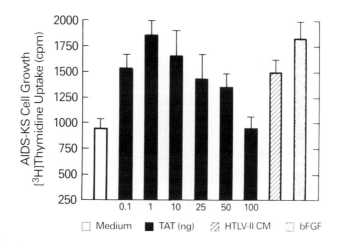

Fig. 5. Proliferation of AIDS-KS cells ([³H]-Thymidine incorporation) with recombinant purified Tat. bFGF and HTLV-II CM were used as positive controls. Reprinted from Ensoli et al. J Virol 1993; 67: 277-287.

COOPERATIVE EFFECTS OF INFLAMMATORY CYTOKINES AND HIV-1 TAT PROTEIN ON VASCULAR CELL PROPERTIES

Vascular cell growth

The previous results indicated that Tat may represent the link between HIV-1 infection and the development of KS. However, normal vascular cells (endothelial and smooth muscle cells), potential progenitors of KS spindle cells (4,5) did not proliferate with Tat (2). In addition, these results did not explain the reasons for the very high risk of KS development in homosexuals men among all HIV-1-infected individuals. Previous data indicated that homosexuals men are more immunostimulated than other groups at risk for AIDS (12,13). In addition, HIV-1-infected homosexual men can develop KS prior to any immunodeficiency (14,15). Inflammatory cytokines normally released by activated immune cells such as tumor necrosis factor (TNF), interleukin-1 (IL-1) and gamma-interferon (gamma-IFN) have been found to be increased in HIV-1-infected individuals (16,17). These observations suggested that immune activation may affect, via the production of cytokines, the onset of KS in homosexual men. To investigate this hypothesis normal vascular cells (H-UVEC and AA-SMC) were cultivated in the presence of CM from activated T cells (PHA-T CM or HTLV-II CM). After exposure to inflammatory cytokines both cell types acquired a spindle cell morphology similar to AIDS-KS cells and they proliferated in response to Tat (6) (Fig. 6).

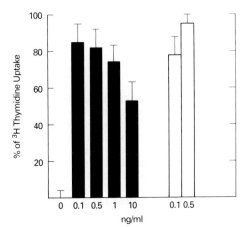

Fig. 6. Tat protein (■) induces the growth ([³H]-Thymidine incorporation) of normal endothelial cells (H-UVE) cultured with inflammatory cytokines. bFGF (□) was used as a positive control. Similar results were obtained with AA-SM cells.

Angiogenic properties of extracellular Tat

The vascular cell growth promoting activity of Tat and the fact that <u>tat</u>-transgenic male mice develop dermal lesions characterized by intense angiogenesis suggested that Tat may have the properties of angiogenic growth factors.

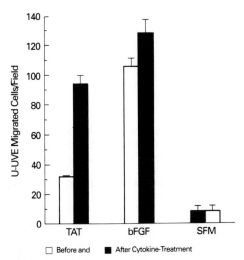

Fig. 7. H-UVE cell migration to Tat is increased by previous cell exposure to inflammatory cytokines. Tat (20 ng/ml), bFGF (20 ng/ml), SFM (serum free medium). Tat-induced cell migration was blocked by anti-Tat antibodies but not by bFGF antibodies. Reprinted from Albini et al. Proc Natl Acad Sci USA 1993; in press.

356

To determine whether Tat has angiogenic properties, experiments
of migration and invasion were performed with Tat and AIDS-KS
cells or normal vascular cell types (18). Tat induced AIDS-KS
cell migration at the same range of concentration promoting cell
growth (2,3,6) with a peak of activity at 10-20 ng/ml. On the
contrary, without prior exposure to inflammatory cytokines, Tat
induced low levels of H-UVE cell migration compared to bFGF
(Fig.7). However, after culture with cytokines, Tat-induced H-UVE
cell migration reached levels similar to bFGF (Fig. 7) (18). Cell
migration to Tat or bFGF was blocked by specific antibodies. Tat
also promoted the invasiveness of AIDS-KS and cytokine-treated
normal endothelial cells at the same concentration inducing
maximal migration of the same cell types (10-20 ng/ml). Thus, Tat
induces endothelial cells to migrate (Fig.7), to cross the
basement membrane (18) and to proliferate (Fig. 6) (6). These are
the events required for angiogenesis (19).

Adhesive interaction of HIV-1 Tat

Tat has been shown to induce adhesion of human lymphoid and rat
skeletal muscle cells when immobilized on culture plates (20).
The finding that cytokines induce endothelial and SM cells to
proliferate in response to Tat (6) suggested that inducible Tat-
receptor(s) mediates the cell growth promoting effect of the
protein. To evaluate whether Tat was capable of a specific cell

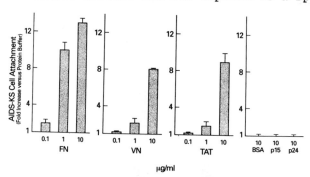

Fig. 8. Immobilized Tat induces attachment of AIDS-KS cells.
AIDS-KS cells were seeded on plates coated with serial dilutions
of FN, VN, and Tat. The number of adherent cells is expressed as
fold of cell attachment compared to the adhesion seen with the
protein buffer. Reprinted from Barillari et al. Proc Natl Acad
Sci USA 1993; 90: 7941-7945

surface interaction with AIDS-KS cells, cell adhesion experiments
were performed with Tat. Tat induced the adhesion of AIDS-KS
cells in a dose-dependent manner, mimicking the effects of
fibronectin (FN) and vitronectin (VN) (Fig.8). To the contrary,
no cell adhesion was observed with BSA, or HIV-1 p15, p24 and Rev
(21). Prior to exposure to inflammatory cytokines, immobilized
Tat protein promoted attachment of SM cells, but had little or no
effect on the adhesion of H-UVE cells (Fig. 9). After culture in
the presence of inflammatory cytokines, SM cell adhesion to the

protein was greatly increased and H-UVE cells became adherent to Tat reaching similar values to those observed with AIDS-KS cells (21). The same cytokines also increased vascular cell adhesion to FN and VN (Fig. 9).

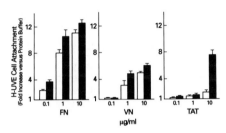

Fig. 9. Attachment of H-UVE cells to Tat is induced by inflammatory cytokines. H-UVE cell were cultured without (□) or with (■) CM from activated T cells. Similar results were obtained with SM cells. Reprinted from Barillari et al. Proc Natl Acad Sci USA 1993; 90: 7941-7944.

To determine the protein domain mediating the attachment of AIDS-KS cells and cytokine-treated vascular cells to Tat, initial experiments were carried out with AIDS-KS cells and two Tat peptides spanning most of the sequences of the products of <u>tat</u> exon 1 ([32-72], containing the basic region) and exon 2 ([65-80] containing the RGD region). Tat and FN were used as the positive controls. Both peptides induced about 2-3 fold increase of cell adhesion, suggesting that both portion of Tat may be involved in the cell attachment to the protein. When the same experiments were repeated by performing the adhesion assays at 4 C, cell adhesion to [32-72]Tat peptide was increased up to 6-fold, while

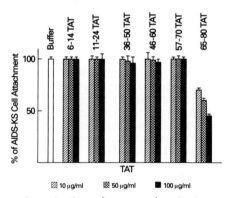

Fig. 10. Mapping of Tat domain required for cell attachment by competition of cell adhesion with overlapping Tat peptides. AIDS-KS cell were preincubated with serial dilutions the peptides or with the peptide buffer. Reprinted from Barillari et al. Proc Natl Acad Sci USA 1993; 90: 7941-7945.

cell adhesion to Tat, FN and [65-80]Tat was reduced or lost. This suggested that the basic region of Tat induces cell attachment in a non-energy-dependent fashion, likely through electric charge interactions with the cell membrane, while the portion of Tat containing the RGD sequence mediates cell adhesion via a receptor. To verify this, competition experiments were performed by preincubating AIDS-KS cells with serial dilutions of overlapping Tat peptides (Fig.10). Only the Tat peptides containing the RGD sequence ([65-80]Tat, [65-85]Tat and [72-86]Tat) inhibited the adhesion of AIDS-KS cells to Tat in a dose-dependent fashion. To the contrary, Tat-adhesion was not inhibited by any of the other peptides, including those containing the basic sequence (residues 49-57) (21). In addition, competition experiments with the peptide GRGDSP, or the mutated

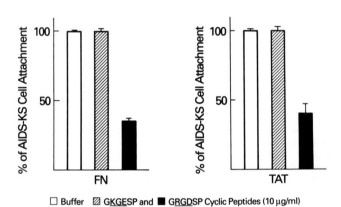

☐ Buffer ▨ GKGESP and ■ GRGDSP Cyclic Peptides (10 µg/ml)

Fig. 11. AIDS-KS cells were preincubated with the cyclic peptides GRGDSP or GKGESP (10 ug/ml) or with protein buffer and seeded on plates coated with Tat or FN (10 ug/ml). Similar results were obtained with cytokine-treated H-UVE and SM cells. Reprinted from Barillari et al. Proc Natl Acad Sci USA 1993; 90:7941-7945.

peptide GKGESP, indicated that the RGD sequence is the major vascular cell attachment domain of Tat (Fig. 11) (21).

As the RGD sequence of extracellular matrix (ECM) molecules is specifically recognized by receptors of the integrin family (22) and the α5ß1 and the αvß3 are RGD-recognizing integrins widely distributed on cells of mesenchymal origin (23), their level of expression was analyzed in AIDS-KS and normal vascular cell types. AIDS-KS cells expressed high levels of both α5ß1 and αvß3, while the expression of these receptors was lower in normal vascular cells cultured in the absence of cytokines. However, after cell exposure to inflammatory cytokines, the level of integrin expression reached that detected with AIDS-KS cells (21). As antibodies directed against α5ß1 or αvß3 inhibited cell adhesion to immobilized Tat (Fig. 12), these results demonstrated that both α5ß1 and αvß3 mediate the cell attachment effect of Tat.

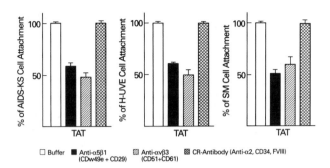

Fig. 12. The attachment of AIDS-KS, SM, and H-UVE cells to immobilized Tat is specifically inhibited by antibodies directed against α5ß1 and αvß3 but not by control antibodies (CR). When AIDS-KS cells were preincubated with anti α5ß1 and αvß3 antibodies combined, cell adhesion to Tat or FN was inhibited by 75% and 60%, respectively. Reprinted from Barillari et al. Proc Natl Acad Sci USA 1993; 90: 7941-7945.

CONCLUSION

The HIV-1 Tat protein is a transactivator of viral gene expression. However, Tat possesses other activities on cellular functions. Early during HIV-1 acute infection of T cells, Tat is released into the extracellular media and promotes growth of AIDS-KS cells and cytokine-activated normal vascular cells (progenitor of AIDS-KS cells). Extracellular Tat exerts other effects on vascular cells which may explain the angiogenic lesions observed in tat-transgenic mice. Tat promotes endothelial cells to migrate, to degrade and cross the basement membrane and to organize in tube-like structures. Each effect of Tat on vascular cells requires previous cell exposure to inflammatory cytokines (TNF, IL-1, and gamma-IFN). As these cytokines are increased in HIV-1 infected individuals and induce normal vascular cells to acquire features of the KS cell phenotype, this suggests that extracellular Tat and cytokines cooperate in the induction of AIDS-KS. Extracellular Tat has also adhesive effects on the same cell types. Cell attachment to Tat is induced by previous exposure of these cells to the same inflammatory cytokines. Adhesion is associated with the amino acid sequence RGD of Tat through a specific interaction with the integrins α5ß1 and αvß3. The expression of both integrins on vascular cells is increased to levels detected in AIDS-KS cells by the same cytokines which promote these cell types to acquire spindle cell morphology and to become responsive to the adhesion and angiogenic affects of Tat. These results indicate that Tat has properties which mimic the ECM molecules and they suggest that integrins also mediate the angiogenic effect of the protein.

REFERENCES

1   Wong-Staal F, Sadaie MR. In Franza R, Cullen B and Wong-Staal, eds. Cold Spring Harbor. Laboratory Press. Cold Spring Harbor, 1988; 1.
2   Ensoli B, Barillari G, Salahuddin SZ, Gallo RC et al. Nature 1990; 345: 84-86.
3   Ensoli B, Buonaguro L, Barillari G, Fiorelli V et al. J Virol 1993; 67: 277-287.
4   Rugters JL, Wieczorek R, Bonetti F, Kaplan KL et al. Am J Pathol 1986; 122: 493-499.
5   Weich HA, Salahuddin SZ, Gill P, Nakamura S et al. Am J Pathol 1991; 139: 1251-1258.
6   Barillari G, Buonaguro L, Fiorelli V, Hoffman J et al. J Immunol 1992; 149: 3727-3734.
7   Vogel J, Hinrichs SH, Reynolds RK, Luciw PA et al. Nature 1988; 335: 606-611.
8   Ensoli B, Barillari G, Gallo RC. Hematol Oncol Clin North Am 1991; 5: 281.
9   Nakamura S, Salahuddin SZ, Biberfeld P, Ensoli B et al. Science 1988; 242: 426-430.
10 Salahuddin SZ, Nakamura S, Biberfeld P, Kaplan MH et al. Science 1988; 242: 430-433.
11 Ensoli B, Nakamura S, Salahuddin SZ, Biberfeld P et al. Science 1989; 243: 223-226.
12 Rabkin CS, Goedert JJ, Biggar RJ, Yellin F et al. Acquir Immun Defic Syndr 1990; 3: s38.
13 Levy JA, Zigler JL Lancet 1983; 2: 78.
14 Ballard HS. Arch Intern Med 1985; 145: 547.
15 Rubinstein P, Rothman WM, Friedman-kien A. Antibiot Chemother 1984; 32: 87.
16 Hober D, Haque A, Wratte P, Beaucaire G et al. Clin Exp Immunol 1989; 78: 329.
17 Honda M, Kitamura K, Mizutani Y, Oishi M et al. J Immunol 1990; 145: 4059.
18 Albini A, Barillari G, Gallo RC, Ensoli B. Proc Natl Acad Sci 1993; in press.
19 Folkman J, Klagsbrun M. Science 1987; 235: 442-447.
20 Brake DA, Debouk C, Biesecker G. J Cell Biol 1990; 111: 1275-1281.
21 Barillari G, Gendelman R, Gallo RC, Ensoli B. Proc Natl Acad Sci USA 1993; 90: 7941-7945.
22 Hynes RO. Cell 1992; 69: 11-24.
23 Vlodavsky I, Gospodarowicz D. Nature 1981; 289: 304-306.

INDEX OF AUTHORS

Aberle, H., 249
Albrecht, R.M., 273

Bar-Shavit, R., 209
Barbas, C.F., 325
Barillari, G., 351
Beckman, A., 343
Behrendt, N., 289
Benezra, M., 209
Bennet, W., 83
Berkenpas, M.B., 303
Bläser, J., 331
Boer, M., 193
Brunner, G., 217
Buddecke, E., 201
Bunch, T., 187

Calafat, J., 193
Cheresh, D.A., 135, 177
Cherny, R.C., 39
Clement, A., 229
Cooper, S., 83

Dahlbäck, K., 91
Danø, K., 289
De Boer, H.C., 103
De Groot, Ph., 103
De Groot, Ph.G., 121
Diefenbach, B., 149
Dumler, I., 163

Ehrlich, H.J., 59
Ensoli, B., 351
Eskohjido, Y., 209

Faissner, A., 229
Felding-Habermann, B., 135, 149
Fessler, J.H., 187
Fessler, L.I., 187
Figdor, C., 193
Fiorelli, V., 351
Fitz, L.J., 45
Fitzgerald, M., 45
Flock, J.-I., 257
Fogerty, F.J., 187

Gallo, R.C., 351
Gechtman, Z., 311
Giannotti, J., 45
Ginsburg, D., 303
Gissler, H.M., 295
Götz, B., 229
Graeff, H., 331
Groß, W., 343
Gullberg, D., 187

Hayashi, I., 13
Hayashi, M., 3
Herrmann, F., 249
Herrmann, M., 273
Hertwig, B., 343
Hess, S., 21
Hilkens, J., 193
Himmelsbach, F., 157
Høyer-Hansen, G., 289

Jacobs, K., 45
Jänicke, F., 331
Joester, A., 229
Jonczyk, A., 149

Kelleher, K., 45
Kirschning, C., 249
Knopf, H.-P., 249
Kouns, W.C., 141
Kramer, M., 331
Kramer, M.D., 295
Kreizman, T., 311
Kriz, R., 45

Lamping, N., 249
Lawrence, D.A., 303
Liebenhoff, U., 171
Linder, D., 55
Litvinov, S.L., 193
Ljungh, A., 257
Loskutoff, D.J., 75, 325

Mandl, C., 229
Marguerie, G., 127
März, W., 343
McKeown-Longo, P.J., 111
Merberg, D.M., 45
Morgenstern, E., 99
Mosher, D.F., 13, 273, 303
Mueller, B.M., 135
Müller, Th.H., 157
Murtha, P., 45

Naski, M.C., 303
Nelson, R.E., 187
Nemerow, G.R., 177
Nguyen, H., 217
Niederländer, C., 229
Nieuwenhuis, H.K., 121

Pache, L., 331
Palaniappan, S., 303
Panetti, T.S., 111
Pannekoek, H., 325
Pera, M.F., 83
Peterson, C.B., 67
Petri, T., 163
Podack, E.R., 237

Podor, T.J., 75, 303
Preissner, K.T., 21, 45, 55,
   59, 99, 103, 295
Presek, P., 171
Proctor, R.A., 273

Richter, B., 59
Rippmann, F., 149
Roach, S., 83
Rønne, E., 289
Rydén, C., 265

Sanders, L.C., 135
Sane, D.C., 31
Sappino, A.-P., 283
Scaltreto, H., 45
Scharnagl, H., 343
Schleuning, W.-D., 163
Schmidt, A., 201
Schmitt, M., 331
Schnädelbach, O., 229
Schneider, W., 343
Scholze, A., 229
Schumann, R.R., 249
Schvartz, I., 311
Seewaldt-Becker, E., 157
Seiffert, D., 75
Shaltiel, S., 311
Siekmeier, R., 343
Sims, P.J., 243
Sixma, J.J., 121
Solberg, H., 289
Steiner, B., 141
Stockmann, A., 21

Temple, P., 45
Teschemacher, H., 55
Thiagarajan, P., 39
Thomssen, C., 331
Tomasini-Johansson, B.R., 223
Tschesche, H., 331
Turner, K., 45

Uzan, G., 127

Van de Wiel van Kemenade, E., 193
Van der Valk, S., 193
Van Meijer, M., 325
Van Zonneveld, A.J., 325
Vassalli, J.-D., 283
Vischer, P., 201
Vlodavsky, I., 209
Völker, W., 21, 201
Von der Ahe, D., 59
Vos, H.L., 193

Wadström, T., 257
Weidenthaler-Barth, B., 295
Weidle, U., 331
Weller, T., 141
Wesseling, J., 193
Wickham, T.J., 177
Wilhelm, O., 331
Wöhner, S., 55

Yacoub, A., 265

Zhao, Y., 31

KEYWORD INDEX

α IIb/β3 receptor, 157
α-granules, 99
α_2-macroglobulin receptor, 343
αv-integrins, 135
adenovirus, 177
adhesion, 121, 209
AIDS-Kaposi's sarcoma, 351
amino acid sequencing, 55
amyloid, 91
angiogenesis, 351
animal vitronectins, 3
antagonists, 141
anti-adhesion, 193
antiproliferative activity, 201
antiproteases, 283
antithrombin, 103
apoptosis, 91
arterial smooth muscle cells, 201

β-endorphin binding domain, 55
bacterial adhesion, 273
baculovirus, 31
balloon injury, 201
basic fibroblast growth factor, 217
binding, 21
binding proteins, 75
binding sites, 149
biomaterial, 257
biosynthesis, 75
blood platelet, 121
body fluids, 75
bone matrix, 265
bone marrow, 217
bone sialoprotein, 265
boundary, 229
breast cancer, 331
bullous pemphigoid, 295

cAMP dependent protein kinase (PKA),
 311
cathepsins, 331
CD59, 243

cell adhesion, 39, 157, 351
chondroitin sulfate proteoglycan, 229
collagen, 121
collagen IV, 187
combinatorial cloning, 325
complement, 243
conformation, 21, 303
cross-linking, 55
cytokine induction, 249
cytotoxic T cells, 193

denaturation, 67
development, 83
dithionitrobenzoic acid (DTNB), 13
DNA-degradation, 237
drosophila, 187

endocytosis, 111
endoproteinase ArgC, 55
endothelial cells, 21, 103, 209
endotoxin-recognition, 249
episialin, 193
error-prone PCR, 325
extracellular matrix, 111, 121, 209
extracellular matrix protein, 257

fas, 237
fibrin, 223
fibrin fibres, 99
fibrinogen, 273
fibrinolysis, 325
fibroblast, 343
fibronectin, 257

germ cell, 83
gp IIb/IIIa receptor, 157
GPI anchor, 217
GPIIb gene, 127
gramnegative sepsis, 249
guanidine, 67

hemopexin, 45, 59

heparan sulfate oligosaccharides, 201
heparan sulfate proteoglycan, 217
heparin, 13
heparin laminin, 257
HIV-1 TAT, 351
homology, 59
human melanoma, 135

immunolabeling, 99
infection, 273
inflammation, 223
integrin specificity, 39
integrins, 127, 141, 157, 177
invasion, 193

kinetics, 303

laminin, 187
LDL receptor, 343
lesional skin, 295
LII6β3-integrin, 273
LIIbβ3-fibrinogen binding, 149
lipoprotein (a), 343

macrophage activation, 249
megakaryocyte, 45, 127
membrane attack complex, 243
membrane glycoprotein, 243
metabolism, 103
metalloproteases, 331
monoclonal antibodies, 289
mouse vitronectin, 59
MUC1, 193
multimeric vitronectin, 21
mutagenesis, 31
mutational simulation, 325

nephrosclerosis, 91
neuritogenesis, 229
neuron-glia interactions, 229

PAI-1, 303, 311, 331
pComb3, 325
penton base, 177
peptide inhibitors, 149
perforin, 237
peroxidase, 187

phospholipase C, 217
physarum vitronectin, 3
plant vitronectins, 3
plasmin, 283, 311
plasmin(ogen), 295, 343
plasminogen activation, 311
plasminogen activator inhibitor-1 (PAI-1),
 31
plasminogen activators, 163
platelet surface, 99
platelets, 171, 273
prognosis, 331
proteases, 283, 331
protein expression, 31
protein kinase C inhibitor, 171
protein tyrosine kinase, 171
proteoheparan sulfate, 201
purification, 3, 13

receptor selectivity, 141
repulsion/inhibition, 229
reversible unfolding, 67
RGD, 209
rheumatoid arthritis, 223

second calcium binding domain of LIIb,
 149
signal transduction, 163
skin, 91
smooth muscle cells, 209
somatomedin B, 45, 59
staphylococcal infection, 265
staphylococcal interaction, 265
staphylococcus aureus, 273
sulfhydryl, 13

tenascin, 229
thrombin, 103, 111, 303
thrombosis, 141
thrombospondin, 273
tiggrin, 187
tissue-type plasminogen activator, 283
transcription, 127
transcytosis, 21
translocation, 171
tumor growth, 135
tyrosin phosphorylation, 163

U-937 cells, 163
uPA, 331
uPA cleavage, 289
uPAR variants, 289
urea, 67
urokinase, 163, 283
urokinase plasminogen activator (uPA),
 289
urokinase receptor (uPAR), 289
urokinase-type plasminogen activator

receptor, 163

virus receptors, 177
vitronectin, 13, 31, 39, 45, 67, 83, 91,
 103, 111, 135, 223, 295, 303, 311
vitronectin-β-endorphin conjugate, 55
vitronectin binding protein, 257
vitronectin redistribution, 99
Von Willebrand factor, 121